Ecotoxicology and Environmental Chemistry

Ecotoxicology and Environmental Chemistry

Edited by **Giselle Tang**

SYRAWOOD
PUBLISHING HOUSE
New York

Published by Syrawood Publishing House,
750 Third Avenue, 9th Floor,
New York, NY 10017, USA
www.syrawoodpublishinghouse.com

Ecotoxicology and Environmental Chemistry
Edited by Giselle Tang

International Standard Book Number: 978-1-68286-168-4 (Hardback)

Contents

Preface

In my initial years as a student, I used to run to the library at every possible instance to grab a book and learn something new. Books were my primary source of knowledge and I would not have come such a long way without all that I learnt from them. Thus, when I was approached to edit this book; I became understandably nostalgic. It was an absolute honor to be considered worthy of guiding the current generation as well as those to come. I put all my knowledge and hard work into making this book most beneficial for its readers.

Ecotoxicology is an umbrella science that integrates concepts from various disciplines like molecular chemistry, earth sciences, environmental engineering and environmental chemistry in order to understand the interactions of toxics with the ecosystem and develop strategies to reduce the negative impacts. Heavy metal contamination, ocean acidification, soil poisoning are some of the crucial factors that fueled the research in this field. This book includes some of the vital pieces of work being conducted across the world, on various topics like standard toxicity tests, ecotoxicological evaluation and biosensors. It is a complete source of knowledge on the present status of this important field.

I wish to thank my publisher for supporting me at every step. I would also like to thank all the authors who have contributed their researches in this book. I hope this book will be a valuable contribution to the progress of the field.

                                                                                          Editor

Uptake and distribution of hexavalent chromium in tissues (gill, skin and muscle) of a freshwater fish, Tilapia, *Oreochromis aureus*

Ayşe Bahar Yılmaz[1]*, Cemal Turan[2] and Tahsin Toker[1]

[1]Water Pollution Laboratory, Faculty of Fisheries, Mustafa Kemal University, 31200, İskenderun-Hatay, Turkey.
[2]Fisheries Genetics Laboratory, Faculty of Fisheries, Mustafa Kemal University, 31200, İskenderun-Hatay, Turkey.

Water pollution by heavy metals, especially chromium pollution from industrial sources can affect aquatic life, all ecosystems and human health directly or through food chain. This study aims to investigate the uptake of hexavalent chromium by a freshwater fish, (Tilapia, *Oreochromis aureus*). Short-term acute toxicity tests were performed over a period of 96 h providing the medium with various concentrations of potassium dichromate. Then the 96 h LC_{50} value was found to be 91.51 mg l^{-1} (Cr^{6+} as 32.35 mg l^{-1}). Five different concentrations of Cr^{6+} varying between 10, 15, 20, 25 and 30 mg l^{-1} were implemented for the uptake of this metal. The experiment was carried on for 28 days, meanwhile sampling fish weekly. With continued exposure, the accumulations were increased and fish progressively lost their ability to respond to this increase in exposure period. The chromium concentration in different organs was in the following order gill > skin > muscles tissues (least). The concentration of Cr in the gill range from 3.11 - 45.23 µg g^{-1} w.w, while the concentration accumulated in the muscle tissue of fish ranged from 0.86 to 12.34 µg g^{-1} w.w.

Key words: Heavy metal, Cr toxicity, fish, Tilapia (*Oreochromis aureus*), accumulation.

INTRODUCTION

Heavy metals are introduced into the environment by a wide spectrum of natural sources such as volcanic activeties, erosion and anthropogenic ones including industrial wastes as well as a leakage. Some of these metals including lead, nickel, cadmium, mercury are toxic to living organisms even at quite low concentrations, while others such as copper, iron, zinc and manganese are biologically essential and natural constituents of the aquatic ecosystems and become toxic only at very high concentrations (Cohen et al., 2001; Storelli et al., 2006; Karadede and Ünlü, 2007).

Chromium is considered as a heavy metal and pollutant as well as an essential micronutrient. Wastewater pollution by chromium originating from electroplating, dyeing, tannery, hard-alloy steel and stainless steel manufacture, has affected the life on earth. Chromium is also used as a catalyst and coating material (Idachaba et al., 2004). Welding, grinding and polishing of stainless steel are among principal ways of introducing chromium into the land environment while other ways of introducing chromium into air and water environments include the burning of fossil fuels and waste incineration (WHO, 1988). This pollution could affect all ecosystems and human health directly or through food chain (Yılmaz et al., 2009).

Chromium exists in different oxidation states which have distinct biological effects (Richard, 1991). Hexavalent chromium [(Cr VI)] is a well known carcinogen metal form for animals and human beings. Cr (VI) compounds readily penetrate into cell membranes via anion transport systems. It was clear from previous studies that Cr (VI) itself was highly active and carcinogen should it arrive as Cr (VI) inclusion to the target (Costa, 1997; Ding and Shi, 2002). Cr (VI) has the ability to lend electrons and be reduced to trivalent chromium [Cr (III)]. In contrast to Cr (VI), the trivalent form of Chromium is 500 to 1000 times less active against living cells because of its poor uptake (Alexander and Aaseth, 1995). Cohen et al. (2001) suggested that Cr (III) was much less active than Cr (VI) to a cell but when

---

*Corresponding author. E-mail: aybahar@ yahoo.com or abyilmaz@ mku.edu.tr.

Cr (III) enters the cell, it could cause toxicity on a basis comparable to Cr (VI).

Fishes are often at the top of the aquatic food chain and may concentrate large amounts of certain metals from water (Mansour and Sidky, 2002). In addition to adsorption on tissue and membrane surface, fish may assimilate metals by ingestion of particulate material or food in water, or ion-exchange of dissolved metals through lipophilic membranes, e.g. the gills (Mendil et al., 2005). Metal distribution between different tissues depends on the mode of exposure, that is, dietary and/or aqueous exposure, and can serve as a pollution indicator (Alam et al., 2002). The objective of our study was to investigate the impact of different chromium (10, 15, 20, 25 and 30 mg l^{-1} Cr^{6+}) pollutions on the survival of juvenile *Oreochromis auras* (Tilapia) and to compare the chromium concentrations in fish tissue (muscle, skin and gill).

MATERIALS AND METHODS

Fish

O. aureus (Tilapia), a tropical freshwater and important culture fish, is a specie commonly found in brackish water in estuaries all over the world and responds promptly to environmental alterations (Vijayan et al., 1996; Almeida et al., 2002; Turan, 2006). Juvenile specimens of Tilapia were captured from ponds at Çukurova University Aquaculture Research Centre and transferred to the laboratory where the experiments were conducted. Holding and acclimatization took place in tanks containing fresh water at temperature of 20 ± 2°C and oxygen content 80%. The samples were placed in aquarium for one week to allow for adaptation of the fish to the new conditions. Tap water used for the experiment had a pH value of 7.6 ± 0.3 and total hardness of 135 ± 5.8 mg $CaCO_3$ l^{-1}. The aquariums were aerated with air stones for proper oxygen saturation (8.7 mg O_2 l^{-1}). Fish were fed once a day with artificial feed meal. Acclimatized fish were moved at random into test aquaria (40 x 40 x 80 cm) as well as into control aquaria, in four aquaria for each concentration and control, containing twelve fishes.

Chemicals

All reagents were analytical grade. Required concentrations of potassium dichromate (Merck) were prepared by adding aliquots of 1% stock solution in double distilled water. A Cr (VI) stock solution was delivered to twenty test aquariums via automatic pipettes. The toxicant solution in the aquarium was replaced with fresh solution of the same concentration every 24 h. Renewal bioassays were conducted using five concentrations of potassium dichromate (Cr^{6+} as 10, 15, 20, 25 and 30 l^{-1}). Controls without toxicant were also run simultaneously.

Experimental exposure

Treatment water was monitored every day for dissolved oxygen, pH and Cr concentrations. Exposure period was 28 days, during which samples were taken on day 7, 14, 21 and 28.
Experimental units were checked daily for mortality and behavioral changes and any dead fish was immediately removed from the aquariums. For measurement purpose, 3 fishes were taken from each replicate aquarium and their lengths were measured (cm) and weighed (g) before dissection with cleansed tools.

Water and fish tissue analysis

Samples of 100 ml of water from each treatment were filtered through a 0.45 µm micropore membrane filter and kept at -20°C until analysis process. Each filtered sample was transferred to a pre-cleaned polyethylene bottle, acidified to 1% HNO_3 and analyzed with Inductively Coupled Plasma Atomic Emission Spectrometry (ICP-AES, Varian model- Liberty Series II). The absorption wavelength and instrument detection limit were 283.553 nm and 0.007 µg l^{-1}. For the analyses, the gill, a part of skin and approximately 5 g of the epaxial muscle on the dorsal surface of each fish were dissected, washed with distilled water, weighed, packed in polyethylene bags and stored at -20°C prior to analysis. The digestion of the fish tissues was in accordance with methods described by Yilmaz (2003) with concentrated HNO_3/HCl (1:3 v/v). Metal concentrations were determined with ICP-AES. Standard reference materials were as follows: Multi-4 Merck for fish, SRM-143d for water acc. to National Institute of Standards and Technology. The results indicated good agreement between the certified and analytical values with recovery rates of Cr between 92 and 103% for fish, 94 and 102% for water. The concentrations were expressed as micrograms per gram wet weight (µg g^{-1} wet wt.) of tissue in organisms.

Statistical evaluation

For the survival tests, Statistical Analysis of data was carried out with SPSS statistical package program. A value of $p < 0.05$ was considered to be significant. For the accumulation tests, the experiments were repeated four times and only the arithmetic mean of the four experiments at each concentration was taken to express the results.

RESULTS

Test organisms (n = 300) were juvenile specimens of Tilapia of mean lengths 10.2 ± 1.3 cm and weights 18.39 ± 2.09 g. There was no statistical difference between the study groups and control group regarding the size of the fish ($p > 0.05$).

Physico-chemical parameters of fresh water quality including pH value, temperature and the amount of dissolved oxygen of the test aquariums were determined as 7.06 ± 1.24, 21.8 ± 2.02°C and 8.18 ± 1.46 mg l^{-1} respectively during the experiment. Higher metal concentrations caused a significant decrease in pH values, however, in all cases; pH values did not reach the acidic range, which could affect the organisms' survival.

The concentrations of Cr measured in the aquariums throughout the experiment were ± 10% of the nominal concentrations (Table 1). Therefore, nominal concentrations from now on will take this value as the basis. The general agreement of Cr^{6+} with nominal concentrations confirmed that virtually all of the Cr remained in the Cr^{6+} form throughout the experiment (Farag et al., 2006).

Table 1 summarizes the survival of the fish during a 4-week-exposure period for juvenile specimens. Survival of control fish was nearly 100% throughout the experiment. Chromium concentrations (25 and 30 mg l^{-1} Cr^{6+}) reduced

Table 1. Mean of measured chromium concentration in water (Cr^{6+} µg l^{-1}), standard deviation (SD) and mean percent survival of juvenile Tilapia (*O. aureus*) exposed to Cr^{6+} as $K_2Cr_2O_7$ during 28 days.

Theorical chromium concentration (Cr^{6+} µg l^{-1}) in water	Mean of measured chromium concentration in water (Cr^{6+} µg l^{-1}) (SD)	Initial number of Tilapia per aquarium	Percent survival in exposure days (d)			
			7th d	14th d	21th d	28th d
Control (0)	1 (0.1)	12	100[a]	100[a]	95.8[a]	95.8[a]
10	9.7 (0.7)	12	95.8[a]	97.2[a]	95.7[a]	90[a]
15	13.9 (0.5)	12	95.8[a]	94.1[a]	85[a]	80[b]
20	19.8 (0.8)	12	93.9[a]	93.9[a]	89.5[a]	80[b]
25	24.4 (0.9)	12	89.6[a]	87.1[a]	80[b]	62.5[b]
30	28.9 (1.2)	12	87.5[a]	76.7[b]	71.4[b]	54.5[c]

Different superscript letters (a, b and c) designates difference at $p < 0.05$ within a sample day.

fish survival beginning from the first week of experiment, where high mortality rate of Cr-exposed fish occurred within 14 to 21 days. Survival rate decreased to 54.5% in fish exposed to 30 mgl^{-1} Cr^{6+} dose from days 21 to 28. The behavioural changes of the control group and Tilapia exposed to various doses of K_2CrO_4 were compared with each other during the experiments. The control group displayed normal behavior during the test period. The lowest concentrations (10 and 15 mg l^{-1} Cr^{6+}) had similar behavior to that of the control group. From the dose 20 mg l^{-1} Cr^{6+}, the fish started to show behavioral disorders such as loss of equilibrium, sudden startling and respiratory difficulties. 25 mg l^{-1} Cr^{6+} and at the highest concentration (30 mg l^{-1} Cr^{6+}) onwards, there was shivering, rather high respiratory disorder and swimming in capsized position.

The accumulations of chromium (µg/g w. wt) in the gill, skin and muscle tissues of Tilapia during the exposure period are shown in Figures 1a - e. The initial Cr^{6+} concentrations were: 1a, 10 mg l^{-1}; 1b, 15 mgl^{-1}; 1c, 20 mgl^{-1}; 1d, 25 mg l^{-1}; 1e, 30 mg l^{-1}. The chromium concentrations in the tissues of control fish were below the detection limit of the instrument (< 0.007µg l^{-1} Cr^{6+}) throughout the experiments. The concentrations (µg g^{-1} wet weight) of Cr in the organs (gill, skin and muscle) of fish increased when they were exposed to Cr in the water (Figure 1a - e). Chromium accumulation in the muscle tissue of fish ranged from 0.86 to 12.34 µg g^{-1} w.w., while concentrations in the gill were within the range of 3.11 - 45.23 µg g^{-1} w.w. (Figure 1a - e).

As can be seen in Figure 1a, maximum level of Cr^{6+} was 1.30 µg g^{-1} w.w on fish muscle having lived 28 days at 10 µg l^{-1} Cr^{6+} initial concentrations. On the other hand, higher concentration of Cr^{6+} (12.34 µg g^{-1} w.w.) was observed on fish muscle after 28 days at 30 µg l^{-1} Cr^{6+} (Figure 1e). Maximum levels of Cr^{6+} concentrations on the muscle tissues at 15 (after 28 days) in Figure 1b, 20 (after 28 days) in Figure 1c and 25 (after 28 days) µg l^{-1} Cr^{6+} treatments in Figure 1d were 2.43, 9.03 and 10.81 µg g^{-1} w.w., respectively. The concentrations of Cr in gills were almost twice, three and four times as high as in the

muscle at 20, 25 and 30 µg l^{-1} Cr^{6+} mediums, respectively. The Cr accumulation on the skin of Tilapia for all experimental concentrations was nearly twice as much as in the muscle of samples at the end of experiment.

DISCUSSION

It is widely known that metal toxicity is more accurately measured in fresh water than in sea water, because metal appear to a great extent as complex compounds in sea water and this reduces the toxicity of metal ions. The reaction and survival of aquatic animals depend not only the biological state of the animals and physico-chemical characteristics of water (such as pH, temperature, hardness and the amount of dissolved oxygen) but also on the kind, toxicity, type and the duration of exposure to the toxicant (Martin et al., 1981; Mays, 1996; Vutukuru, 2005). In the present study, the mortality increased with an increase in concentration and also the duration of the exposure. This may be rather significant, because smaller fish being generally more active than larger ones, metal uptake and elimination of metal could also occur in higher rates in these smaller ones (Heath, 1987; Larsson et al., 1985; Canli and Furness, 1993).

A comparison of toxicity values for 10 µg l^{-1} Cr^{6+} to 30 µg l^{-1} Cr^{6+} experiments demonstrated a decline in survival rate following longer exposure periods. Even thought the magnitude of this decline varied between the different chromium concentrations, lower (10 to 20 µg l^{-1} Cr^{6+}) mediums were about 10 to 20% mortality while higher (25 and 30 µg l^{-1} Cr^{6+}) mediums were 38 to 45% mortality at the end of experiments. Nearly no mortality was recorded in the experiment controls.

The World Health Organization (WHO), U.S. EPA, as well as nearly every state agency (Turkish Environmental Guidelines, Class I-II, 1998; TSE 266) has set the drinking water standard for Cr (VI) at 50 or 100 ppb. These parameters (Cr (VI)) are not a small amount of Cr (VI) since it would be equated to 1 to 2 µM Cr (VI) (Costa, 2003). The Cr concentrations [10 to 30 µg l^{-1} (ppb) Cr (VI)]

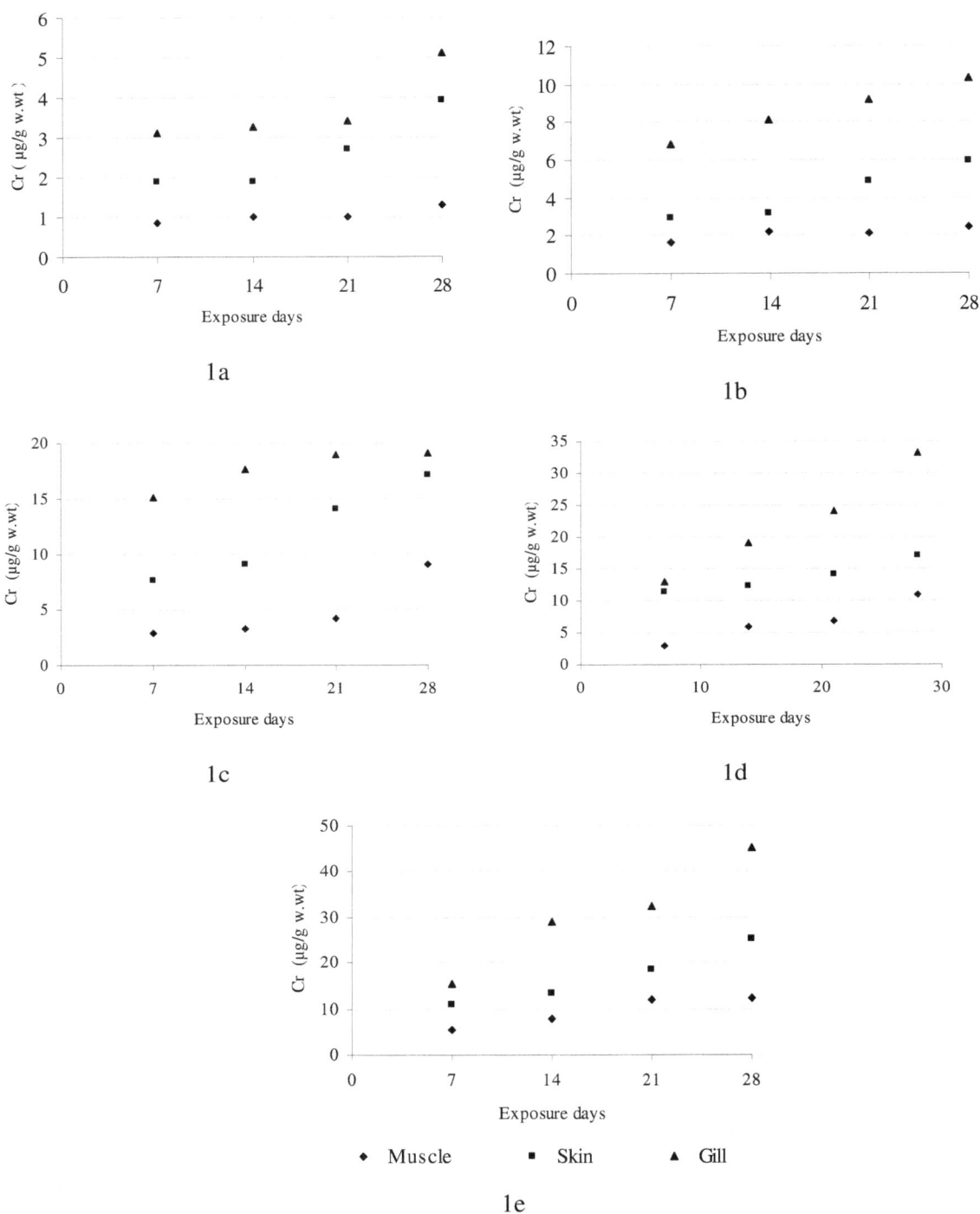

Figure 1. Time-dependent bioconcentration levels of chromium μg per gram wet weight (μg/g w. wt.) in muscle (♦), skin (■) and gill (▲) tissues of Tilapia. The initial Cr^{6+} concentrations: 10 mg l^{-1} (a), 15 mg l^{-1} (b), 20 mg l^{-1} (c), 25 mg l^{-1} (d) and 30 mg l^{-1} (e).

used in the present study has been proved not to be sublethal for juvenile Tilapia and have been considered as representatives of environmental exposure. However, high Cr concentrations indicate that effects on survival were related to exposure concentrations and duration. At physiological pH, hexavalent Cr exists as an oxyanion and in this form, it resembles oxyanions such as sulfates and phosphates, which are used extensively in various biochemical processes. The individual cells of the body, which have active transport systems for these nutrients, take up sulfates and phosphates (Costa, 2003). In this way Cr (VI) can enter the body. Vutukuru (2005) showed that chromium induced alterations at the biochemical level, more pronounced changes occurring at the end of 96 h and thus it was time-dependent. Furthermore, metal induced alterations in the protein content may probably affect the enzyme mediated bio defence mechanisms of the fish.

In the present investigation, the tendency of the body surface to acquire a dark color appears to be the first symptom of toxication in these fish just like in the previous study for cadmium (Hilmy et al., 1985). At the end of the experiment, fish showed externally abdominal distention and hemorhagic (reddening) anal-uregenital pours. Erosion and fusion were also detected in lamellar epithelium of gills.

The effects of hexavalent chromium on aquatic living things have been evaluated by several studies. Some data exist on the effects of Cr (VI) on salmon (Buhl and Hamilton, 1991; Farag et al., 2006). Other researcher studied heavy metal accumulation in tissues of Tilapia, a freshwater fish (Ay et al., 1999; Wepener et al., 2001; Aleya et al., 2005; Zirong and Shijun, 2007). However, no study has yet assessed chromium exposure and accumulation on juvenile Tilapia. Previous studies did not investigate accumulation of this metal on skin and gill as a result of Cr (VI) exposure (Arillo and Melodia, 1988; Buhl and Hamilton, 1991; Outridge and Scheuhammer, 1993; Farag et al., 2006).

It was noted that Cr accumulation in the tissues showed the following sequence: gills > skin > muscle tissues (least). As in the present study, Storelli et al. (2006) found that the lowest concentration of Cr was detected in the muscle tissue and skin and the gills showed similar levels (p > 0.05). Some researchers found metal accumulation in the tissues of rainbow trout *Salmo gairdneri* in the following organs: lead in the bone, spleen and kidney; chromium in the spleen, muscle and gills; copper in the kidney; zinc in the gills (Camusso et al., 1995). Yılmaz (2003) indicated that concentrations of heavy metals on wild fish were higher in all of the skin samples than in the muscles.

The reason for high metal concentrations in the skin could be due to the metal complexion with the mucus that is impossible to remove completely from the tissue before analysis. In general, higher metal concentrations in gills reflect metal concentrations of the water where the fish live. Gills are vital respiratory and osmoregulatory organs and cellular damage induced by the metal might impair the respiratory function of the fish by reducing the respiratory surface area (Vutukuru, 2005). Gills are regarded as an important part for direct metal uptake from the water whereas the body surface is generally assumed to play a minor role (Pourang, 1995). It was demonstrated by Wepener et al., (2001) that gill tissue in banded tilapia was the initial site of accumulation of water-borne metals. The tendency of the metal to bind to the external gill surface was via ionic bonds, and to gill cytosolic compounds is via covalent bonds. It was also observed that the heavy metals in muscle tissue were at low levels compared with other organs (Sağlamtimur et al., 2003; Karadede and Ünlü, 2007). The results also show that chromium is more accumulated in the samples of gill than in the skin or muscles.

For the Cr metal, the European legislation has not established maximum level and therefore, an evaluation of the chemical quality of this fish is only possible utilizing dietary standards fixed in other countries. For example, the Western Australian Food and Drug Regulation List limit the level for Cr 5.5 μg g^{-1} (Usero et al., 2003). As it was observed that fish could live even at higher levels than those mentioned in EPA standards, it is safe to say that fish can cope with a higher accumulation rate of heavy metals in their bodies, which itself is not safe for human consumption. Therefore, it would be advisable to take into consideration the duration of contact with toxic substances and the fish species in addition to concentration rate before establishing a basis for permissible levels in environmental studies.

Conclusion

The results in the current study showed that the exposure of Tilapia to waterborne Cr (VI) caused significant accumulations in their organs and the accumulations were associated with the exposure period, concentrations of chromium and different tissues. Even studied levels of Cr (VI) were below drinking water standard (50 or 100 ppb) that the accumulations on gill, skin and also muscles were high. With longer exposure, the accumulations were increased and fish progressively lose their ability to respond with the increase in exposure period. These malfunctions are particularly important because they are associated with changes in survival, which can be related to effects at the accumulation levels. Future research should focus on the effect of chromium toxicity on bio defence mechanisms of Tilapia at sub-lethal levels.

ACKNOWLEDGMENTS

This study was supported by Mustafa Kemal University, Research Fund by project 05 E 0102.

REFERENCES

Alam MGM, Tanaka A, Allison G, Laurenson LJB, Stagnitti F, Snow E (2002). A comparison of trace element concentrations in cultured and wild carp (*Cyprinus carpio*) of lake Kasumigaura, Japan. Ecotoxicol. Environ. Saf. 53: 348 -354.

Aleya B, Md. Nurul A, Satoshi K, Kiyohisa O (2005). Selected elemental composition of the muscle tissue of three species of fish, *Tilapia nilotica, Cirrhina mrigala* and *Clarius batrachus,* from the fresh water Dhanmondi Lake in Bangladesh. Food Chem. 93: 439- 443.

Alexander J, Aaseth J (1995). Uptake of chromate in human red blood cells and isolated rat liver cells: the role of the anion carrier. Analyst 120: 931-933.

Almeida JA, Diniz YS, Marques SFG, Faine LA, Ribas BO, Burneiko RC, Novelli ELB (2002). The use of oxidative stress responses as biomarkers in Nile Tilapia (Oreochromis niloticus) exposed to in vivo cadmium contamination. Environ. Int. 27: 673-679.

Arillo A, Melodia R (1988). Effects of hexavalent chromium on trout mitochondria. Toxicol. Lett. 44: 71-76.

Ay Ö, Kalay L, Tamer M, Canli M (1999). Copper and lead accumulation in tissues of a freshwater fish Tilapia zilli and its effects on the branchial Na, K-ATPase activity. Bull. Environ. Contam. Toxicol. 62: 160-168.

Buhl KJ, Hamilton SJ (1991). Relative sensitivity of early life stages of arctic grayling, coho salmon, and rainbow trout to nine inorganics. Ecotoxicol. Environ. Saf. 22: 184 -197.

Camusso M, Vigano L, Balestrini R (1995). Bioconcentration of trace metals in rainbow trout: A field study. Ecotoxicol. Environ. Saf. 31: 133 -141.

Canli M, Furness RW (1993). Toxicity of heavy metals dissolved in sea water and influences of sex and size on metal accumulation and tissues distribution in the Norway lobster Nephrops norvegicus. Mar. Environ. Res. 36: 217-236.

Cohen T, Hee S, Ambrose R (2001). Trace metals in fish and invertebrates of three California Coastal Wetlands. Mar. Pollut. Bull. 42: 224-232.

Costa M (1997). Toxicity and carcinogenicity of Cr (VI) in animal models and humans. Crit. Rev. Toxicol. 27: 431-442.

Costa M (2003). Potential hazards of hexavalent chromate in our drinking water. Toxicol. Appl. Pharmacol. 188: 1-5.

Ding M, Shi X (2002). Molecular mechanisms of Cr (VI)-induced carcinogenesis. Mol. Cell. Biochem. 234(235): 293-300.

Farag AM, Mayb T, Marty GD, Easton M, Harper DD, Little EE, Cleveland L (2006). The effect of chronic chromium exposure on the health of Chinook salmon (Oncorhynchus tshawytscha). Aquat. Toxicol. 76: 246-257.

Heath AG (1987). Water pollution and fish physiology. CRC Pres Inc. Florida, USA. pp. 245.

Hilmy AM, Shabana MB, Daabees AY (1985). Bioaccumulation of cadmium: Toxicity in Mugil cephalus. Comp. Biochem. Physiol. 81(1): 139 -143.

Idachaba MA, Nyavor K, Egiebor NO (2004). The leaching of chromium from cementbased waste form via a predominantly biological mechanism. Adv. Environ. Sci. Res. 8: 483-491.

Karadede-Akin H, Ünlü E (2007). Heavy metal concentrations in water, sediment, fish and some benthic organisms from Tigris River, Turkey. Environ. Monit. Assess. 131: 323-337.

Larsson A, Haux C, Sjöbeck M (1985). Fish physiology and metal pollution: Results and experiences from laboratory and field studies. Ecotoxicol. Environ. Saf. 9: 251-281.

Mansour SA, Sidky MM (2002). Ecotoxicologcal Studies. 3. Heavy metals contaminating water and fish from Fayoum Governorate, Egypt. Food Chem. 78: 15-22.

Martin M, Osborn KF, Billing P, Glickstein N (1981). Toxicities of ten metals to Crassostre gigas and Mytilus edulis embryos and Cancer magister larvae. Mar. Pollut. Bull. 12(9): 305-308.

Mays LW (1996). Water Resources Handbook, McGraw-Hill. ISBN 0-07-114517-6.

Mendil D, Uluozlu OD, Hasdemir E, Tuzen M, Sari H, Suiçmez M (2005). Determination of trace metal levels in seven fish species in lakes in Tokat, Turkey. Food Chem. 90: 175-179.

Outridge PM, Scheuhammer AM (1993). Bioaccumulation and toxicology of chromium: implications for wildlife. Rev. Environ. Contam. Toxicol. 130: 31-77.

Pourang N (1995). Heavy metal bioaccumulation in different tissues of two fish species with regards to their feeding habits and trophic levels. Environ. Monit. Assess. 35: 207-219.

Richard CF, Bourg CMA (1991). Aqueous geochemistry of chromium: A review. Water Res. 25(7): 807-816.

Sağlamtimur B, Cicik B, Erdem C (2003). Effects of different concentrations of copper alone and copper+cadmium mixture on the accumulation of copper in the gill, liver, kidney and muscle tissues of Oreochromis niloticus (L). Turk. J. Vet. Anim. Sci. 27: 813-820.

Storelli MM, Barone G, Storelli A, Marcotrigiano GO (2006). Trace Metals in Tissues of Mugilids (Mugil auratus, Mugil capito, and Mugil labrosus) from the Mediterranean Sea. Bull. Environ. Contam. Toxicol. 77: 43-50.

Turan F (2006). Improvement of growth performance in Tilapia (Oreochromis aureus, Linnaeus) by supplementation of red clover (Trifolium pratense) in diets. Bamidgeh, Isr. J. Aquac. 58(1): 34-38.

Turkish Environmental Guidelines (TSE-266) (1998). Publications of Turkish Foundation of Environment. Institution of Turkish Standards, Annual Progress Report.

Usero J, Izquierdo C, Morillo J, Gracia I (2003). Heavy metals in fish (Solea vulgaris, Anguilla anguilla and Liza aurata) from salt marshes on the southern Atlantic coast of Spain. Environ. Int. 29: 949-956.

Vijayan MM, Morgan JD, Sakamota T, Grau EG, Iwama GK (1996). Food deprivation affects seawater acclimation in tilapia: hormonal and metabolic changes. J. Exp. Biol. 199: 2467-2475.

Vutukuru SS (2005). Acute effects of hexavalent chromium on survival, oxygen consumption, hematological parameters and some biochemical profiles of the Indian Major Carp, Labeo rohita. Int. J. Environ. Res. Public Health 2(3): 456-462.

Wepener W, Vuren van JHJ, Preez du HH (2001). Uptake and distribution of a copper, iron and zinc mixture in gill, live rand plasma of a freshwater teleost, Tilapia sparrmanii. Water SA. 27: 99-108.

WHO (1988). Chromium. In: Environmental Health Criteria, 61. World Health Organization, Geneva. pp. 1-197.

Yılmaz AB (2003). Levels of heavy metals (Fe, Cu, Ni, Cr, Pb and Zn) in tissue of Mugil cephalus and Trachurus mediterraneus from Iskenderun Bay, Turkey. Environ. Res. 92: 277-281.

Yılmaz S, Türe M, Sadıkoğlu M, Duran A (2009). Determination of total Cr in wastewaters of Cr electroplating factories in the I. Organize industry region (Kayseri, Turkey) by ICP-AES. Environ. Monit. Assess. DOI 10.1007/s10661-009-1045-z.

Zirong X, Shijun B (2007). Effects of waterborne Cd exposure on glutathione metabolism in Nile tilapia (Oreochromis niloticus) liver. Ecotoxicol. Environ. Saf. 67: 89-94.

Evaluation of heavy metals in roadside soils of major streets in Jos metropolis, Nigeria

E. S. Abechi[1]*, O. J. Okunola[2], S. M. J. Zubairu[1], A. A. Usman[3] and E. Apene[4]

[1]Department of Chemistry, Ahmadu Bello University, Zaria, Kaduna State, Nigeria.
[2]National Research Institute for Chemical Technology, Basawa, Zaria, Kaduna State, Nigeria.
[3]Salem University, Lokoja, Kogi State, Nigeria.
[4]Federal College of Forestry Mechanization P. M. B. 2273, Afaka, Kaduna, Kaduna State, Nigeria.

A study of heavy metals in roadside soils is critical in assessing the potential environmental impacts of automobile emission on the soil. The soil samples were collected and analyzed for the levels of Pb, Zn, Mn, Cu, Ni, Cd, Co and Fe using AAS. Results indicate the decreasing order of the average total metal content for the studied metals: Fe > Zn > Mn > Pb > Cd > Cu. Except for Cd, all metals are lower than the levels of those reported in other studies. The absence of Co and Ni indicate no pollution due to these metals. Correlation analysis between metals and the traffic volume (V) indicates significant positive correlation ($p < 0.05$) between Pb, Cd and Mn, and V. This further indicates that the metal pollution in the soil is mostly originated from vehicular emissions e. g. motor vehicles. Therefore, this study provides a practical approach to monitor the level of these metals.

Key words: Heavy metals, roadside, Jos, metropolis.

INTRODUCTION

Roads are important infrastructure that plays a major role in stimulating social and economic activities. However, road construction has also resulted in heavy environmental pollution (Bai et al., 2008).

Several researchers have indicated the need for a better understanding of trace metal pollution of roadside soils (De Kimple and Morel, 2000; Manta et al., 2002). Trace metals in roadside soils may come from various human activities, such as industrial and energy production, construction, vehicle exhaust, waste disposal, as well as coal and fuel combustion (Li et al., 2001). According to Adefolalu (1980) and Mabogunje (1980), in developing countries like Nigeria, improved road accessibility creates a variety of ancillary employment which range from vehicle repairs, vulcanizer and welders to auto-electricians, battery chargers and dealers in other facilitators of motor transportation. These activities send trace metals into the air and the metals subsequently are deposited into nearby soils, which are absorbed by plants on such soils. Sakagami et al. (1982) reported that there was a close relationship between trace metal concentration in roadside soil and those in the dust falls. Trace metals in the soils can also generate airborne particles and dusts, which may affect the air quality (Gray et al., 2003). Among the numerous environmental pollutants, an important role is ascribable to heavy metals whose concentration in soils, water and air are continuously increasing in consequence of anthropogenic activity.

According to Lagerwerff and Specht (1970), while many studies have been made on lead, little attention has been focus on the contamination of other trace metals in the roadside environment. Metals such as iron Fe, Cu, Zn are essential component of many alloy, pipe, wire and tyre in motor vehicles and are released into the roadside environment as a result of mechanical abrasion and normal wear and tear (Harrison et al., 1981). Soil tends to accumulate metals on a relatively long term basis since many metals in the soil are so mobile. This explains the overall higher contamination level of metals in the soil and why, in sampling, the top layer of soil should be taken (Ho and Tai, 1988). Although, there have been considerable number of studies on the concentrations of heavy metals in roadside soils, vast majority have been carried out in developed countries with long histories of industrialization and extensive use of leaded gasoline since 1935 (Otte et al.,

*Corresponding author. E-mail: abeshus@yahoo.com / okunolaoj@gmail.com.

Table 1. Traffic volume of the sampling roads.

Location	Site code	Average traffic volume per day
Murtala Muhammed way	A	1,521
Yakubu Gowon way 2	B	1,340
Yakubu Gowon way 1	C	1,320
Tudun Wada road	D	1,001
Pam road	E	965
Rukuba road	F	930
Off Yakubu Pam road	G	603

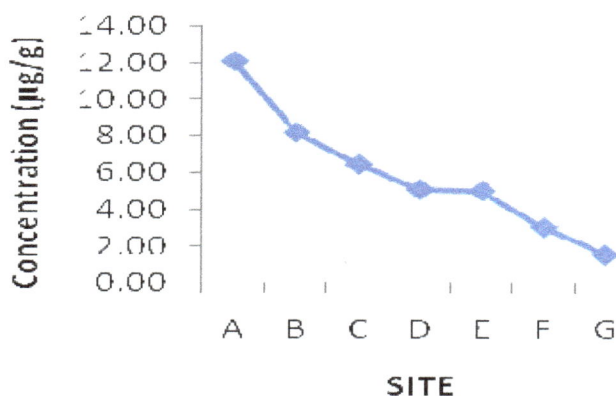

Figure 1. Concentration of Pb across the sampling site.

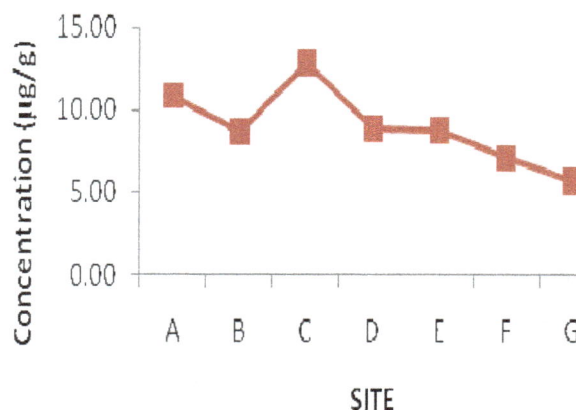

Figure 2. Concentration of Zn across the sampling site.

1991; Mateu et al., 1995). Very few studies were carried out in developing countries such as Nigeria and data on pollutant metal concentrations and distribution in such country is extremely scarce.

Hence, this research work was carried out to ascertain the heavy metal concentrations in roadside soils of major roads of Jos Metropolis as related to traffic volume.

MATERIALS AND METHODS

Sampling description

The sampling locations were chosen to span a wide range of traffic density and to give a good geographical coverage in the Jos city. Seven sites were selected for study within the Jos environment and their locations are as designated below in decreasing order of traffic volume. The traffic density was determined by counting the number of motor vehicles passing the sampling sites over a period of fifteen hours starting 6.00 a.m. to 6.00 p.m. each day for three days (Grace, 2004). The average number of vehicles passing the site per day was then calculated, as shown in Table 1.

Sample collection

Field collection were made in February (2008), about five months into the dry season such that possible wash away or leaching of the heavy metals where avoided. The samples were also collected within two consecutive days such as to minimize temporal changes.

At each site, three replicated samples were taken from the surface soil, composite and represented samples preserved in an acid pre-washed cleaned polyethene bags and subsequently treated and analyzed separately.

Sample digestion

All reagent used were of analytical grade and double distilled water was used in all preparation except otherwise stated. The method of Ho and Tai (1988) was used for sample digestions. Samples were sealed in polythene bag and air dried. The samples were grinded using an acid pre-washed mortar and pestle sieved by passing them through a 1 mm mesh. 1 g of soil of each of the samples was accurately weighed and treated with 10 ml aliquots of high purity conc. HNO_3. The mixture was on a hot plate until the sample is almost dry and then cooled. This procedure was repeated with another 10 ml conc. HNO_3 followed by 10 ml of 2 M HCl. The digested soil samples were then warmed in 20 ml of 2 M HCl to redissolved the metal salts. Extract were filtered through filter papers and the volume was then adjusted to 25 ml with doubled distilled water. Metal concentrations were determined by UNICAM SOLAR 32 Data station V7.15 AAS model.

RESULTS AND DISCUSSIONS

The average of triplicate determinations of metal concentrations (Pb, Zn, Cu, Cd, Mn and Fe) in soils of the various sampling sites is presented in µg/g in Figures 1- 6.

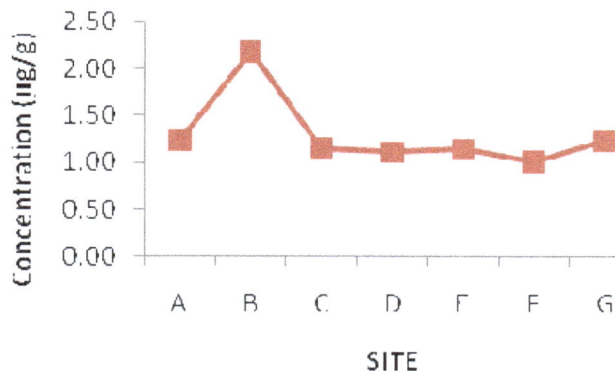

Figure 3. Concentration of Cu across the sampling site.

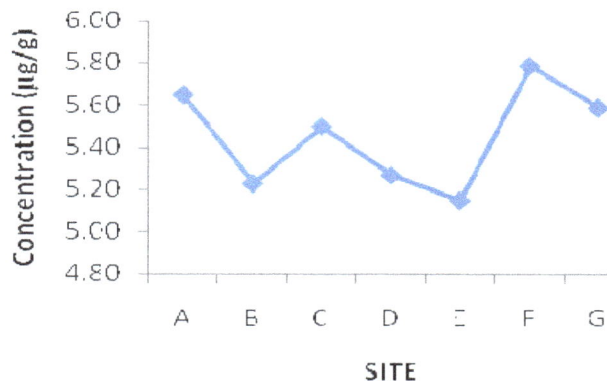

Figure 4. Concentration of Cd across the sampling site.

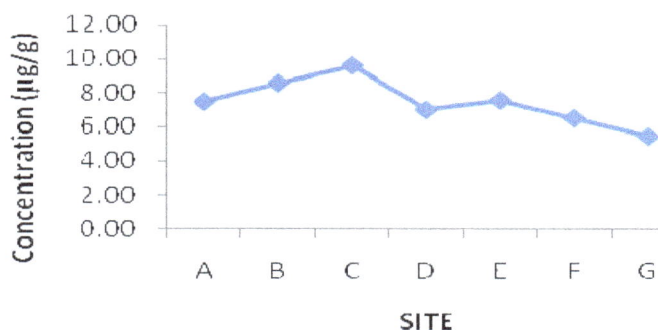

Figure 5. Concentration of Mn across the sampling site.

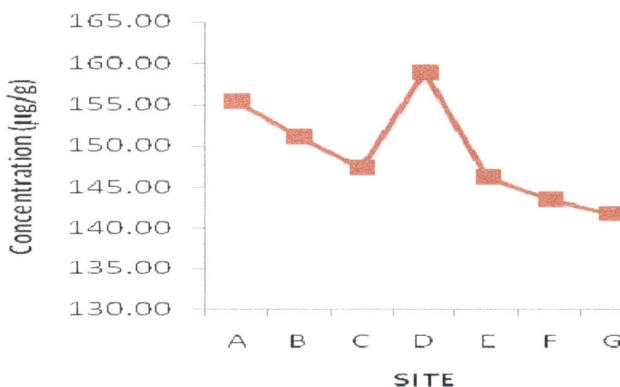

Figure 6. Concentration of Fe across the sampling site.

Analysis of variation of metals among the site indicated significant variations ($p < 0.05$) for all metals except Cd and Cu. The soil pH is generally slightly alkaline across the sites with exception of Site A that is slightly acid.

The roadside soil lead level ranges from a very low concentration of 1.59 (Site G, low traffic volume) to a high concentration of 12.10 µg/g (Site A, high traffic volume). The high mean value of the concentrations attested to the overall high level of contaminations of this metal in the roadside environments. This is in agreement with the report of Lagerwerff and Specht (1970). The high concentrations of lead observed could be attributed to lead particle from gasoline combustion which consequently settles on roadside soils. The case of Site A being more predominant could be attributed to the heavy traffic of the route. At the same way, vehicles are often moving slowly as a result of the heavy traffic jam in this area and this may account for the high level of lead. This is in line with Francek (1992) report that traffic junction and cross roads, records higher levels of metals. Also, it is therefore not surprising that high level of Pb is associated with sites B and C junctions which serve as a mini garage for heavy trucks. This is in addition to the auto mechanic work that dominates the business in the area. Conversely, Site G has the least concentration value of Pb (1.59 µg/g) and this could be expected, because of the least volume of

vehicles recorded on the road. Though, the high mean soil lead level confirms that the roadside environment is generally lead enriched despite a relatively low traffic volume compared to other studies (Ho et al., 1987). The mean lead level observed is far lower than that reported by Ho et al. (1987) and Francek (1992).

Zn ranged from 5.67 (Site G) to 12.88 µg/g (Site C). This value is small compared with many other studies (Jaradat and Momani, 1999; Bai et al., 2008). In this study, the Pb/Zn ratio in soil was less than unity with exception of Site A, which is contrary to report that soil-lead pollution may be caused by automobiles (Jaradat and Momani, 1999). However, other reports found a ratio of less than unity, which was related to the local conditions (Davies, 1984; Hewitt and Candy, 1990). Since no major industry exists in the study areas such as smelting operations, we may assume that the primary sources of Zn are probably the attrition of motor vehicle tire rubber exacerbated by poor road surfaces, and the lubricating oils in which Zn is found as part of many additives such as zinc dithiophosphates.

Also, the mobility of the metal depends on soil pH and also depends on the organic matter and granulometric

Table 2. Correlation analysis among traffic volume and metals

Parameter	Traffic volume (V)	Pb	Zn	Mn	Cu	Cd	Fe
Traffic volume (V)	1.000						
Pb	.939**	1.000					
Zn	.808*	.674	1.000				
Mn	.758*	.544	.860*	1.000			
Cu	.365	.352	-.029	.349	1.000		
Cd	-.059	-.081	-.110	-.351	-.419	1.000	
Fe	.572	.648	.430	.264	.154	-.344	1.000

** Correlation is significant at 0.01, * Correlation is significant at 0.05.

composition of the soil. Acidic pH makes easier the solubilization of the Zn compounds, although the soils in this study are alkaline with exception of Site A that is slightly acidic. An indication that Zn and other metals remain in soils for a longer.

The Jos city roadside copper level ranges from 1.01 µg/g at Site F to 2.19 µg/g at Site B, along Yakubu Gowon road which serves as one of the major roads with associated ancillary vehicle workshops located along it. Compared to other studies, the value of Cu is less than 27, 61, 24 and 29.7 µg/g reported in literature (Ward et al., 1977; Ndiokwere, 1984; Davies et al., 1985; Jaradat and Momani, 1999), though comparable to 2.78 µg/g obtained along Sixao highway in Southwest China (Bai et al., 2008).

Cd on the hand, was obtained from all the sites, range of 5.15 (Site E) - 5.79 µg/g (Site F) was found. Soils in this study exhibited higher levels of contamination than 0.75 µg/g (Jaradat and Momani, 1999), 2.11 µg/g (Amusan et al., 2003) and 0.88 µg/g (Bai et al., 2008). However, the level of Cd in this study is comparable to 6.8 µg/g reported in North Wale (Davies, et al., 1985) and about 5 times that reported by Ndiokwere (1984). In absence of any major industry in the sampling sites, the levels of Cd could be due to lubricating oils and/or old tires that are frequently used on the rough surfaces of the roads which increase he wearing of tires.

Mn and Fe obtained from this study ranged from 5.51 (Site G) - 9.61 µg/g (Site C) and 141.80 (Site G) - 159.00 µg/g (Site D), respectively. Both Fe and Mn form the composition of soils in northern Nigeria, there availability in a trace amount as obtained in this study could be due to local condition of soil weathering.

Correlation calculations, among concentrations of the heavy metals in surface soil and traffic volume (V) are shown in Table 2. This coefficient measures the strength of a linear relationship between any two variables on a scale of -1 (perfect inverse relation) through 0 (no relation) to +1 (perfect sympathetic relation). In this study, the raw geochemical data was used in calculating the correlation coefficient using the SPSS (Statistical

Program for the Social Sciences) computer software package (SPSS Inc., version 13). The results showed significant positive correlations ($p < 0.05$) are present between V: Pb, Zn and Mn, and Mn and Zn. Though mild positive and negative correlation were observed between metals, negative correlation between Cd and V indicate that other sources other than automobile emission could contribute to concentration of this metals in the soils. Significant positive correlations between metals and V could indicate possible contamination of the soils by automobile emissions.

Conclusion

The result of this study generally revealed the presence of all metals with exception of Co and Ni in the roadside soils. The concentrations of the metals in the soils are in the order of Fe > Zn > Pb > Mn > Cd > Cu. The level of Cd at the moment is high compared to other studies in Nigeria. Hence, possible accumulation in the soil and transfer to plants growing along the edge of the highway could occur as a result continual usage of the road by automobile. This can also lead to accumulation of the metal in the tissues of organisms that feed on the plant and other plants growing along the highway. This can be transferred to other consumers in the food chain (Akinola and Adedeji, 2007).

REFERENCES

Adefolalu AA (1980). Transport and rural integrated development In: proceedings of the National Conference on: Integrated Rural Dev. Women Dev., 1: 294-299.

Akinola MO, Adedeji OA (2007). Assessment of lead concentration in panicum maximum growing along the Lagos-Ibadan expressway, Nigeria. Afr. J. Sci. Technol., 8(2): 97-102.

Amusan AA, Bada SB, Salami AT (2003). Effects of traffic density on heavy metal content of soil and vegetation along roadsides in Osun State, Nigeria. West Afr. J. App. Ecol., 4: 107-114.

Bai J, Cui B, Wang Q, Gao H Ding Q (2008). Assessment of heavy metal contamination of roadside soils in Southwest China. Stoch. Environ. Res. Risk Ass. DOI 10.1007/s00477-008-0219-5.

Davies BD, Elwood PC, Gallacher J, Ginnver RC (1985). Environ.

Pollut., 9: 255-266.

Davies BE (1984). Urban Ecol., 6: 285-294.

De Kimple CR, Morel JF (2000). Urban soil management: a growing concern. Soil Sci., 165: 31-40.

Francek MA (1992). Soil lead levels in a small environment: a case study from Mt. pleasant, Michigan. Environ. Pollut. J., 76: 251-257.

Grace N (2004). Assessment of heavy metal contamination of food crops and vegetables from motor vehicle emissions in Kampala City, Uganda. Department of Botany Makerere University, Kampala. A technical report submitted to IDRC-AGROPOLIS .

Gray CW, Mclaren RG, Roberts AHC (2003). Atmospheric accessions of heavy metals to some New Zealand pastoral soils. Sci. Total Environ., 305: 105-115.

Harrison RM, Laxen DPH, Wilson SJ (1981). Chemical Association of Lead, Cadmium, Copper, and Zinc in Street Dusts and roadside soils. Environ. Sci. Technol., 15: 1378-1383.

Hewitt CN, Candy GB (1990). Soil and street dust heavy metal concentrations in and around Cuenca, Equador. Environ. pollut., 63: 129-136.

Ho YB, Tai KM (1988). Elevated levels of lead and other. Metals in roadside soils and grass and their use to monitor aerial metal depositions in Hong Kong. Environ. Pollut. J., 49: 37-51.

Jaradat QM, Momani KA (1999). Contamination of roadside soil, plants and air with heavy metals in Jordan, a comparative study. Turk. J. Chem., 23: 209-220.

Lagerwerff JV. Specht AW (1970). Contamination of roadside soil and vegetation with Cadmium, Nickel, Lead and Zinc. Environ. Sci. Technol., 4:583-586.

Li XD, Poon CS, Pui SL (2001). Heavy metal contamination of urban soils and street dusts in Hong Kong. App. Geochem.. 16: 1361-1368.

Mabogunje AL (1980). "Development process-a spatial perspective". Hutchinson and Co publishers Ltd. pp. 234-244.

Manta DS, Angelone M, Bellanca A, Neri R, Sprovieri M (2002). Heavy metal in urban soils: A case study from the city of Palermo (Sicily), Italy. Sci. Total Environ., 300: 229-243.

Mateu J, Forteza R, Cerda V, Colom-altes M (1995). Comparison of various methods for the determination of inorganic species in airborne atmospheric particulates. Water, Air, Soil Pollut., 84: 61-79.

Ndiokwere CL. (1984). A study of heavy metal pollution from motor Vehicle emission and its effect on vegetation and crops in Nigeria. Environ. Pollut., (7): 35-42.

Otte ML, Bestebroar SJ, Linden Vander JM, Rozema J, Broekman RA (1991). A survey of zinc, copper and cadmium concentrations in salt marsh plants along the Dutch coast. Environ. Pollut., 72: 175-189.

Sakagami KI, Eamada R, Kurobe T (1982). Heavy metal contents in dust fall and soil of the National Park for Nature Study in Tokoyo. Mitteeilungen der Deutschen Bodenkundlichen Gessellscaft., 33: 59-66.

Ward NI, Brook RR, Roberts E, Boswell C (1977). Environ. Sci. Technol., 11: 917-920.

3

Evaluation on rare earth elements of Brazilian agricultural supplies

C. Turra*, E. A. N. Fernandes and M. A. Bacchi

Centro de Energia Nuclear na Agricultura, Universidade de São Paulo, Piracicaba, SP, Brazil.

This work focuses on the determination of rare earth elements (REE) in Brazilian agricultural supplies by instrumental neutron activation analysis (INAA). The results obtained have shown that La, Ce, Nd, Sm, Eu, Tb, Yb, Lu and Sc are present within a large range of mass fractions in the agricultural supplies analysed. The thermophosphate and single superphosphate showed the highest mass fractions of REE. Considering the recommended dose and long-term use, NPK fertilizers, single superphosphate and thermophosphate can significantly increase the content of REE in soil and may cause harmful effects to environment and humans.

Key words: Agricultural supplies, rare earth elements, lanthanoids, phosphate fertilizer, INAA.

INTRODUCTION

The rare earth elements (REE) are in Group IIIB of the periodic table, including Sc, Y and the lanthanoids (La, Ce, Pr, Nd, Pm, Sm, Eu, Gd, Tb, Dy, Ho, Er, Tm, Yb, Lu) (IUPAC, 2005). REE have similar physico-chemical properties (Evans, 1990) and, despite the denomination, they cannot all be considered rare in nature (Hu et al., 2004), since the abundance of Ce in soil is similar to Cu and Zn (Tyler, 2004). Despite being present in plants, REE are not considered essential. Most of the uptake occurs by roots (Wyttenbach et al., 1998), but it can also occur through the surface of leaves after spraying or atmospheric deposition (Chua, 1998). Anyhow, REE have been used in agriculture applied to leaves, seeds and roots (Diatloff et al., 1996).

The application of fertilizers containing REE has being claimed to increase growth and productivity of plants (Challaraj et al., 2010), however this effect is not yet clear (Tyler, 2004; Shi et al., 2006). On the other hand, the accumulation of REE in the soil may have a toxic effect on macro fauna (Li et al., 2010) and micro fauna (Xu and Wang, 2001; Chu et al., 2001). Moreover, excessive application in agriculture may cause harmful effects to humans by bioaccumulation along the food chain.

Studies conducted in areas rich in REE reported that, constant exposure can cause damages in the circulatory, immunologic (Zhang et al., 2000), digestive (Zu-Yi and Xu-Dong, 2009; Li et al., 2010), respiratory (Censi et al., 2011), and nervous systems (Zhu et al., 2005), as well as can decrease intelligence quotient in children (Fan et al., 2004), and can start development of arteriosclerosis and pneumoconiosis (Sabbioni et al., 1992).

Mixture of REE in fertilizers has been used in Chinese agriculture to improve the nutrition of plants for more than 25 years. In 1986, the first commercial fertilizer with REE was registered in China under the name "Changle", having in its composition La_2O_3 (25 to 28%), CeO_2 (49 to 51%) and Nd_2O_3 (15 to 17%) (Hu et al., 2004). Studies have reported both positive (Hu et al., 2004; Wang et al., 2001) and negative effects (Diatloff et al., 1995; Barry and Meehan, 2000; Babula et al., 2008) in plants after application of REE fertilizers. Two rice areas receiving REE fertilizers presented mass fractions of La in soil (42.1 and 83.3 mg kg^{-1}) at critical levels in terms of environmental safety (Zeng et al., 2006). In Brazil, as in many other countries, there is no recommendation to add REE to fertilizers.

Nevertheless, besides providing nutrients for plants, fertilizers can also have impurities as metals and other elements. There is little information regarding the presence of REE in agricultural supplies available worldwide.

---

*Corresponding author. E-mail: christian.turra@gmail.com.

Table 1. Agricultural supplies collected for this study, main chemical composition and recommended use.

Supplies	Chemical composition	Recommended use
Agro-silicon	36% CaO, 9% MgO, 23% SiO_2	Corrective
Calcium	22 % Ca	Fertilizer
Calcium nitrate	24% Ca, 17% N, 59% O	Fertilizer
Chicken manure	2-3.5 % N, 2-4% P_2O_5, 1- 2% K_2O	Fertilizer
Copper oxychloride	48 – 52 % Cu	Fertilizer and fungicide
Cow manure	1-2% N, 0.8-1.4% P_2O_5, 1-1.8% K_2O	Fertilizer
Cromcitrus	2 – 4% Mn, 3 – 5% Zn	Fertilizer
Dolomite lime 1	25 – 35% CaO, 12 – 15% MgO	Corrective
Dolomite lime 2	25 – 35% CaO, 12 – 15% MgO	Corrective
Hydrated lime	40 % Ca, 17% Mg	Fertilizer and fungicide
Iron	6 % Fe	Fertilizer
Lime sulphur powder	3.5% Ca (total) and S	Fertilizer and insecticide
Magnesium nitrate	10.7% N, 15.5% Mg	Fertilizer
Magnesium sulphate	9.4% Mg, 12% S	Fertilizer
Monoammonium phosphate (MAP)	48% P_2O_5, 9% N	Fertilizer
NPK	%N, %P_2O_5, %K_2O variable	Fertilizer
Reactive phosphate	24 – 27% total P_2O_5	Fertilizer
Single superphosphate	18% P_2O_5, 18-20% Ca, 12-20% S	Fertilizer
Sulphur	30 – 40 % S	Fertilizer
Thermophosphate	16-18% P, 16-18% Ca, 6-7%Mg	Fertilizer

Therefore, it is important to assess the contents of REE in these supplies in order to regulate their correct use in agriculture. This work focuses on the determination of REE by instrumental neutron activation analysis (INAA) in some agricultural supplies used in Brazil. The determination will allow better use of agricultural supplies and the identification of possible sources of REE contamination.

MATERIALS AND METHODS

Agricultural supplies were sampled in different farms located in the state of São Paulo, Brazil. Samples are listed in Table 1, together with the main chemical composition and the recommended use of each product. At Centro de Energia Nuclear na Agricultura (CENA), in Piracicaba, materials were oven-dried and the particle size was reduced to less than 0.5 mm. Test portions of 300 mg were directly weighed into high purity polyethylene vials. For quality control purposes, certified reference materials of geological matrices (IAEA-Soil 7 and SRM-2710 Montana Soil) were added to the analytical series.

Empty vials were also included for blank correction. Irradiation was carried out for 8 h at a thermal neutron flux of 8.5×10^{12} cm^{-2} s^{-1} in the nuclear research reactor IEA-R1 of the Instituto de Pesquisas Energéticas e Nucleares, Comissão Nacional de Energia Nuclear (IPEN/CNEN), São Paulo. The material was transported back to CENA, where the induced radioactivity was measured at four decay periods, that is 4, 7, 15 and 40 days after irradiation. Germanium detectors with 50 and 55% relative efficiencies in the 1332 keV photopeak from ^{60}Co were used for measurements. Mass fractions were calculated by the k_0-method using the in-house software package Quantu (Bacchi and Fernandes, 2003).

RESULTS AND DISSCUSSION

Data obtained for the certified reference materials (Table 2) demonstrated that the analytical procedure was adequate for the determination of nine REE, that is, La, Ce, Nd, Sm, Eu, Tb, Yb, Lu and Sc, since there is a good agreement with the values provided in the certificates. For some samples, it was necessary to correct the results of La, Ce, Nd and Sm due to uranium interference, which has to be considered when the concentration of U exceeds 5 mg kg^{-1} (Kuleff and Djingova, 1990). The highest values of U were found in thermophosphate 2 (49.3 ± 1.7 mg kg^{-1}), thermophosphate 146.2 ± 1.4 mg kg^{-1}), NPK 4-20-20 (39.4 ± 1.1 mg kg^{-1}), reactive phosphate (39.3 ± 1.4 mg kg^{-1}) and single superphosphate (38.2 ± 1.7 mg kg^{-1}).

The largest relative interference was observed for Sm (16%) in hydrated lime. For Nd, the largest interference was 8% (in hydrated lime), the same maximum value found for Ce (in lime and limestone 2). For La, the relative interference was lower than 2% in all samples. Considering these results, it can be assumed that, after suitable correction, the uranium interference did not impair the quality of results.

Results showed that La, Ce, Nd, Sm, Eu, Tb, Yb, Lu and Sc are present at variable levels in the agricultural supplies (Table 3), being higher mass fractions found in phosphates and NPK fertilizers. The two limes analyzed showed higher mass fractions of La, Ce and Nd and lower of Eu compared to the values obtained in Florida,

Table 2. Mass fractions and standard deviations (mg kg^{-1}) of REE obtained for the reference materials IAEA Soil 7 and SRM 2711 Montana Soil compared to data provided in the respective certificates

Element	IAEA Soil 7		SRM 2711 Montana Soil	
	Obtained value	Confidence internal	Obtained value	Information value
La	27.7 ± 0.6	27 – 29	40.7 ± 0.8	40
Ce	60.5 ± 1.1	50 – 63	71.5 ± 1.2	69
Nd	31.1 ± 0.6	22 - 34	32.0 ± 0.7	31
Sm	5.0 ± 0.2	4.8 – 5.5	6.1 ± 0.2	5.9
Eu	1.02 ± 0.03	0.9 – 1.3	1.08 ± 0.04	1.1
Tb	0.63 ± 0.04	0.5 – 0.9		-
Yb	2.37 ± 0.06	1.9 – 2.6		-
Lu	0.32 ± 0.02	0.1 – 0.4		-
Sc	8.33 ± 0.04	6.9 – 9.0	9.07 ± 0.05	9.0

Table 3. Mass fractions and standard deviations (mg kg^{-1}) of REE determined in agricultural supplies, also including values from literature for NPK fertilizers or lime

	La	Ce	Nd	Sm	Eu	Tb	Yb	Lu	Sc
Agrosilicon	9.3	77.8	<11	1.74	0.34	0.06	1.13	0.17	3.55
	0.2	1.6		0.04	0.01	0.005	0.007	0.044	0.001
Calcium nitrate	2.17	3.10	<1.95	0.18	0.033	<0.031	<0.077	<0.005	<0.004
	0.001	0.02		0.0025	0.001				
Calcium	2.09	2.69	<1.68	0.13	0.031	<0.029	<0.06	<0.012	<0.008
	0.02	0.10		0.0017	0.0001				
Chicken manure	1.59	4.5	2.8	0.31	0.094	0.065	0.18	0.085	0.48
	0.07	0.02	0.26	0.004	0.0007	0.004	0.015	0.067	0.002
Copper oxychloride	0.50	0.61	<2.26	0.08	0.02	<0.04	<0.2	<0.018	0.069
	0.02	0.011		0.0008	0.002				0.0004
Cow manure	6.76	23.8	6.57	0.92	0.28	0.25	0.66	0.16	3.02
	0.38	0.59	1.16	0.09	0.013	0.017	0.001	0.009	0.037
Cromcitrus	<0.25	<7.0	<30	<0.26	<0.42	<0.74	<0.56	<0.23	<0.05
Dolomite lime 1	7.77	17.1	8.0	1.35	0.22	0.25	0.67	0.14	2.01
	0.078	0.28	0.25	0.007	0.007	0.046	0.007	0.006	0.057
Dolomite lime 2	6.78	13.1	7.14	1.35	0.24	0.15	0.55	0.12	0.72
	0.071	0.28	0.028	0.064	0.001	0.005	0.003	0.002	0.004
Hydrated lime	2.5	3.4	2.36	0.35	0.07	<0.028	0.146	<0.017	0.47
	0.04	0.25	0.37	0.01	0.001		0.001		0.005
Iron	<0.48	<0.68	<1.2	<0.022	<0.007	<0.036	<0.135	<0.023	0.042
									0.0008
Magnesium sulphate	<0.61	<0.55	<1.57	<0.19	<0.007	<0.026	<0.020	<0.011	<0.01
Magnesium nitrate	<0.31	<0.22	<1.8	<0.02	<0.008	<0.02	<0.023	<0.04	<0.04

Table 3. Contd.

Monoammonium phosphate (MAP)	0.97 0.10	7.3 0.43	<2.4	<0.21	<0.008	<0.03	<0.025	<0.028	0.69 0.01
NPK 10 10 10	372 3	770 9	360 2.14	47 0.86	10.6 0.14	3.04 0.064	1.78 0.071	0.53 0.024	8.18 0.071
NPK 20 0 10	0.31 0.02	0.45 0.06	<5.8	0.05 0.0009	0.008 0.0003	<0.04	<0.06	<0.03	0.04 0.001
NPK 12 6 12	237 4	538 13	250 3	32 2	7.00 0.12	1.82 0.04	1.08 0.01	0.26 0.002	5.41 0.004
NPK 4 20 20	421 0.02	875 2	443 7	70 1	18.3 0.43	6.32 0.03	5.21 0.06	1.54 0.04	23.1 0.002
NPK 4 14 8	534 1	1181 7	571 0.01	77 1.5	17.1 0.01	4.57 0.03	2.03 0.13	0.55 0.04	15.2 0.37
NPK 25 5 20	91 0.78	211 1.4	101 12.3	13 1.1	2.73 0.2	0.72 0.017	<0.09	0.12 0.046	3.54 0.28
NPK 20 5 15	368 8.6	839 1.4	379 11.5	45 0.9	10.8 0.072	2.45 0.05	1.71 0.11	0.12 0.02	6.8 0.043
Single superphosphate	673 7	1499 29	770 16	122 2	32.5 0.15	6.53 0.19	8.8 0.08	1.84 0.07	24.6 0.005
Sulphur	<0.50	<0.27	<2.8	<0.06	<0.029	<0.027	<0.06	<0.052	<0.010
Reactive phosphate	99 0.1	164 0.7	110 0.7	17 1.7	3.59 0.04	2.38 0.01	6.1 0.6	1.4 0.07	3.7 0.02
Thermophosphate 1	755 4.3	1575 7	748 31	105 3	24.5 0.07	8.03 0.04	10.4 0.07	1.8 0.09	23.9 0.14
Thermophosphate 2	673 8	1430 14	687 11	102 3	24.5 0.28	8.54 0.02	10.8 0.07	2.2 0.17	26.5 0.07
NPK[25]	<0.5-619	<3 -744	<5 - 214	<0.1-42	0.2 - 12	0.5-3.2	0.2-5.5	0.16-0.85	-

United Sates (Wutscher and Perkins, 1993). Sc is known as tracer of geological material (Fernandes, 1992). Typical values for Sc in soils lie between 0.5 and 45 mg kg^{-1} (Markert, 1998). The agricultural supplies with geological origin presented values within this range. The mass fractions of REE in NPK fertilizers analyzed in Spain showed values varying in a similar range compared to the NPK fertilizers evaluated here (Table 3). NPK is considered a compound fertilizer because it contains two or more primary nutrients (Otero et al., 2005), and the content of REE is somehow related to the amount of P in the fertilizer. Total REE in NPK and phosphate fertilizers are shown in Table 4. The results evidenced that the highest amounts of REE are normally found in fertilizers containing phosphorous in the composition. For the NPK fertilizers without phosphorous ("zero" of P in NPK formulation), the mass fractions of REE were significantly lower than for NPK fertilizers with

Table 4. Total REE found in fertilizers ordered according to the content of P_2O_5.

Fertilizer	% P_2O_5	Σ REE (mg kg^{-1})
NPK 20 0 10	0	3
NPK 25 5 20	5	420
NPK 20 5 15	5	1640
NPK 12 6 12	6	1060
NPK 10 10 10	10	1550
NPK 4 14 8	14	2370
Single superphosphate	18	3070
Thermophosphate 1	18	3190
Thermophosphate 2	18	2895
NPK 4 20 20	20	1810
Reactive phosphate	27	390
Monoammonium phosphate (MAP)	60	12

phosphorus. The results suggest that, the amount of REE in NPK fertilizers depends on P contents, but is most probably also related to industrial process and P source used in the formulation. For instance, monoammonium phosphate presented a low level of REE because its industrial process involves purification steps.

Brazil is the fourth world consumer of fertilizers, after China, India and the United States (ANDA, 2006). The country imports part of the phosphate applied in agriculture, mostly coming from Morocco (51%), Algeria (18%), Israel (17%) and also from Togo and Tunisia (13% together). The crops that most use fertilizers with P_2O_5 in the country are soybean, corn and sugar cane. The distribution pattern of REE in phosphates and NPK fertilizers is shown in Figure 1. The mass fraction of Ce was higher than La for all samples analyzed. In general, the samples showed the same REE distribution pattern of chondrite. However, Tb and Yb presented a different behavior in NPK fertilizers. The value of Yb (0.248 mg kg^{-1}) in chondrite is higher than Tb (0.058 mg kg^{-1}) and in the NPK fertilizers the values of Tb were higher than Yb (Table 4).

A study of phosphorite deposits in Pakistan showed enrichment of light REE (Javied et al., 2010). Some samples of NPK fertilizers analyzed in Spain also showed different distribution patterns for Tb and Yb, while others have La higher than Ce (Otero et al., 2005). The mass fractions of REE determined by INAA in Egyptian phosphate fertilizer ingredients were in descending order La>Ce>Sc>Eu (Abdel-Haleem et al., 2001), a different distribution pattern compared to that found in this study. According to Asher et al. (1990), the recommended dose for REE fertilizer (e.g. Changle) is 600 to 675 g/ha, which represents an application of about 150 to 170 g of REE per hectare. The recommended application of REE is dependent on their bioavailability in soils, which in turn is mainly dependent on the exchangeable fraction of REE, strongly affected by the physico-chemical properties of

soils (Liang et al., 2005). Considering that the recommended dose of thermophosphate ranges from 300 to 1500 kg/ha (Mitsui fertilizers, 2006), the amount of Ce applied would vary from 0.47 to 2.4 kg/ha and the total amount of REE coming from fertilizer would vary between 1 and 5 kg/ha. The average REE content in NPK fertilizers with phosphorous in the formulation was 1450 mg kg^{-1}. For a recommended dose between 100 and 400 kg/ha, the input of all REE is between 0.15 and 0.58 kg/ha for each application. Considering the large consumption of NPK fertilizers, that is 8.9 million tons per year in Brazil (ANDA, 2009), significant amounts of REE have been applied via this fertilizer.

Due to a low risk of REE leaching in groundwater (Hu et al., 2006) and the low REE mobility in soil (Kabata-Pendias and Mukherjee, 2007), the continuous application may lead to accumulation of REE in agricultural soils. In general, phosphate fertilizers are rich in REE (Tyler, 2004), however such elements are usually not considered as nutrients. Therefore, the possible effects of REE in plants and environment are not taken into account.

Conclusions

The amount of REE in the agricultural supplies was largely variable. The highest mass fractions of REE were found in fertilizers containing phosphate. In general, the REE distribution pattern in the supplies was similar to that of chondrite, except for Tb and Yb in NPK fertilizers. Considering the recommended dose and long-term use, NPK fertilizers, single superphosphate and thermophosphate can provide more REE than the recommended doses found in literature. Such high inputs can significantly increase the content of REE in soil, which may cause harmful effects to environment and humans.

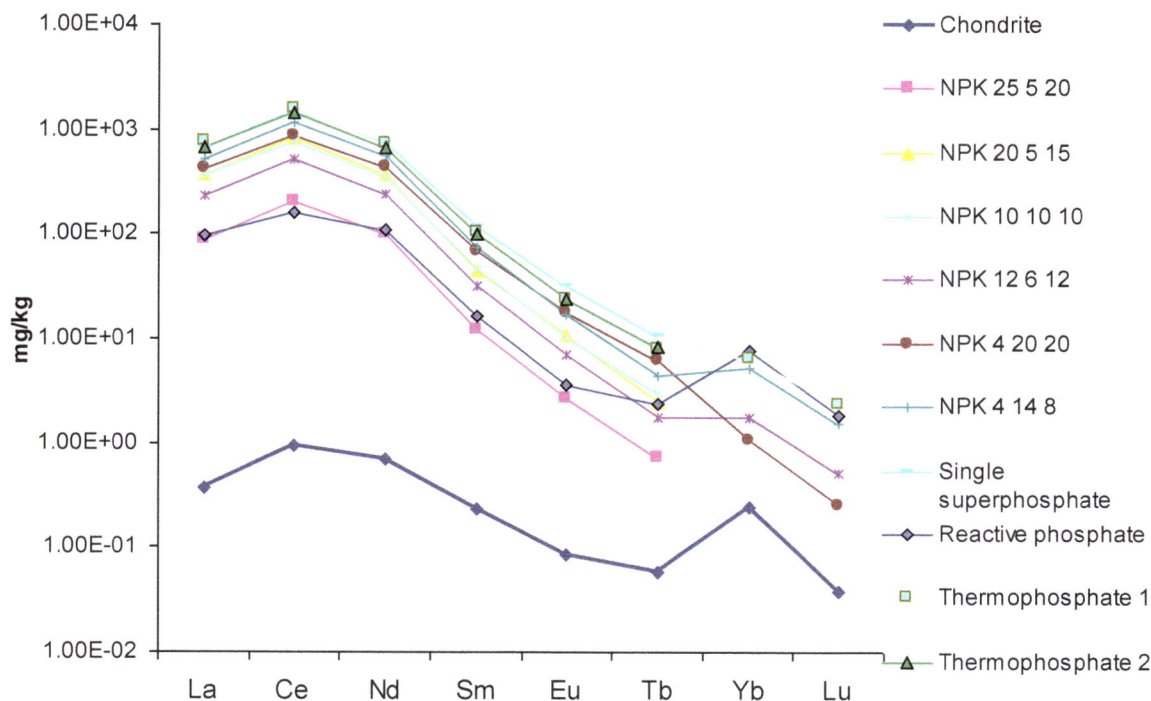

Figure 1. Distribution pattern of REE in NPK and phosphate fertilizers.

REFERENCES

Abdel-Haleem AS, Sroor A, El-Bahi SM, Zohny E (2001). Heavy metals and rare earth elements in phosphate fertilizers components using instrumental neutron activation analysis. Appl. Radiat. Isotopes, 55: 569–573.

ANDA (Agência Nacional de Difusão dos Adubos) (2006) Available at http://www.anda.org.br/estatisticas.aspx Accessed 12 April 2010

Asher CJ, Reghenzani JR, Robards KH, Tribe DE (1990). Rare earths in Chinese agriculture. In 'Report of an Australian Mission which visited China from 31 March–15 April 1990'. Aust. Acad. Technol. Sci. Eng., p. 54.

Babula P, Adam V, Opatrilova R, Zehnalek J, Havel L, Kizek R (2008) Uncommon heavy metals, metalloids and their plant toxicity: a review. Environ. Chem Lett., 6: 189-213.

Bacchi MA, Fernandes EAN (2003). Quantu – design and development of a software package dedicated to k0-standardized NAA. J. Radioanal. Nucl. Chem., pp.257-577.

Barry MJ, Meehan BJ (2000). The acute and chronic toxicity of lanthanum to Daphnia carinata. .IUPAC. Recommendations 2005. Cambridge CB4 0WF, UK Chemosphere, 41: 1669-1674.

Censi P, Tamburo E, Speziale S, Zuddas P, Randazzo LA, Punturo R, Cuttitta A, Aricò P (2011). Yttrium and lanthanides in human lung fluids, probing the exposure to atmospheric fallout. J. Hazard. Mater., 186(2-3): 1103-1110

Challaraj EES, Vignesh V, Anandkumar B, Maruthamuthu S (2010). Bioaccumulation and physiological impact of rare earth elements on wheat (triticum aestivum) Indian J. Plant. Physiol., 15(2): 177-180

Chu HY, Zhu JG, Xie ZB, Cao ZH, Li ZG, Zeng Q (2001). Effects of lanthanum on soil microbial biomass carbon and nitrogen in red soil. J. Rare Earths, 19: 63 – 66

Chua H (1998). Bio-accumulation of environmental residues of rare earth elements in aquatic flora Eichhornia crassipes (Mart) Solms in Guangdong Province of China. Sci. Total Environ., 214:79-85.

Diatloff E, Smith FM, Asher CJ (1995). Rare earth elements and plant growth. Effects of lanthanum and cerium on root elongation of corn and mungbean. J. Plant Nutr., 18: 1963–1976.

Diatloff E, Asher CJ, Smith FW (1996). Concentrations of rare earth elements in some Australian soils. Aust. J. Soil Res., 34:735-747.

Evans CH (1990). Biogeochemistry of the lanthanides. Plenum Press. New York. 8: 444

Fan GQ, Yuan ZK, Zheng HL, Liu ZJ (2004). Study on the effects of exposure to rare earth elements and health-responses in children aged 7– 10 years, J. Hyg. Res., 33:23– 28.

Fernandes EAN (1992). Scandium as tracer in the sugar and alcohol agroindustry. J. Radioanal. Nucl. Chem., pp.168-41.

Hu Z, Richter H, Sparovek G, Schnug E (2004). Physiological and biochemical effects of rare earth elements on plants and their agricultural significance: a review. J. Plant Nutr., 27: 183–220.

Hu Z, Haneklaus S, Sparovek G, Schnug E (2006). Rare earth elements in soils. Commun. Soil Sci. Plant Anal., 37(9): 1381-1420.

Javied S, Waheed S, Siddique N, Shakoor R, Tufail M (2010). Measurement of rare earths elements in Kakul phosphorite deposits of Pakistan using instrumental neutron activation analysis J Radioanal. Nucl. Chem., 284: 397–403

Kabata-Pendias A, Mukherjee AB (2007). Trace elements from soil to human. New York: Springer, 450 p.

Li J, Hong M, Yin X, Liu J (2010). Effects of the accumulation of the rare earth elements on soil macrofauna community. J. Rare Earths 28:957-964

Li XB, Yu XL, Wang TG, Su SY, Zhao K, Zhang XJ (2010). Effect of rare earth La2O3 on digestive system of the red-eared slider. Adv. Mater. Res., 108-111: 866

Liang T, Zhang S, Wang L, Kung H, Wang Y, Hu A, Ding S (2005). Environmental biogeochemical behaviors of rare earth elements in soil plant systems. Environ.Geochem. Health, 27: 301- 311

Markert B (1998). Distribution and biogeochemistry of inorganic chemical in the environment. In: G. Schüürmann and B. Markert (eds.), Ecotoxicology, John Wiley & Sons, Inc. and Spektrum Akademischer Verlag, Weinheim. pp. 165–222.

Mitsui Fertilizer Company (2006). Available at http://www.fertilizantesmitsui.com.br/template.php?page=produtos/produtos_yoorin Accessed 22 November 2009

Otero N, Vitòria L, Soler A, Canals A (2005). Fertiliser characterisation:

major, trace and rare earth elements. Appl. Geochem., 20: 1473–1488.

Sabbioni E, Pietra R, Gaglione P, Vocaturo G, Colombo F, Zanoni M, Rodi F (1992). Long-term occupational risk of rare-earth pneumoconiosis. A case report as investigated by neutron activation analysis. Sci. Total Environ., 26: 19-32.

Shi P, Huang ZW, Chen GC (2006). Influence of lanthanum on the accumulation of trace elements in chloroplasts of cucumber seedling leaves. Biol Trace Elem. Res., 109: 181–188

Tyler G (2004). Rare earth elements in soil and plant systems – a review. Plant Soil 267: 191–206.

Wang Z, Liu D, Lu P, Wang C (2001). Accumulation of rare earth elements in corn after agricultural application. J. Environ. Qual., 30: 37-45.

Wyttenbach A, Furrer V, Schleppi P, Tobler L. (1998). Rare earth elements in soil and in soil-grown plants. Plant Soil. 199: 267–273.

Wutscher HK, Perkins RE (1993). Acid extractable rare-earth elements in Florida citrus soils and tREE. Commun. Soil Sci. Plan., 24: 2059-2068.

Xu X, Wang Z (2001). Effects of lanthanum and mixtures of rare earths on ammonium oxidation and mineralization of nitrogen in soil. Eur J Soil Sci., 52: 323 – 329

Zeng Q, Zhu JG, Cheng HL, Xie ZB, Chu HY (2006). Phytotoxicity of lanthanum in rice in haplic acrisols and cambisols. Ecotoxicol. Environ. Saf., 64: 226-233.

Zhang H, Feng J, Zhu WF, Liu C, Xu S, Shao P, Wu D, Yang W, Gu J (2000) Chronic toxicity of rare-earth elements on human beings - Implications of blood biochemical indices in REE-high regions, South Jiangxi. Biol. Trace Elem. Res., 73:1-17.

Zhu W, Xu S, Shao P, Zhang H, Wu D, Yang W, Feng J, Feng L (2005) Investigation on liver function among population in high background of rare earth area in south China. Biol. Trace Elem. Res., 104:1-7.

Zu-Yi C, Xu-Dong Z (2009). Accumulation and toxicity of rare earth elements in liver. Acta Ecology of Domestic Animal, 04. Available at:http://en.cnki.com.cn/Article_en/CJFDTOTAL-JCST200904028.htm

4

Effect of concentration and contact time on heavy metal uptake by three bacterial isolates

L. O. Odokuma[1]* and E. Akponah[2]

[1]University of Port-Harcourt, Port-Harcourt, Rivers State, Nigeria.
[2]Delta State University, Abraka, Delta State, Nigeria.

The effect of heavy metal concentration and contact time (exposure period) on heavy metal up take by pure cultures of three bacteria (*Pseudomonas*, *Bacillus* and *Aeromonas*) isolated from a crude oil impacted brackish aquatic system in the Niger Delta were investigated. Heavy metals employed included metals found in this Bonny light crude oil (Fe, Zn, Cd, Cu, Ni and Pb). Accumulation of these metals was gradual and the amount increased in direct proportion to initial metal concentration up to an extent that ranged from 1 - 100 (mg/l) after which uptake remained either constant or declined. Maximum uptake of Fe, Pb, Cd, Cu, Zn and Ni were obtained at initial concentrations of 10, 10, 10, 100, 100 and 10 (mg/l), respectively and the values were 8.75, 0.01, 1.3, 0.06, 4.2 and 0.001 milligram per gram dry weight (mg/g dry wt) of *Bacillus* cells. For *Pseudomonas* sp. initial metal concentration that resulted in maximum uptake were 10 mg/l (Fe), 1 mg/l (Zn,) 10 mg/l (Pb), 0.1 mg/l (Cd), 10 mg/l (Cu) and 10 mg/l (Ni). Values accumulated at these concentrations were 14, 0.7, 1.0, 0.08, 0.3 and 0.11 (mg/g dry wt), respectively. Whereas, maximum amounts accumulated by *Aeromonas* sp. were 1.65, 0.1, 0.001, 0.9, 0.2 and 0.013 (mg/g dry wt) respectively. The respective initial concentration that yielded these uptake values was 100, 10, 1.0, 10, 10 and 10 (mg/l). Contact duration increased the amount of metal bioconcentrated by each test organism. At all tested concentrations maximum uptake of Fe, Zn, Cu, Cd, Pb and Ni by *Bacillus* were at the 8th, 8th, 8th, 4th, 4th and 4th hours of exposure respectively. Slight decreases in uptake were noticed on further incubation beyond these durations. Maximum accumulation of Fe, Zn, Cu, Cd, Pb and Ni by *Pseudomonas* sp. were obtained at incubation durations of 8th, 12th, 12th, 8th, 2nd and 24th hours. 12th, 12th, 12th, 4th, 8th and 4th h were the incubation periods that resulted in maximum bioconcentration of Fe, Zn, Cu, Cd, Pb and Ni by *Aeromonas* sp. The three test organisms presented distinct uptake capacities which decreased thus: *Pseudomonas* sp. > *Bacillus* sp. ≥ *Aeromonas* sp. Affinities of *Bacillus* sp., *Pseudomonas* sp. and *Aeromonas* sp. for the various heavy metals followed the pattern Fe > Zn > Cd > Cu > Ni > Pb, Fe > Pb ≥ Zn > Cu > Ni ≥ Cd and Fe > Cd > Cu ≥ Zn > Ni > Pb respectively. Results showed that heavy metal concentrations between 10 – 100 mg/l and exposure periods of between 4 - 12 h depending on the metal and the test organism rapidly promoted accumulation in heavy metal polluted sites.

Key words: Accumulation bioconcentration, heavy metal, toxicity, contact time, bacteria.

INTRODUCTION

The threat of heavy metal pollution to public health and the ecosystem (Davies et al., 2006; Mallampoti et al., 2007; Vinodhini and Narayanam, 2008) has led to an increased interest in developing systems that can remove or neutralize its toxic effects in soil, sediments and wastewater (Kotrba et al., 1999; Gonzalez et al., 2005). Unlike organic contaminants, which can be degraded to harmless chemical species, heavy metals can neither be degraded nor destroyed. Remediation of the pollution they cause can therefore only be brought about mainly by bioaccumulation, biosorption or their re-speciation into less toxic forms (Abou-Shanab et al., 2007; Adenipekun and Isikhuemhen, 2008).

Microorganisms can physically remove heavy metals

*Corresponding author. E-mail: luckyodokuma@yahoo.co.in.

from solution through either bioaccumulation or bio-sorption. Bioaccumulation is the retention and concentration of a substance by an organism (Odiete, 1999). Metals are transported from the outside of microbial cell, through the cellular membrane and into the cell cytoplasm, where the metal is sequestered and therefore immobile bioaccumulation consume cellular energy while biosorption does not (Bull et al., 1981; Mullen et al., 1989; Strandberg et al., 1995).

A number of factors may affect the rate at which heavy metals are accumulated in bacterial cells. Several authors such as Galun and Gelun (1987), Asku et al. (1992) and Fourest and Roux (1992) have shown that biosorption rates are greatly influenced by factors such as temperature, pH, metal concentration, contact time as well as concentration of biomass. Asku et al. (1992) reported that temperature ranging from 20 - 33 °C seems not to influence the biosorption performance but at higher temperatures, biosorption increases. Galum and Gelun, 1987 have shown that pH affects the solution chemistry of the metals, the activity of the functional group in the biomass and the competition of metallic ion. Fourest and Roux (1992) reported that biomass concentration in solution seems to influence the specific uptake. They also attributed the interference between the binding sites resulting from an increase in biomass concentration to metal concentration shortage in solution. Bioaccumulation increased with increase in contact time for low concentration of metals (0.001 - 1.0 mg/l) of CdS, ZnO and Fe_2O_3 until equilibrium is achieved (Odokuma and Emedolu, 2005). However, at higher concentration of 10.0 - 100 mg/l of these metal salts, bioaccumulation decreased with increasing contact time (Odokuma and Emedolu, 2005). Bioaccumulation varies between individual organisms as well as between individual species. Odokuma and Emedolu (2005) have shown that bioaccumulation of CdS at 0.1, 11.0, 10.0 100 mg/l was 80, 64, 52 and 15% for *Bacillus* and 94, 57, 73 and 13% for *Aeromonas*. Some chemicals bind to specific site whereas, others move freely in and out. Those im-mediately eliminated do not bioaccumulate. Odokuma and Emedolu (2005) showed that the bioaccumulation of heavy metals by *Bacillus* and *Aeromonas* followed the trend Fe > Zn ≥ Cd > Pb

The bioconcentration potentials of heavy metals (Cd, Pb, Zn and Fe) associated with crude oil by *Bacillus*, *Chromobacteiium*, *Staphylococcus* and *Aeromonas* have been investigated by Odokuma and Emedolu (2005). The authors observed that both *Bacillus* and *Aeromonas* were tolerant to the salts of these metals at 0.001, 0.1, 1.0 10, 100 and 1000 and thus bioconcentrated these metals. The objective of this study was to enhance the bio-concentration capabilities of these three organisms by determining the optimum concentrations and contact time required for bioconcentration of these metals. Armed with this information, one could suggest the inclusion of these organisms in the protocol for bioreme diation of heavy metal contaminated environments in the Niger Delta.

MATERIALS AND METHODS

Source of samples

River water used in this study was collected from the New Calabar River located in Choba, River State of Nigeria. The New Calabar River is a short coastal river about 200 km in length. It is under the influence of tidal cycles and consists of brackish water due to marine water influx during the tidal cycles.

Water sample collection

Composite samples of water (0 - 30 cm depth) were collected from the river about 1 km southwest of the University of Port- Harcourt with 100 ml sterile plastic containers. Composite sampling was performed by collecting ten samples about 1 m apart and pooling them together.

Digestion of crude oil

Crude oil used was Bonny light. This was obtained from the Nigerian National Petroleum Corporation (NNPC), Port Harcourt. Crude oil sample was digested using the wet oxidation method employing a mixture of concentrated nitric acid, perchloric acid and sulphuric acid (APHA, 1998). The heavy metals in the crude oil were determined using model AA320 atomic absorption spectrophotometer.

Chemical reagents

All chemical reagents employed in this study were products of Aldrich chemical Co, Milwaukee, USA, BDH chemicals, Poole, England and Sigma chemical company St. Louis Missouri, USA.

Preparation of stock solution of heavy metal salts

The heavy metal salts employed in this study include: Nickel tetraoxosulphate (vi) salt ($NiSO_4$), copper (ii) tetraoxosulphate (vi) salt ($CuSO_4$), lead trioxonitrate (v) salt ($PbNO_3)_2$, iron (ii) tetraoxosulphate (vi) salt ($FeSO_4$), cadmium tetraoxosulphate (vi) salt ($CdSO_4$) and zinc tetraoxosulphate (vi) $ZnSO_4$ salt. A weight of each of these heavy metal salts that gave a 1 g of each of the respective heavy metal (metal without the salt) was weighed and dissolved in 1000 ml of deionised water. These were left to stand for 30 min to obtain complete dissolution. This was followed by sterilization by membrane filtration (0.2 μm pore size Aerodisc).

Isolation of heavy metal resistant bacteria from the river water

Heavy metal resistant bacteria were isolated from the river water. An amount (0.1 ml) of a 10^{-4} dilution of the river water sample was inoculated onto the surface of freshly prepared nutrient agar plates using the spread plate technique (APHA, 1998). The plates were incubated at 37 °C for 24 h. Isolated colonies were purified by two subsequent single colony transfers. Pure colonies were specifically transferred into nutrient agar slants. The slants were incubated at 37 °C for 18 - 24 h. These served as the stock cultures and were stored at 4 °C in the refrigerator. Pure bacterial isolates were characterized and identified using criteria as in Holt et al. (1994).

Nine predominant bacterial genera; *Achromobacter*, *Alcaligenes*, *Aeromonas*, *Bacillus*, *Chromobacterium*, *Corynebacterium*, *Micrococcus*, *Pseudomonas* and *Serratia* were identified.

Preparation of standard inoculum of isolates

A loopful of cells from the respective stock cultures were incubated into 100 ml sterile nutrient broth contained in 250 ml Erlenmeyer flasks. The flasks were incubated at 37°C for 24 h with intermittent shaking. At the end of the incubation period, cells were harvested by centrifugation at 4000 rpm for 30 min and re-suspended in 100 ml sterile physiological saline. The total viable counts were carried out to estimate the number of viable organisms. During this process, the cultures were subjected to serial dilutions up to 10^6 dilutions. An amount (0.1 ml) from each dilution was inoculated by spread plate technique into freshly prepared nutrient agar plates, which were incubated at 37°C for 24 h. The dilutions produced between 30 - 300 colonies were chosen and served as inoculum for preliminary screening experiments.

Preliminary screening test

This was carried out to determine the isolates that possess resistance to all the heavy metals associated with the crude oil. One hundred millilitres of 1 mg/l of the respective heavy metal solutions were prepared as earlier described. Nine millilitres were dispensed into test tubes and sterilized. Controls contained 9 ml of physiological saline. One millilitre of respective standardized isolates' inoculum was then added and incubation followed immediately at a temperature of 25°C ± 2 for duration of 24 h. At the end of the incubation period, 0.1 ml were withdrawn and plated onto the surface of freshly prepared nutrient agar plates using the spread plate technique as described by APHA (1998). Incubation followed immediately at 25°C ± 2 for 18 - 24 h. Colonies formed were counted and percent log survival were calculated according to Williamson and Johnson (1981).

$$\% \log survival = \frac{\log of\ count\ in\ toxicant\ concentration}{Log of\ count\ in\ control} \times 100$$

Based on the results, *Pseudomonas*, *Bacillus* and *Aeromonas* were chosen for further studies.

Toxicity test

The preliminary range finding test was carried out to determine the lowest observed effect concentration (that is, the lowest concentration tested that had a significant effect) and the highest observed effect concentration (highest concentration tested beyond which inhibition was 100%). Lethal response (or percentage log survival) was used as an index of toxicity.

Preparation of toxicant concentration

Based on the result of the preliminary range finding test, various toxicant (metal) concentrations (0.001, 0.01, 0.1, 1.0, 10 and 100.0 mgl[-1]) of each heavy metal salts were prepared from the stock solution (1 g equivalent of heavy metal-determined from the heavy metal salt in 1000 ml of deionised water) of the heavy metals. These concentrations were employed for the percentage log survival test.

Percentage log survival test

The test was carried out to determine the lethal effect of the heavy metal salts on *Pseudomonas*, *Bacillus* and *Aeromonas* by monitoring the total viable count of the bacterial isolates with time, when exposed to varying concentrations of the various salts.

The experimental set up for each of the six heavy metal salt is as follows:

1. Ninety millilitres of the various toxicant concentration (0.001, 0.01, 0.1, 1.0, 10 and 100 mgl[-1]) contained in each of six 250-ml Erlenmeyer flask in duplicates were inoculated with 10 ml of standard inoculum of *Pseudomonas*, *Bacillus* and *Aeromonas*, respectively. Control consisted of a duplicate set of flask containing sterile normal saline without any toxicant added.
2. The flask was shaken and at exposure times of 0, 2, 4, 12, 24, 48 and 96 h, 1 ml was aseptically withdrawn from each flask and serially diluted (10[-1] - 10[-4]) using the ten-fold dilution technique. Using the spread plate method, each dilution was plated out in duplicate set of plates containing sterile nutrient agar medium. The plates were incubated at 37°C for 18 - 24 h and discrete colonies that developed were counted. Results were expressed as colony forming units per millilitre (cfu/ml). The same procedure was carried out for the controls.
3. Percentage log survival was calculated by dividing the log of counts in each toxicant concentration by the log of count in the control (zero toxicant concentration) and multiplying by 100 (Williamson and Johnson, 1981).

$$\text{Percentage log survival} = \frac{\log C}{\log c} \times 100$$

where C = count in each toxicant concentration
c = count in the control.

Metal uptake experiment

The methods of Kurek et al. (1982), Bauda and Block (1985) and Boularbah et al. (1992) were adopted with some modifications for the heavy metal uptake test.

Growth and preparation of test isolates

The biomass of the respective bacterial isolates were developed by growing each isolate in 250-ml Erlenmeyer flasks containing 100 ml of freshly prepared nutrient broth (pH 7.0) at 37°C for 18 - 24 h, under shaken conditions (120 rpm). Cells were harvested by centrifugation at 4000 rpm for 30 min. Harvested cells (biomass) were washed thrice with sterile phosphate buffered saline. The cells were resuspended in 100 ml-deionized water. This served as inoculum for the various bioaccumulation tests. To access viability of cells, 0.1 ml was plated onto surface of nutrient agar plates.

Metal solution

Stock solutions of the various heavy metal salts and their various concentrations were prepared. Solutions were adjusted to desired pH values (7) with 0.1 M sodium hydroxide and 0.1 M trioxonitrate (V) acid. The initial metal concentrations at the beginning of all experiments carried out were as calculated from the dilution.

Effect of initial heavy metal concentration on bioconcentration

The experiment was conducted to ascertain the effect of varying concentrations of heavy metal salts on bioconcentration potentials of the test organisms. The experimental set up for each of the heavy metal was as follows: From the various concentrations of the

Table 1. Response of Isolates to the toxicity of the various heavy metals.

Isolates	Fe	Zn	Cd	Cu	Ni	Pb	Control
Alcaligenes	++	++	-	-	-	-	+++
Aeromonas	+++	+++	++	+++	++	++	+++
Bacillus	+++	+++	+++	+++	+++	+++	+++
Achromobacter	+++	++	-	-	+	-	+++
Chromobacterium	++	+	-	+	-	-	+++
Corynebacterium	++	++	-	+	+	-	+++
Micrococcus	+++	++	+	++	+	-	+++
Pseudomonas	+++	+++	+++	+++	++	++	+++
Serratia	+++	+	-	+	+	-	+++

Key: +++ = > 70% log survival, ++ = 50-69% log survival, + = 30-49% log survival, - = < 29% log survival.

respective heavy metal salts, 40 ml were withdrawn using sterile pipette into duplicate set of 100-ml Erlenmeyer flask. 10 ml of each of the standard inoculum were then added. The control however contained 40 ml of sterile normal saline and 10 ml of the respective isolate. All flasks were incubated at 25 °C ± 2 for 24 h. At the end of the incubation period, cells were harvested by centrifugation at 4000 rpm for 30 min (using 800 D centrifuge), washed thrice in sterile PBS, dried, weighed, digested and analysed for heavy metal uptake. Metal uptake was expressed as mg metal/gram dry weight of cells (Vol1esky and May-Philips, 1995). All glasswares were rinsed with 0.1 M hydrochloric acid before and after each experiment to prevent sorption (precipitation) of metals on glass surface.

Effect of contact time

This experiment was done to determine the shortest exposure time for the respective isolate to maximally accumulate each of the various heavy metals (that is, to determine the equilibrium point/residence time of the metals in each test isolate). The experimental set up is as described as: Ten millilitres of the standard inoculum was contacted with 40 mls aliquots of varying concentrations of heavy metal solution in 100-ml Erlenmeyer flasks. These were then incubated at 25 °C ± 2 for different time intervals ranging from 1, 2, 4, 8, 12 and 24 h. At each of the exposure periods, cells were harvested, dried, weighed, digested and analysed for heavy metal uptake using AAS.

Statistical analyses

Analyses of variance (ANOVA) and Rank correlation coefficient (Finney, 1978) were employed to determine the existence of significant statistical variations in the results

RESULTS

The results of the preliminary screening for the resistance of isolates to the toxicity of the various heavy metals are presented in Table 1. The isolate obtained from the river water sample were *Achromobacter*, *Alkaligens*, *Aeromonas*, *Bacillus*, *Corynebacterium*, *Chromobacterium*,

Micrococus, *Pseudomonas* and *Serratia*. Three of the test isolates (*Aeromonas*, *Bacillus* and *Pseudomonas*) showed resistance to the six heavy metal salts; hence these three were selected for further experiments.

The effect of heavy metal concentration on the percentage (%) survival of organisms is presented in Figures 1a - 3f. Results showed that there was a general decrease in percentage survival with increase in concentration and contact time of heavy metals to all three organisms. Results showed that Fe and Zn were the least toxic to all three organisms while Pb, Cd and Ni were the most toxic to the organisms.

Results from the two-way analyses of variance showed a significant difference between % log survival and heavy metal concentration for *Bacillus*. The calculated F value (F_{cal} was 13.8 while the tabulated F value (F_{tab}) was 2.5. For *Pseudomonas* F_{cal} was 4.6 while F_{tab} was 2.5. For *Aeromonas* F_{cal} was 3.8 while F_{tab} was 2.5. There was also a strong positive correlation between metal concentration and % log survival

Figures 4a - f shows the effect of initial metal concentration on metal uptake by *Bacillus*. The data showed that high concentrations (10 - 100 mg/l) of heavy metals favoured the increase of metal uptake as against low concentrations (0.1 - 1.0 mg/l). Only the low concentration of Pb (1.0) promoted metal up-take.

In Figures 5a - f, the effect of initial heavy metal concentration on metal up-take by *Pseudomonas* is presented. Results showed that heavy metal concentrations (10 – 100 mg/l) favoured metal up-take. However, low concentrations (1.0 mg/l) of Pb, Cd and Fe also favoured accumulation of these metals by *Pseudomonas*.

In Figure 6a - f the effect of initial metal concentration on metal uptake by *Aeromonas* is presented. High metal concentrations (10 – 100 mg/l) favoured metal up-take in *Aeromonas*.

However, low concentrations (0.1 - 1.0 mg/l) of Pb, Cd and Fe also promoted heavy metal uptake by this

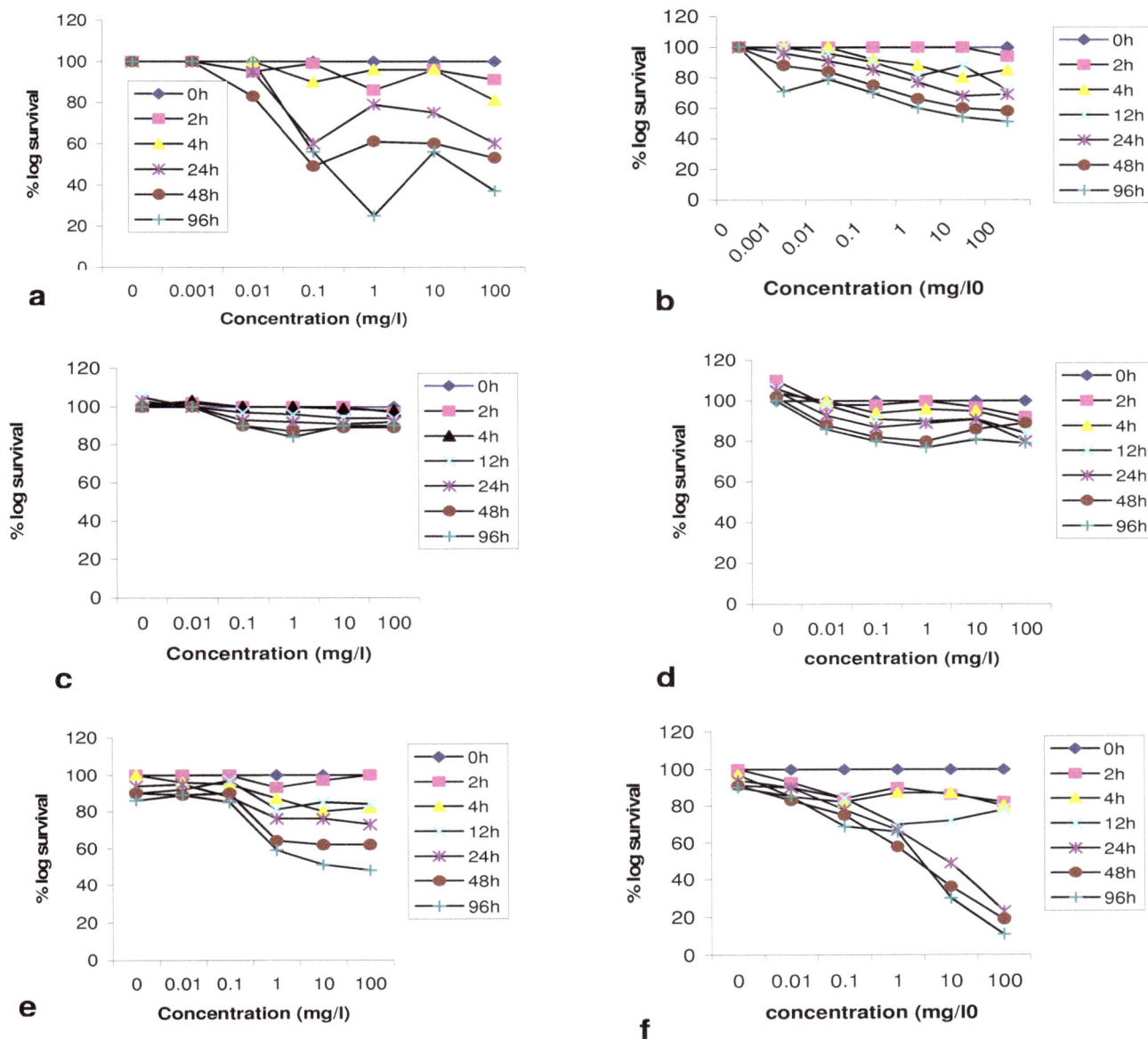

Figure 1. Effects of heavy metals; (a) $CuSO_4$, (b) Ni SO_4, (c) $FeSO_4$, (d) $ZnSO_4$, (e) $Pb(NO_3)_2$ and (f) $CdSO_4$; on the survival of *Bacillus* sp.

organism.

The effect of contact time on metal uptake by *Bacillus* is presented in Figure 7 (a – f). An increase in metal uptake with increase in contact time from 0 - 4 h followed by an equilibrium position till the 8 h then a decrease till the 24 h was observed. Similar results were observed in *Pseudomonas* (Figures 8a - f) and *Aeromonas* (Figures 9a -f) with very slight variations depending on the metal. At all tested initial concentrations, results showed that contact duration affected uptake a great deal.

There was a significant difference between metal uptake and contact time for all three organisms. For iron uptake by *Bacillus* F_{cal} was 13.8 while F_{tab} was 3.2. For

Pseudomonas F_{cal} was 13.2 while F_{tab} was 3.2. For *Aeromonas* F_{cal} was 8.2 while F_{tab} was 3.2. For cadmium uptake by *Bacillus* F_{cal} was 5.5 while F_{tab} was 4.1. For *Pseudomonas* F_{cal} was 5.5 while F_{tab} was 3.2 while for *Aeromonas* F_{cal} was 18.4 while F_{tab} was 4.1. For lead uptake by *Bacillus* F_{cal} was 4.2 while F_{tab} was 3.2. For *Pseudomonas* F_{cal} was 28.9 while F_{tab} was 3.2. For *Aeromonas* F_{cal} was 6.7 while F_{tab} was 3 .2

A strong positive correlation between metal uptake and contact time was observed until the 12th h at the various concentrations of heavy metals used for all organisms. The r value at 0.1.1.0, 10.0 and 100 mg/l of Fe for *Aeromonas* were 0.8811, 0.8712, 0.8276 and 0.8663,

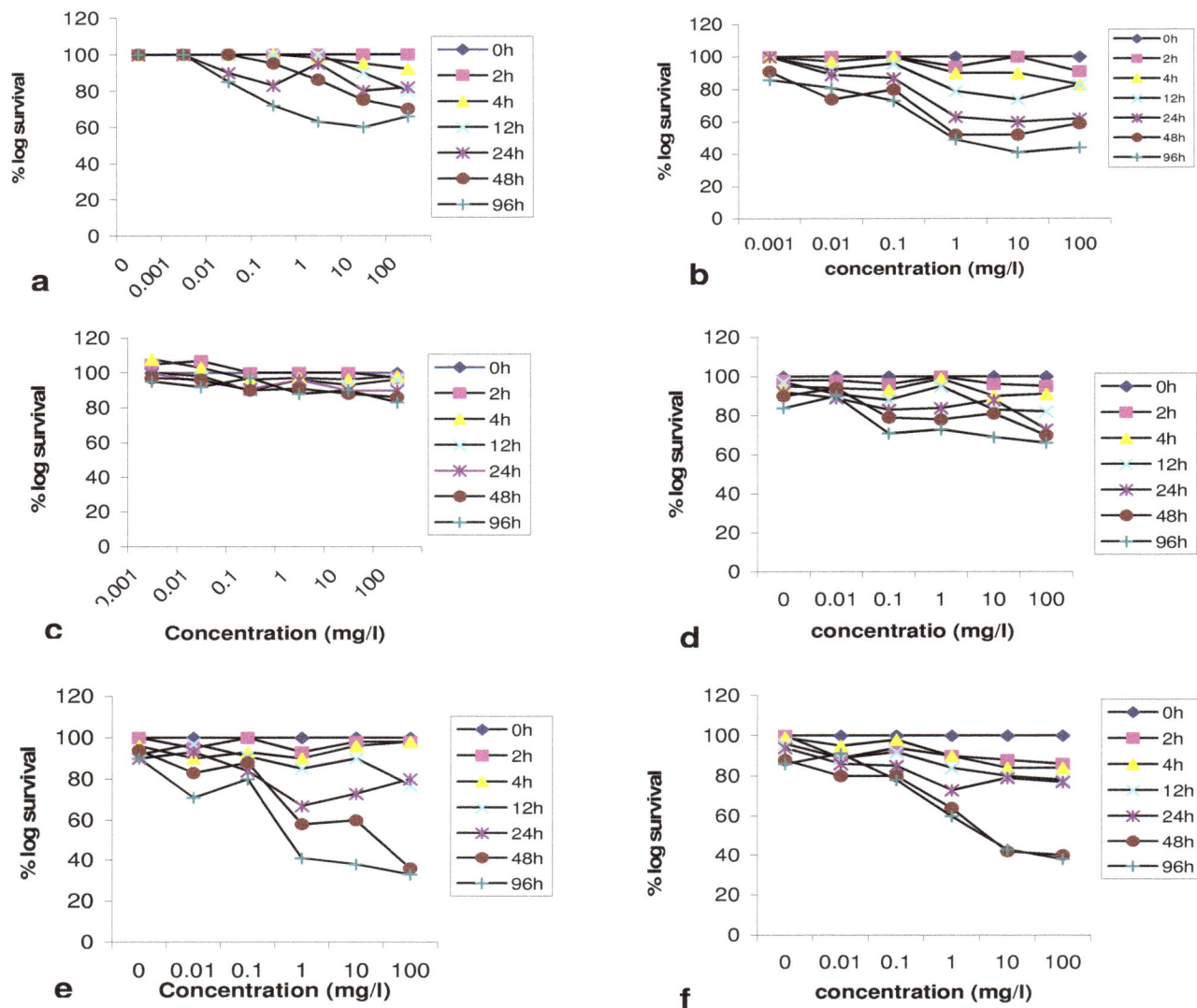

Figure 2. Effects of heavy metal salts; (a) $CuSO_4$, (b) $NiSO_4$, (C) $FeSO_4$, (D) $ZnSO_4$, (e) $Pb(NO_3)_2$ and (f) $CdSO_4$; on the survival of *Pseudomonas* sp.

respectively. The r values at similar concentrations for Zn for *Bacillus* were 0.7848, 0.7856, 0.7821 and 0.7950 respectively. Zinc uptake by *Pseudomonas* within 12 h exposure time showed positive r values of 0.8773, 0.8783, 0.8543, and 0.8594 respectively at metal concentrations of 0.1, 1.0, 10.0 and 100 mg/l. At similar concentrations metal uptake and contact time of Zn by *Aeromonas* showed strong positive concentrations of 0.8890, 0.8550, 0.7625 and 0.8127.

However for Cd, Cu and Pb correlation was less than 0.5 at these concentrations for all test organisms. For *Aeromonas* r values for Cd at these concentrations were 0.2356, 0.1296, 0.1092 and 0.2894. For *Bacillus* r values for Cd at 10.0 and 100 mg/l were 0.4887 and 0.3707 respectively. However at low concentrations of Fe, 0.0 and 1.0 mg/l r values were 0.8095 and 0.7856,

respectively. For *Pseudomonas* r values for Cd at all concentrations were 0.7865, 0.7900, 0.7895 and 0.7878 respectively. Correlation value for Pb uptake by *Bacillus* were negative at all concentrations, -0.41903, -02155, -0.3300 and -0.3498, respectively.

DISCUSSION

The results of the preliminary screening for resistance to the toxicity of the various heavy metals showed that six of the nine genera were not capable of survival when exposed to the various heavy metal salts associated with the crude oil sample within 24 h exposure duration.

Aeromonas, *Bacillus* and *Pseudomonas* were however, resistant to the toxicity of all the heavy metals within

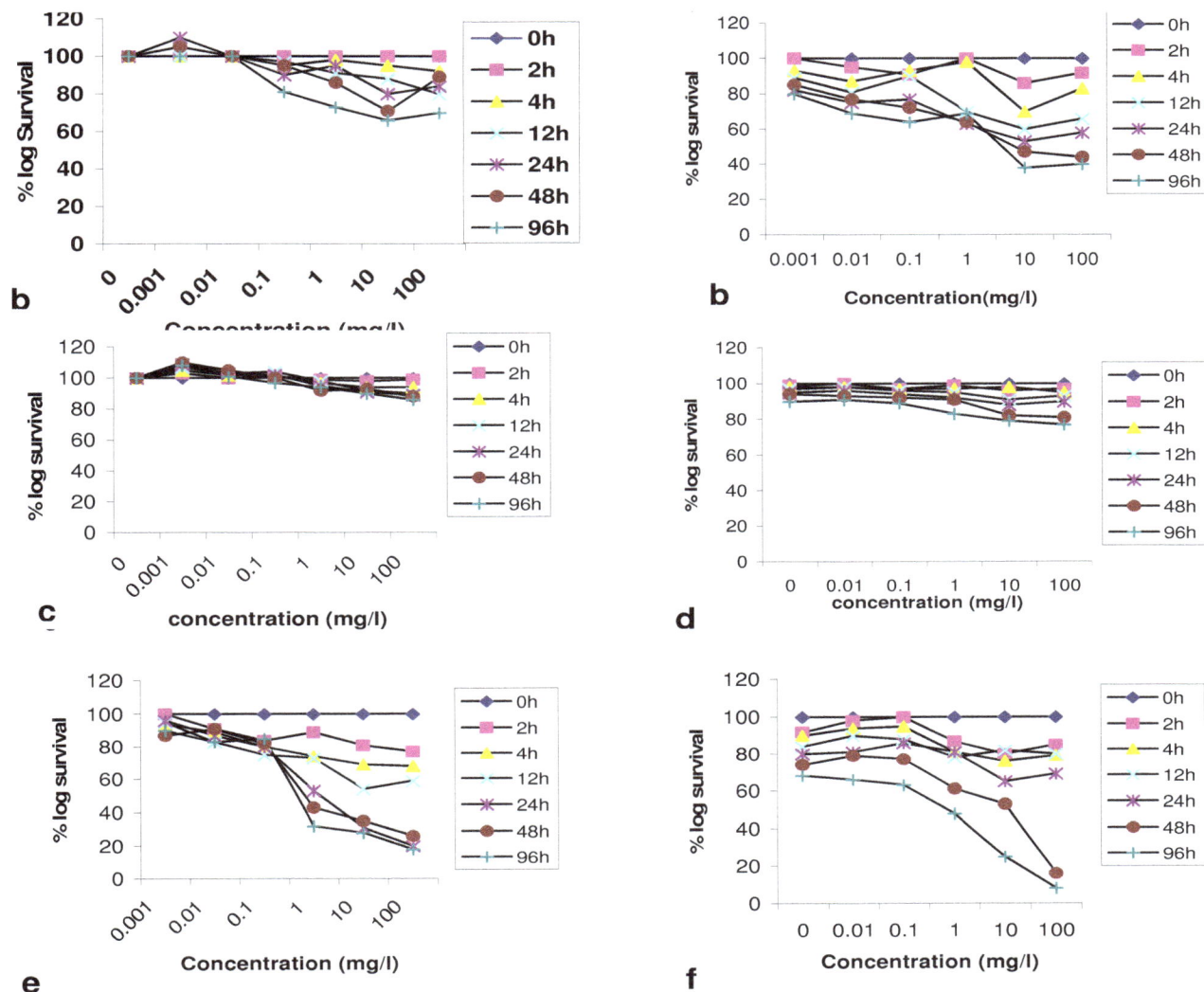

Figure 3. Effects of various heavy metals salts; (a) $CuSO_4$, (b) Ni SO_4, (c) $FeSO_4$, (d) $ZnSO_4$, (e) $Pb(NO_3)_2$, (f) $CdSO_4$; on the survival of *Aeromonas* sp.

same exposure period; hence, they were selected for the studies. Similar results had been reported by Odokuma and Ijeomah (2003), Odokuma and Emedolu (2005). In their reports *Bacillus* sp. and *Aeromonas* sp. were shown to be resistant to the toxicity of heavy metals. The persistence of these isolates in the presence of the respective heavy metals may be as a result of the possession of heavy metal resistant plasmids (Odokuma and Oliwe, 2003). The spore forming ability of *Bacillus* sp. might also, have contributed to its ability to survive when exposed to the various heavy metal salts (Stainer et al., 1982; Dutton et al., 1990). The resistance of *Pseudomonas* sp. to the toxicity of the various salts may be due to its ability to use diverse compounds (organic and inorganic) as sole carbon source (Schlegel, 1997). In addition to its genetic make-up, complexity of its cell wall being a Gram-negative organism may have contributed to

its resistance. Similar reasons can also be advanced for the resistance of *Aeromonas* sp to the toxicity of the various heavy metal salts.

All the nine isolates were observed to be resistant to Fe and Zn at the tested concentrations. This might be due to the fact that these heavy metal salts are important in certain biosynthetic activities in the organisms as well as being components of enzyme (Taylor et al., 1997). Thus, at such a low concentration, they have proved to be useful to the various test isolates. However, Pb Cd, Ni and Cu were observed to be highly toxic to *Alcaligenes* sp., *Achromobacter* sp., *Chromobacterium* sp., *Corynebacterium* sp., *Micrococcus* sp., and *Serratia* sp. as the percent survival of these respective isolates when exposed to 1 mg/l of the various heavy metal salts were less than 30% as shown in Table 1. Similar results had been observed by Odokuma and Ijeomah (2003). Results

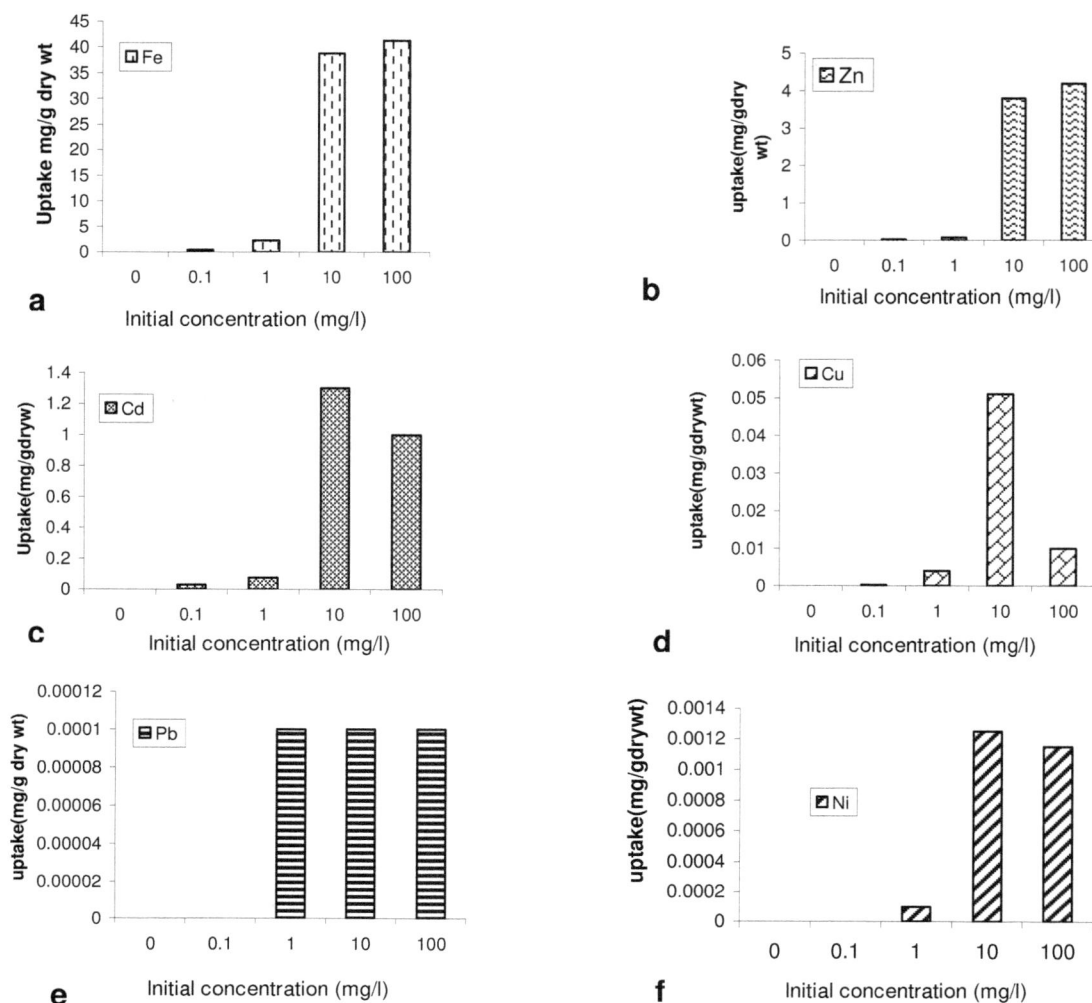

Figure 4. Uptake of various heavy metals; (a) Fe, (b) Zn, (c) Cd, (d) Cu, (e) Pb and (f) Ni; by *Bacillus* sp.

obtained therefore, suggest that the occurrence of these heavy metals in the environments will greatly reduce the population of these organisms and hence the microbial diversity of the affected ecosystem.

Results of the toxicity of the various metals to *Bacillus* sp., *Pseudomonas* sp. and *Aeromonas* sp. as presented in Figures 1a - 3f showed the mechanism of response of the isolates to the various logarithmic concentrations of the toxicants. Percentage survival decreased with increase in contact time as well as concentration (Buikema et al., 1982). The toxicity of the various heavy metals to *Bacillus* sp., *Pseudomonas* sp. and *Aeromonas* sp. respectively, followed the decreasing trend: Cd > Pb > Cu > Ni > Zn > Fe, Cd > Ni > Cu > Pb > Zn > Fe and Cd > Pb > Ni > Cu > Zn ≥ Fe respectively. Beyond the 24th h of exposure, percent log survival greatly decreased indicating that contact time is a crucial factor in establishing the resistance of organisms to the toxic pressure of the

metals. Thus, further experiments were conducted within duration of 24 h.

The initial bioaccumulation experiments carried out in this study revealed that all the three isolates were not only resistant to the toxicity of the various heavy metals within the duration of exposure, but also, had the capability of accumulating these heavy metals. It is well recognized that microorganisms have affinity for metals and can accumulate heavy and toxic metals by a variety of mechanisms (Simmons et al., 1995; Malekzadeh et al., 1995; Gupta and Keegan, 1998; Odokuma and Emedolu, 2005). Several principal sites of metal-complex formation in biological systems have been proposed (Vieira and Volesky, 2000). These include accumulation in the cell wall, carbohydrate or protein polyphosphate complexes, and complexion with carboxyl groups of the peptideglycan in the cell wall or entering into cells via an present study, an increasing uptake pattern was observed in

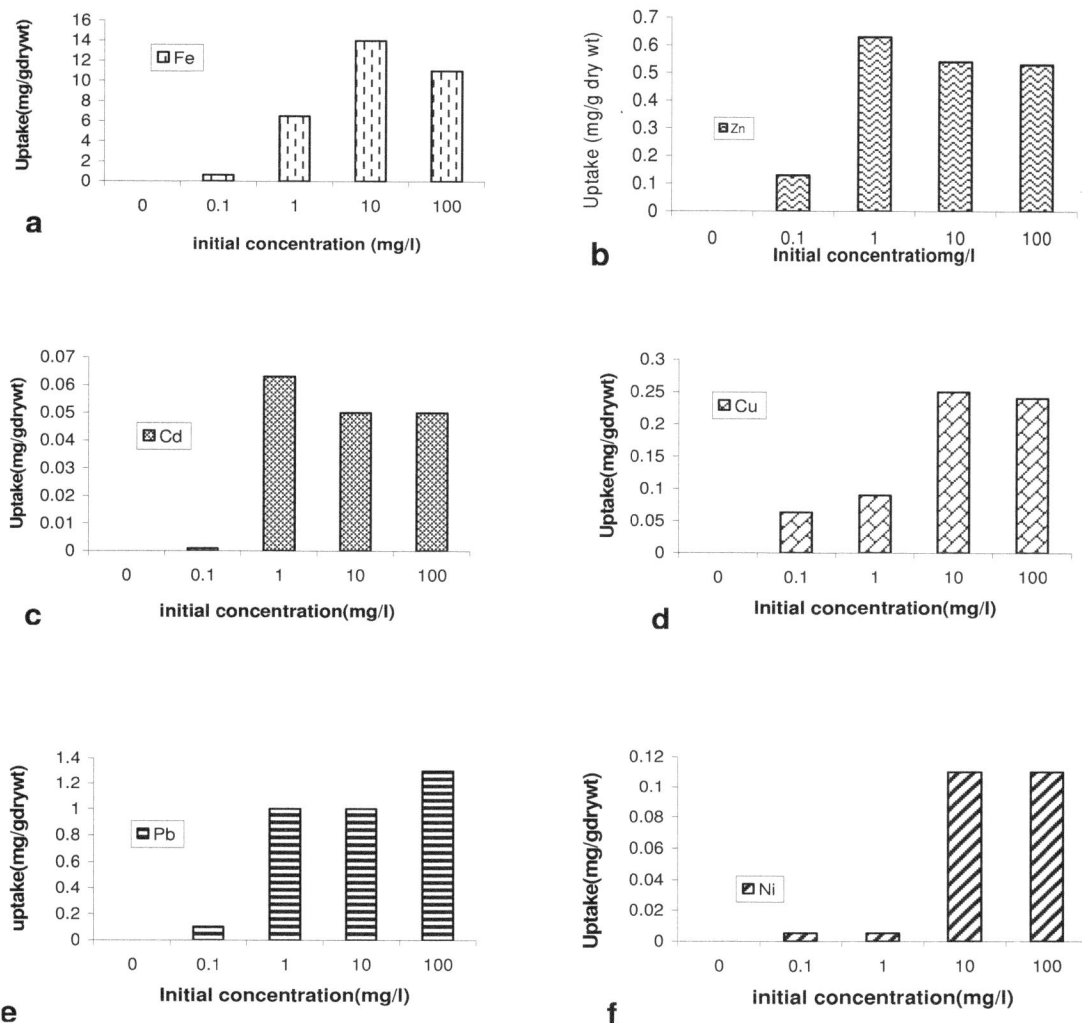

Figure 5. Uptake of various heavy metals by *Pseudomonas* sp. as a function of initial concentration. (a) Fe uptake (b) Zn uptake (c) Cd uptake (d) Cu uptake (e) Pb uptake (f) Ni uptake.

the respective test isolates as the initial concentration of the various heavy metal salts were increased. These obser-vations suggest that metal uptake may involve diffusion phenomenon whereby, metal ions move from regions of high concentrations to low concentrations and the fact that the steeper the concentration gradient, the more ra-pid is the movement of molecules or ions. (Taylor et al., 1997) Investigations indicated that under the conditions of test, maximum uptake was obtained at initial concen-trations that ranged from 1.0 - 10 mg/l after which uptake either remained constant (100.0 mg/l) or dropped to an extent. This might be suggestive of saturation of the test organism by the metal in question while the decrease in uptake at certain high initial concentration [as observed in Pb uptake by *Bacillus* sp. (Figure 4f) might be due to increasing toxic action of the metals with increase in their initial concentrations. Similar observations, had been made by Kaewehai and

Prasertson (2002), Al-Garni (2005) and Odokuma and Emedolu (2005).

Bacillus sp. showed a selective uptake: Fe > Zn > Cd > Cu > Ni > Pb. Iron > Pb >Zn > Cd > Cu > Ni and Fe > Cd > Cu > Zn > Ni > Pb were the trends demonstrated by *Pseudomonas* and *Aeromonas* respectively. All the three isolates accumulated Fe the most. This may be attributable to the low toxicity of this metal to the test organisms as revealed by the results of the toxicity tests. The production of siderophores by organisms especially *Pseudomonas* had been advanced for Fe accumulation (Ford and Mitchel, 1992). These authors also reported that in the absence of Fe, accumulation of its analogs, which include Cu and Cd, could be enhanced by siderophore production. This might have contributed to the uptake of these metals. Results of toxicity test energy-dependent mechanism (Silver, 1991). In the (Figures 1e, 2e and 3e) showed that Pb and Ni were the

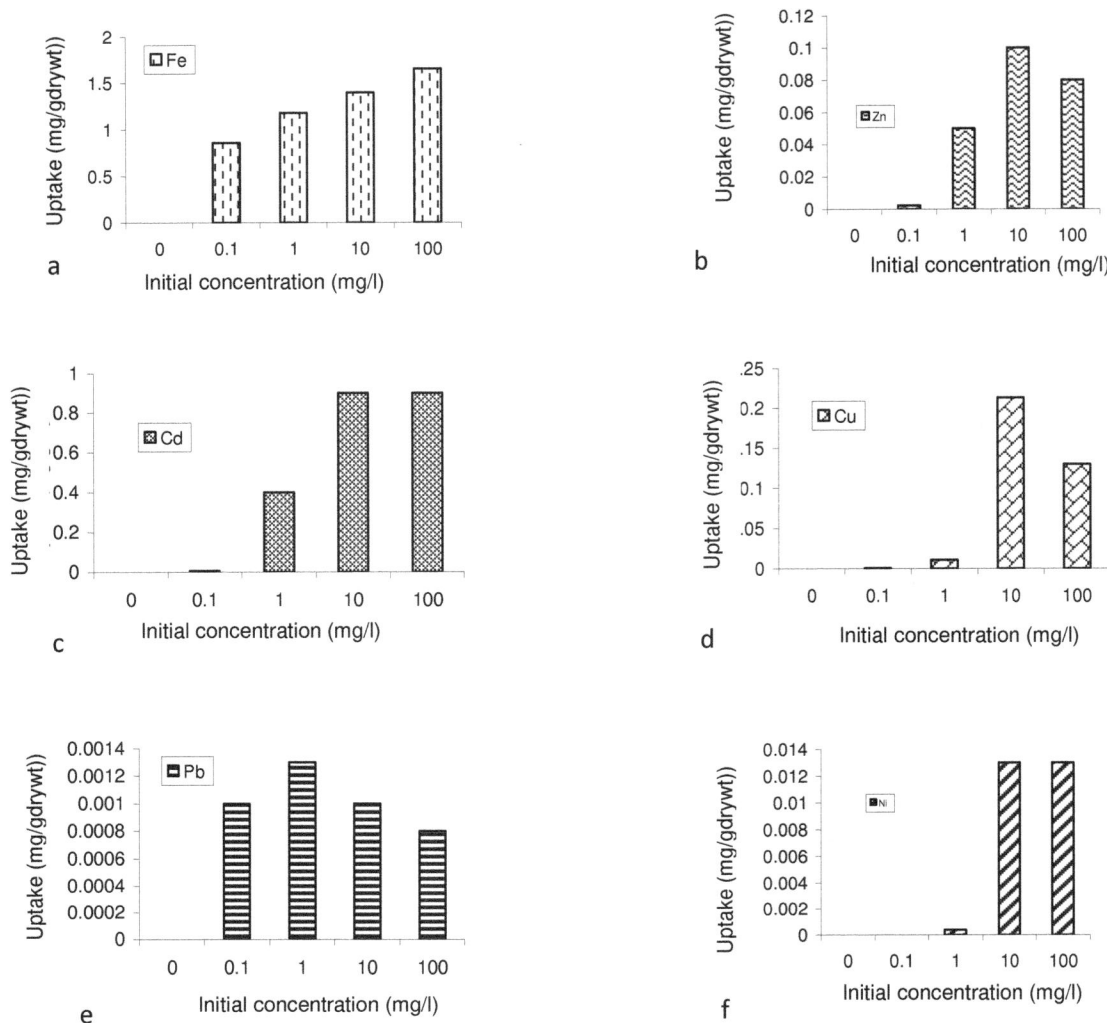

Figure 6. Uptake of various heavy metals; (a) Fe, (b) Zn, (c) Cd, (d) Cu, (e) Pb and (f) Ni; by *Aeromonas* sp. as a function of initial concentration.

most toxic to the test isolates. This might also be responsible the poor capability of *Bacillus* sp. and *Aeromonas* sp. in accumulating these metals. Again the high molecular weight of these metals might have resulted in this limitation observed in their uptake by the isolates. *Pseudomonas* however, demonstrated a good ability of accumulating this metal. A similar observation had been made by Malekzadeh et al. (1995). They reported that a strain of *Pseudomonas* was highly efficient in accumulating highly toxic uranium up to 174 mg/g dry weight of the bacterial biomass which far exceeded its ability to ac-cumulate less toxic metals (Cu and Cd). This is indicative of the fact that an organism may have specificity for a particular metal. Also, it has been reported that toxicity increases cell membrane permeability, hence toxicity can increase uptake if the toxic pressure is not sufficient to kill the organisms

(Malekzadeh et al., 1995). Thus, this probably accounts for the increased influx of Pb ions into *Pseudomonas* sp.

Data obtained on the effect of contact time on the uptake of the various heavy metals by the respective test isolates, showed rapid uptake up to contact time of between 4 and 8 h depending on the heavy metal as well as the test isolates. Time for attaining equilibrium was less than 24 h. The rate of metal uptake is influenced by agitation or shaking (Gadd, 1988) diffusion of metal through a hydrodynamic boundary layer around the biosorbent surface (Weber, 1985) and adsorption of metal ions by active sites of the biomass. In this case of metal accumulation by the tested bacteria, the shaken conditions employed, probably allowed a thorough mixing of solutes and biomass in the system and hence suppressing the kinetic limitations of bulk transport of metal ions across the cell membrane. Thus, equilibrium

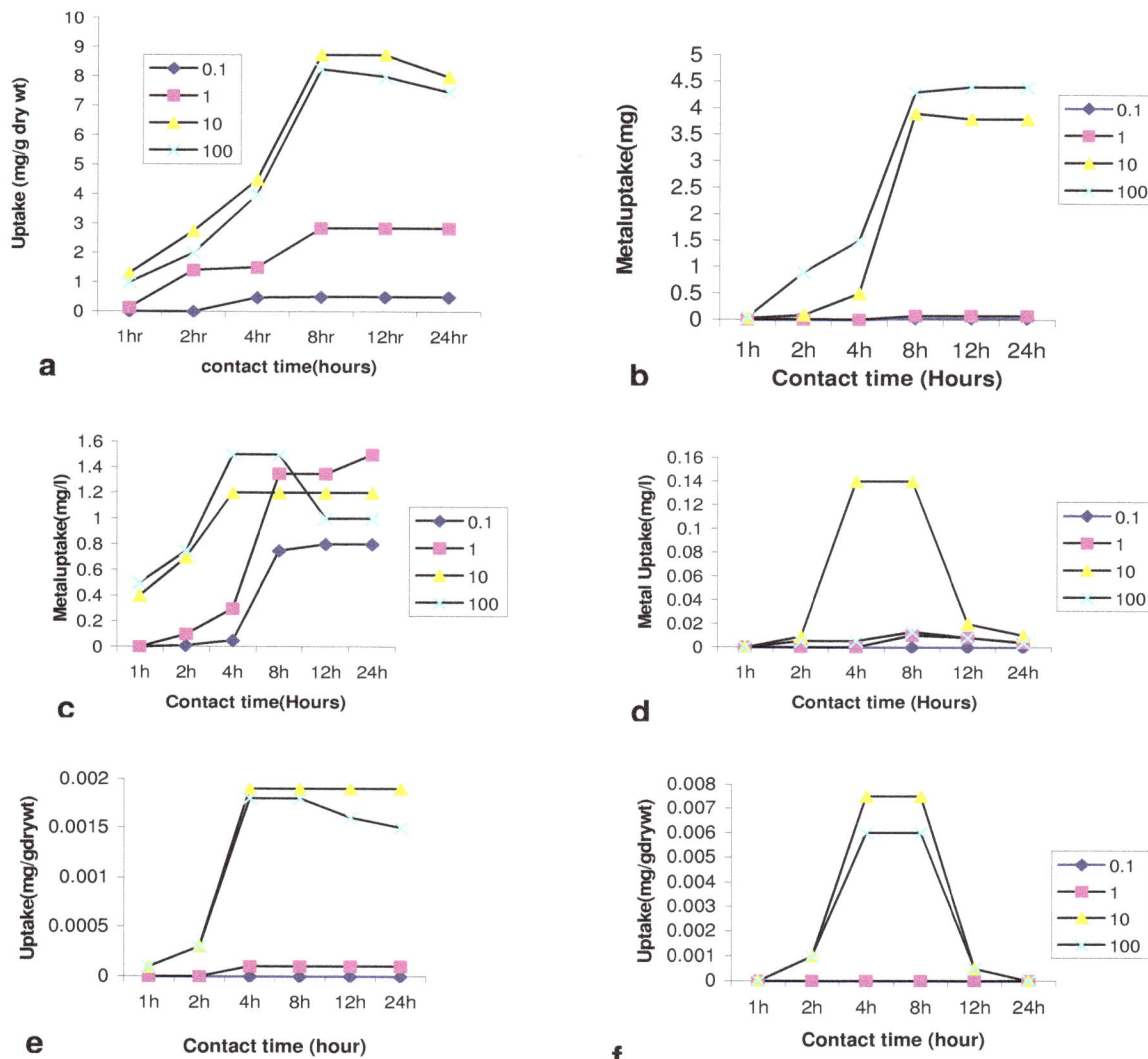

Figure 7. Effects of contact time on bioconcentration of various heavy metals by *Bacillus* sp. (a) Fe (b) Zn (c) Cd (d) Cu (e) Ni (f) Pb.

was attained in less than 24 h. The results of the effect of contact time on uptake indicates that the respective test isolates had an optimum residence time for each heavy metal and once this time elapsed, uptake remained either constant or diminished slightly. This agrees with metal uptake models, where the process can be considered as an equilibrium that involves adsorption and desorption due to saturation (Panchanadikar, 1994). Maximum nickel adsorption at 2 and 6 h by different bacteria in-cluding *Pseudomonas*, *Acinectobacter* and *Enterobacter agglomerans* have been reported (Panehanakikar, 1994; Kaewehai and Prasertson, 2002; Abu Al-Rub et al., 2002). Also, it had been reported that the biosorption of Pb by *Phanerochaete chrysporium* was rapid in the first 15 min and equilibrium was attained after 3 h (Yetis and Ceribass, 2001). The three test isolates showed a similar pattern in accumulating the heavy

metals with time although the uptake of Zn by *Bacillus* and *Aeromonas* were not significantly different with respect to time.

Conclusion

The three test organisms (*Bacillus, Pseudomonas* and *Aeromonas*) had the capability of accumulation of the test metals. Results showed that metal concentrations bet-ween 10 - 100 mg/l promoted rapid metal uptake. There-fore to promote accumulation of heavy metals in polluted aquatic systems, it is necessary to ensure initial metal concentrations of between 10 - 100 mg/l. Metal up-take by the three test isolates increased between 4 - 12 h depending on the metal then reached equilibrium. The rate of attaining equilibrium varied between test organism

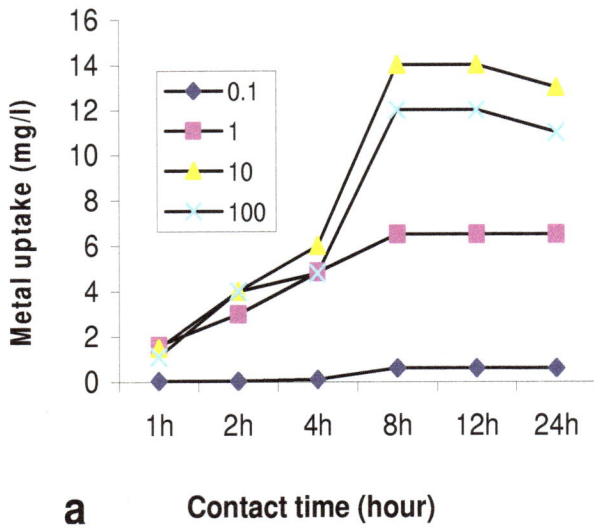

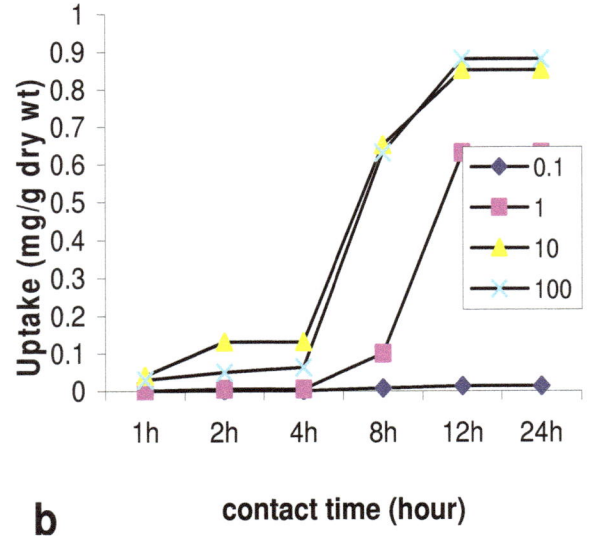

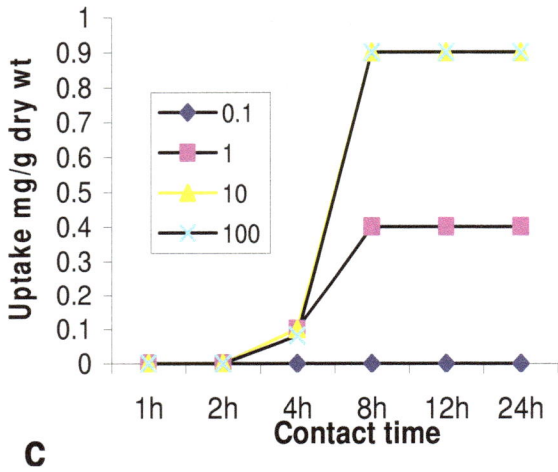

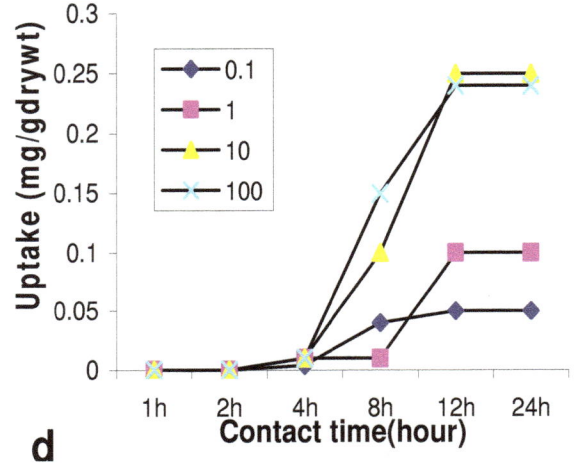

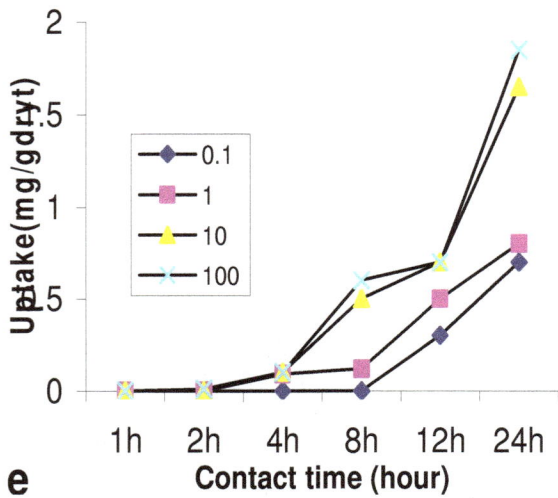

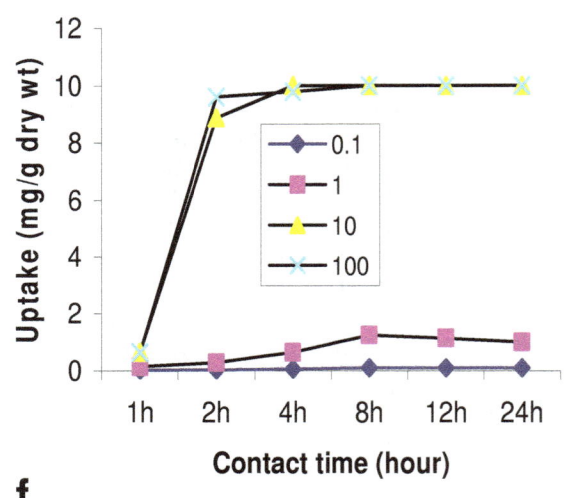

Figure 8. Effects of contact time on bioconcentration of various heavy metals; (a) Fe (b) Zn (c) Cd (d) Cu (e) Ni (f) Pb; by *Pseudomonas* sp.

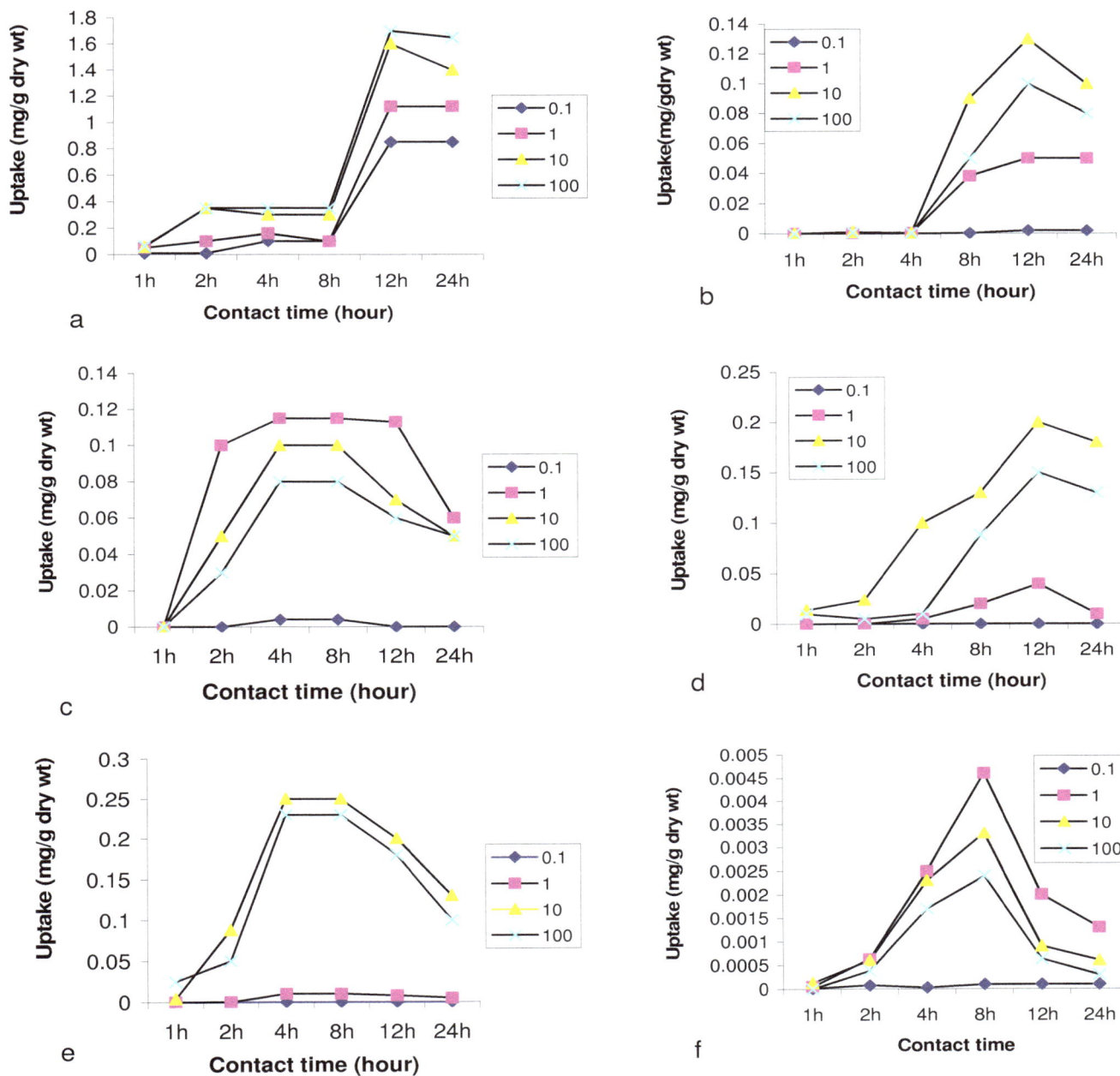

Figure 9. Effects of contact time on bioconcentration of various heavy metals; (a) Fe (b) Zn (c) Cd (d) Cu (e) Ni (f) Pb; by *Aeromonas* sp.

and metal type. Thus exposing test organisms to metal concentrations of above 24 h may not improve accumulation of these metals. Harvesting these organisms within 12 h and introducing fresh organisms will promote further accumulation of the metals. Thus, the combination of maintaining initial metal concentration of between 10 - 100 mg/l at test site and exposure for not more than 12 h followed by subsequent harvest and repeated

introduction of test organism would promote rapid accumulation.

REFERENCES

Abu Al-R, Ashour EAE, Naas MEC, Benyahm F (2002). Adsorption of Nickel on Immobilized Algal Biomass. The Fourth Annual University Research Conference, Al-Yan pp. 17-20.

Al- Garni SM (2005). Biosorption of Lead By Gram –ve Capsulated and non-Capsulated Bacteria. Wat. Sci. Technol., 31(3): 345-349.

American Public Health Association (1998). Standard Methods For The Examination of Water And Wastewater. 20th Ed. Washington, D.C. USA. American Water Works Association Water Pollution Control Federation.

Asku Z (1992). The Biosorption of Copper (II) By C. vulgaris and zamigera. Technol., 113: 579-586.

Bauda P, Block JC (1985). Cadmium Biosorption and Toxicity To Laboratory Grown Bacteria. Environ. Technol., 6: 445-454.

Boularbah A, Morel IL, Bitton G, Guckert A (1992). Cadmium Biosorption and Toxicity To Six Cadmium Resistant Gram Positive Bacteria Isolated From Contaminated Soil. Environ. Toxicol. Wat. Qual. 7: 237-246.

Buikema AL, Neiderleher BR, Cairns J (1982). Biological Monitoring Part IV Toxicity Testing. Wat. Res., 16: 230-262.

Dutton RJ, Bitton G, Koopman E, Agami O (1990). Effects ofEnvironmental Toxicants On Enzymes Biosynthesis. A Comparison of β- Galactosidase, α -Glucosidase and Tryptophanase. Archives to Environmental Contamination and Toxicology. 19: 395-398.

Finney DJ (1978). Statistical Methods in Biological Assay 3rd edition, Charles Griffin London pp.1-252.

Ford T, Mitchel R (1992). Microbiol Transport of Toxic Metals. In: Environmental Microbiology. A. J. Wiley and Sons. Inc. New York, pp. 83-97.

Fourest E, Roux CJ (1992). Heavy Metal Biosorption By Fungal Mycelial By- Products: Mechanisms and Influence of pH. Appl. Microbiol. Biotechnol., 37(3): 399-403.

Gadd GM (1988). Accumulation of Metals by Microorganisms and Algae In: Biotechnology. Weinheim, Germany, pp. 401-433.

Kaewehai S, Prasertson P (2002). Biosorption of Heavy Metals. J. Sci. Technol., 24(3): 422-430.

Kurek E, Zabun JC, Bollag JM (1982). Sorption of Cadmium by Microorganisms in Competition with Other Soil Constituents. Appl. Environ. Microbiol., 42: 1011-1015.

Galun M, Gelun E (1987). Removal of Metal Ions From Aqueous Solution By Penicillin Biomass. Kineitic and Uptake Parameter. Water, Air and Soil Pollution, 3: 359-371.

Gupta G, Keegan B (1998). Biosorption and Bioaccumulation of Lead by Poultry Litter Microorganisms. Poult. Sci. Mar., 77(3): 400-404.

Holt JG, Krieg NR, Sneath PHA, Staley JT, Williams T (1994). Bergey's Manual of Determinative Bacteriology. 9th ed. Williams and Wilkins, Baltimore, USA.

Malekzadeh F, Farazmand A, Ghafourian H, Shahamar M, Levin M, Grim C, Colwell RK (1995). Accumulation of Heavy Metals By a Bacterium Isolated From Electroplating Effluent. Proceedings of the Biotechnology Risk Assessment Symposium Canada, pp. 388-398.

Mullen MD, Wolf DC, Ferris FG, Beveridge TJ, Fleming CA, Boyle GW (1989). Bacterial Sorption of Heavy Metals. Appl. Environ. Biotechnol. 3: 402-410.

Odiete WO (1999). Environmental Physiology of Animals and Pollution. Diversified Resources Ltd. Surelere Lagos. pp. 225-228.

Odokuma LO, Ijeomah SO (2003). Tolerance of Bacteria To Toxicity of Heavy Metals In The New Calabar River. Glob. J. Environ. Sci. 2(2): 128-132.

Odokuma LO, Oliwe SI (2003). Toxicity of Substituted Benzene Derivatives to Four Chemolithotropic Bacteria Isolates from the New Calabar River, Nigeria. Glob. J. Environ. Sci. 2(2): 72-76.

Odokuma LO, Emedolu SN (2005). Bacterial Sorbents of Heavy Metals Associated With Two Nigerian Crude Oil Glob. J. Pur. Appl. Sci. 11(3): 343-351.

Panchanadikar VV (1994). Biosorption Process For Removing Lead (II) Ions From Aqueous Effluents Using Pseudomonas sp Int. J. Environ. Sci., 46: 243-250.

Schlegel HG (1997). General Microbiology. 2nd edn. Sci. Tech. Publishers Madison. p. 1119

Silver S (1991). Bacterial Heavy Metal Resistance Systems and Possibility of Bioremediation. In: Biotechnology, Bridging, Research and Application. Kluwer Academic Publishers, London. pp. 265-287.

Simmons P, Tobin JM, Singleton I (1995). Consideration of the Use of Commercially Available Yeast Biomass For The Treatment of Metal Containing Effluents. J. Ind. Microbiol., 14: 240-246.

Strandberg GW, Shumate SE, Parrol JR (1985). Microbial Cells As Biosorbents For Heavy Metals: Accumulation of Uranium By Saccharomyces cerevisiae and Pseudomonas aeruginosa. Appl. Environ. Microbiol. 41: 237-245.

Stainer RY, Adelberg EA, Ingraham JLI (1982). General Microbiology. .4th edn. Macmillan New York p.1023.

Taylor DJ, Green NPO, Stout GW (1997). Biological Sciences. 3rd ed. Cambridge University Press, p. 983.

Vieira HSF, Volesky B (2000). Biosorption: A Solution To Pollution? Int. Microbiol., 3: 17-24.

Volesky B, May-Philips HA (1995). Biosorption of Heavy Metals By Saccharomyces cerevisiae. Appl. Microbiol. Biotechnol., 42: 797-806.

Weber WJ (1985). Adsorption Theory Concepts and Models. In: Adsorption Technology: A Step By Step Approach to Process Evolution and Application. Marcel Dekker, New York, pp. 1-35.

Williamson KJ, Johnson DG (1981). A Bacterial Bioassay for Assessment of Waste water Toxicity. Wat. Res., 15: 383-390.

Yetis C, Cerribass H (2001). Biosorption of Ni (II) and Pb (II) By Phanerochaete Chrypsosporium from a Binary Metal System. Wat. Sci. Technol., 27: 15-20.

5

Radioactivity concentration and heavy metal assessment of soil and water, in and around Imirigin oil field, Bayelsa state, Nigeria

R. K. Meindinyo[1] and E. O. Agbalagba[2*]

[1]Department of Physics, Bayelsa State College of Education, Okpoama, Brass, Nigeria.
[2]Department of Physics, University of Port Harcourt, Rivers state, Nigeria.

The alpha and beta activity concentration and heavy metal assessment of soil and water in and around Imirigin oil field has been carried out. Study area was subdivided into five locations, Soil and water samples were collected from field undisturbed environment and oil spilled areas. Sample collection and preparation follows standard procedures. *In situ* measurement was conducted for pH and electrical conductivity, heavy metal analysis was carried out using Atomic absorption Spectrophotometer (AAS) and gross alpha and beta activity concentration was by using gas filled proportional counter. Average value for pH and E.C. are 6.5±0.2 and 46.8±1.0 µs/cm respectively for soil and 6.4±0.5 and 406.1±5.2 µs/cm respectively for water. The mean values obtained for ASS analysis for soil are 11.9±1.0, 3.3±0.4, 1.7±0.7, 8.1±0.5, 42.5±1.9, 3.3±0.5, 8.0±0.6, 0.08±0.02 and 79.5±2.2 mg/kg. For Ca, Mg, Zn, Ni, Fe, Cd, Pb, Hg and Cr respectively for water, mean value obtained are 8.3±0.5, 4.2±0.4, 1.6±0.4, 1.5±0.3, 1.3±0.2, 0. 0.06±0.004, 0.07±0.003, 0.05±0.01 and ND mg/l for Ca, Mg, Na, K, Fe, Pb, Cd, Hg and AS respectively. Gross alpha and beta activities mean concentrations for soil are 0.53±0.02 Bq/g and 29.29±0.17 Bq/g respectively, and 4.02±0.01 Bq/l and 54.23±1.76 Bq/l respectively for water. The results show that the level of the various metals obtained differs from location to location. Values obtained in soil are within reported values in the Niger Delta region except Iron level. Heavy metals such as Ca, Fe and Cd exceed the WHO limits for drinking water. The mean values for alpha and beta activity in soil are above reported values in similar environment while mean values obtained in water samples are above WHO recommended maximum permissible limit for drinking water. These values obtained shows that drinking water from sampled locations may pose some long term health hazards to the public users though soil from the area is still safe as construction material for buildings.

Key words: Assessment, radionuclide concentration, heavy metal, oil field.

INTRODUCTION

The advent of oil mineral exploitation and exploration activities resulted in increased pollution of the Niger Delta environment. Some of such problems are increase presence of potentially toxic metals in soil and water bodies of the area and increase in human exposure to ionizing radiation. This contamination and pollution of soil and water has been a great concern to many a person in this region and beyond. This is because if contamination/

pollution occurs, damage may be extensive and effects may be long term and extend over many seasons. Studies have shown that almost all the elements in the periodic table including heavy metals are found in the crude oil matrix (Abison, 2000; Avwiri et al., 2007). Thus, the release of gas through flaring, oil spillage and its derivative has serious radiological and hazardous effects to men and direct impact on the soil (Aroganjo et al., 2004; RRC, 2007). This may also result in surface and ground water pollution (Oyinloye and Jegede, 2004). Moreso, leachate from municipal, oil spill, gas flare, solid, landfills, surface water run-off are potential sources of

*Corresponding author. E-mail: ezek64@yahoo.com.

contamination of both surface and ground water (Odukoye et al., 2002). Pollution of water is on the increase at frightening proportion. Frightening in the region especially in the oil host communities due to gas flare and oil spillage into water bodies and land and other allied factors responsible for environmental degradation (Egila and Techemen, 2004).

The concentration of heavy metals in soil, surface and ground water depends on the human activities distributed through the geological stratification of an area (Bolaji and Tse, 2009). The present of such contaminants in water and soil above recommended standard set by water, soil and environmental regulatory bodies like EPA and WHO may result in serious health hazard (US-EPA, 2002). These perceived consequences of the consumption of foodstuff and vegetables from these polluted soil and the consumption of untreated and unregulated water and the attending radiological burden has triggered various environmental studies especially in the Niger Delta region on soil equality and water aquifer quality and aquatic ecosystem (Akpan and Offen, 1993; Udom et al., 1999; Ekpete, 2002; Oguzie et al., 2007; Egila and Terhemen, 2004; Abam et al., 2007; Nwala et al., 2007; Bolaji and Tse, 2009). But none of these studies have been able to relate the presence of heavy metal (physio-chemical parameters) with natural radioactivity concentration in these studies.

This present study examined the presence heavy metals in the Imiringi Oil field environment which has been characterized by frequent oil spillage due to oil pipe failure and vandalization. The study also evaluates the gross activity concentration and its attending radiological burden on the population living in this environment. Baseline data will be gotten from the analysis since no known similar study have been carried out in the oil field. The study will therefore, be of great benefit to the host communities, operating company and the government.

METHODS

Study area

Imiringi oil and gas field is one of the main oil producing fields onshore of Bayelsa State. It lies within latitudes 04 54N and 05'02N and longitudes 006°ʹ 15E and 006° 24E and is situated in the Southern part of the Niger Delta. Its geomorphological features consist of mainly fresh water swamp and mangroves with table land within the swamp. The study area geology is consistence with the geology of Bayelsa State and that of the Niger Delta Complex as reported elsewhere (Bolaji and Tse, 2009; Manila and Nwachukwu, 2009).

Using lithological and geophysical logs, Etu-Efeotor and Akpokodje (1990) delineate five levels of aquifers in the study area. The region ground water problems include salt water intrusion, pollution of the upper aquifers of the area, high iron content in some of the area horizons, causing encrustation and contamination of surface water pollution with oil spillage as major contributor (Avwiri and Agbalagba, 2009).

Deposition of heavy metals on surface water from gas flare and washing of contaminated soil through surface water run-off into

water bodies are other major environmental problems in the region.

Sample collection and preparation

The sampled Imiringi oil field was subdivided into five locations; A– Imiringi Town, B- Around Imiringi Gas Turbine, C– Imiringi Flare and Flow Station, D- Ede-Epie area, Yenagoa and E– Agudama area, Yenagoa) for effective coverage of the study area, four soil and water samples were collected from each of the selected location with the study field.

The bulk soil samples were collected from old oil spill site and undisturbed, uncultivated grass cover level area and in remote location from man-made structures such as roads and buildings to prevent any external influence on this type of samples results (Senthilkumar et al., 2010). Soil samples were collected double for Atomic Absorption spectrophotometer (ASS) and gross alpha and beta analysis in a black polythene bag from an area of approximately 100 cm^2 and in range of 10– 15 cm depth for grass cover and undisturbed area while 2–10 cm depth in oil spilled areas (Senthilkumar et al., 2010). For the heavy metal analysis samples were initially broken into smaller aggregates, de-watered and dried at ambient temperature (20–30°C) for two weeks. They were subsequently grounded in a mortar passed through a 2 mm sieve and stored in a sealed container until needed for analysis (Onwuka and Uzoukwu, 2008). Samples for gross alpha and beta analysis were air dried at room temperature for ten (10) days. Samples were then transfer into a low temperature (50°C) drying cabinet to help accelerate the drying process without loss of radionuclides from the samples (Onoja et al., 2004). The dried samples were grounded with mortar and pestle and allowed to pass through a 100-mesh sieve. Prepared soil samples were pelleted into counting planchette size using the hydraulic compressing machine. Samples were kept in desiccators and store for four weeks to attain a state of secular equilibrium.

Twenty water samples were collected for the study, four each from the five sampling locations. Samples were collected from the location creeks/River water which has direct bearing of the oil and gas activities. One sample of borehole water was collected from each location for comparison and check. Samples collection was carried out according to prescribed sample collection procedure in pre-wash and clean 300ml screw plastic containers and as reported elsewhere (Apha, 2000; ISO, 9697 and 9698 (E), 1992; Juliet, 2006; Bolaji and Tse, 2009).

Samples analysis

The PH and electrical conductivity were measured and recorded insitu, this is because the chemistry of soil, surface and under-ground water is sensitive to environmental changes. The analytical methods used in the determination of these heavy metals in water have been reported in Nwala et al. (2007), Bolaji and Tse (2009) and were in accordance with ASTM and ALPHA Standard procedures for the heavy metals in soil samples analysis, a high precision variance spectral unicorn 1969 Atomic Absorption Spectrometer (Perkin Elmer Model 2280) was used in acetylene/air flame after sample preparation.

In the gross alpha and beta analysis, prepared samples were analyzed using the gas filled proportional Counted EURISYS MEASUE IN-20 low background eight channels alpha and beta Counter at the Center for Energy Research and Training Ahamdu Bello University, Zaria, Nigeria. The background measurements, calculation for machine efficiency, plateau test were carried out using Standard Methods (ASTM) follows as reported by Avwiri and Agbalagba (2007).

Table 1. Mean physic-chemical parameters concentration in soil sample (±SD).

S/n	Location code	pH	EC μS/cm	Ca cmol/Kg	Mg cmol/Kg	Zn mg/Kg	Ni mg/Kg	Fe mg/Kg	Cd mg/Kg	Pb mg/Kg	Hg Mg/Kg	Cr Mg/Kg
1.	A	6.7 ± 0.1	0.8 ± 0.2	0.12 ± 0.2	1.6±0.2	0.2±0.1	0.2±0.1	40.1±0.1	1.3±0.3	6.2±0.1	0.2±0.00	161.20±2
2.	B	5.3±0.3	41.6±2.6	0.2±O.1	2.6±0.1	1.3V0.5	4.1±1.0	50.4±2.6	0.8±0.1	5.2±0.5	ND	58.4±1.
3.	C	4.5±0.2	32.5±1.5	56.0±4.0	8.66V0.8	1.1±0.8	31.1±0.9	56.1±0.3	0.5±0.2	3.3±0.7	0.01±0.00	79.9±2.
4.	D	7.5±0.5	34.1±0.3	1.3±0.2	2.4±0.5	3.0±0.6	0.3±0.02	45.1±1.5	14.0±2.1	7.2±1.6	0.12±0.01	40.8±
5.	E	8.6±0.1	45.1±0.2	1.2±0.3	1.3±0.3	2.8±1.2	5.0±0.2	21.0±1.5	ND	8.3±0.4	0.01±0.00	57.2±
	Mean	6.5±0.2	46.8±1.0	11.9±1.0	3.3±0.4	1.7±0.7	8.1±0.5	42.5±1.9	3.3±0.5	6.0±0.7	0.07±0.002	79.5±2.2

Table 2. Physico-chemical parameters concentration in water sample (±SD).

S/n	Location code	pH	MS/CM electrical cond. (EC)	Ca Mg/l	Mg Mg/l	Na Mg/l	K Mg/l	Fe Mg/l	Pb Mg/l	Cd Mg/l	Hg Mg/l	AS Mg/l
1.	A	6.8±0.2	290±5.0	7.1±0.3	7.4±0.3	2.3±0.1	1.3±0.2	1.4±0.02	ND	0.01±0.0	0.07±0.001	ND
2.	B	6.1±0.4	300.0±4.6	14.3±0.8	3.8±0.2	1.76±0.2	1.3±0.4	1.3±0.1	0.2±0.0	0.04±0.0	ND	ND
3.	C	7.6±1.0	246.1±3.3	4.4±0.2	4.4±0.3	1.8±0.02	0.9±0.20	0.8±0.2	0.08±0.0	0.3±0.02	0.02±0.002	ND
4.	D	6.4±0.6	570.4±7.0	8.5±0.6	2.57±0.2	0.9±0.1	2.1±0.5	1.0±0.2	ND	ND	ND	ND
5.	E	5.2±0.2	624.7±6.7	7.1±0.7	3.0±0.4	1.3±0.3	1.6±0.5	1.9±1.4	0.03±0.002	ND	ND	ND
	Average Values	6.4±0.5	406.1±5.3	8.3±0.5	4.2±0.4	1.6±1.4	1.5±0.3	1.3±0.2	0.06±0.004	0.07±0.0003	0.054±0.006	0.00
	WHO limit	6.5 – 8.5	500	7.5	200	200		0.30	0.01			0.05

RESULTS AND DISCUSSION

The results of the heavy metal (physcio-chemical) analysis of the soil and water samples are presented in Tables 1 and 2. The pH of the soil samples ranged from 4.5 acidic to 8.6 alkaline with an average value of 6.52 ±0.24. Jones (1998) reported that acidic soil/water results in corrosion of Iron and Steel materials. Electrical conductivity mean value is 46.8 ± 1.0 μS/cm. This indicates that the soil contain moderate ionic content hence, the capacity of the soil to conduct or transmit electric current would be moderate. The mean values of Calcium (Ca), Magnesium (Mg), Zinc (Zn), and Nickel (Ni), Iron (Fe) are 11.86±0.96,

3.32±0.38, 1.68±0.64, 8.14±0.48 and 42.54±1.92 mg/kg respectively. All these metals except Iron are well within the values for uncontaminated soil (FUGRO, 2004). The high concentration of Fe may be peculiar characteristics of the area soil.

For heavy metals, Cadmium concentration ranged from ND to 14.0±2.1 mg/kg with mean value of 3.32±0.54 mg/kg. This shows that studied area is Cadmium polluted. This is true because the mean value obtained exceeded the recommended upper limit of Cadmium concentration in soil (FUGRO, 2004). This values obtained are also higher than previous reported values in the Niger Delta (FUGRO, 2004; Uzoukwu and Onomake, 2005; Onwuka and Uzoukwu, 2008).

This high value may be attributed to the oil spillages in the environment as seen in location of oil spilled site. The lead content of the soil sample ranged from 5.2 to 13.2 mg/kg with mean value of 8.04±0.64 mg/kg. The possible sources of lead in these samples are crude oil and burnt leaded gasoline. Pollution by high levels of lead could be detrimental not only to soil organisms but also to aquatic organism. This is because run-off water from the soil is discharge into a body of water, but the obtained level of lead in the studied area may not be toxic to organisms. However, if there is long time accumulation of lead due to additional pollution, it could result to adverse effect. High concentration of lead (46-218 mg/kg) in bioavailable forms could harm soil and aquatic organism (Onwuka and Uzoukwu, 2008). Lead-induced effects include neurological dysfunction in the central nervous system, altered behavior and immune suppression (FUGRO, 2004). The values obtained in the study area agrees with that obtained elsewhere in the Niger Delta prior to the establishment of Exxon oil and gas facilities (FUGRO, 2004) and those obtained by Uzoukwu and Onomake (2003) (11.9±90 mg/kg) and 6.59±2.96 (mg/kg) obtained by Onwuka and Ozoukwu (2008) at the university of Port Harcourt botanic garden. This obtained value of lead is below the USEPA standard 50 mg/kg value.

Mercury (Table 1) value concentration ranged from N.D in location B to 0.2 mg/kg in location A. with a mean value of 0.07±0.002 mg/kg. This shows that Hg level in the study area is near insignificant compared to the 25 mg/kg maximum recommendation value (Adekola et al., 2002). The concentration of Chromium (Cr) in the digested soil samples ranged from 40.9 mg/kg in location D to 161.20 mg/kg in location A with a mean value of 79.5±2.2 mg/kg. The concentration of Chromium in uncontaminated environment (soil and estuarine) are in the range of 50-100 mg/kg (CNL, 2001).

Accordingly, the Chromium level in the studied area was within the acceptable range for unpolluted soil or sediments. Thus the level observed will not cause any toxic effects on soil micro and macro organism nor on man that can consume any flora that is cultivated on the soil.

The results of the physico-chemical analysis of the water samples are presented in Table 2. The pH of the sampled water ranged from 4.5 to 8.6 with a mean value of 6.4±0.5. This range of values obtained is well within the WHO range of 6.5-8.5 for domestic and portable water. The electrical conductivity (EC) value ranged from 290-624 µs/cm with a mean value of 406.1±5.3 µs/cm. This shows that EC level of locations D and E exceeded the permissible limit of 500µs/cm set by WHO.

EC is an indicator of water quality and soil salinity hence the relatively high value observed in some river water samples show high salinity, thus water may not be very suitable for domestic and agricultural use.

Ca^{2+} and Mg^{2+} values obtained for the water samples ranged from 4.4-14.3 mg/l with a mean value of 8.3±0.5 mg/l for Ca^{2+} and 2.6-7.4 mg/l with a mean value of 4.2±0.4 mg/l for Mg^{2+}. The values obtained for Mg^{2+} are well in agreement with values reported by Bolaji and Tse (2009) in Port Harcourt well water, and FUGRO (2004). But Ca^{2+} mean values obtained, is above the WHO recommended limit of 7.5 mg/l for domestic and drinkable water. These values are in deviation from the values reported by Nwale et al. (2007), but agree with the values reported in the Niger Delta Environmental Survey report (NDES, 2000).

Na^+ concentration in the river water analyzed range from 0.9 to 2.3 mg/l with a mean value of 1.6±0.4 mg/l while K^+ ranged from 0.9 to 2.1 mg/l with a mean value of 1.5±0.3 mg/l. These observed concentrations for Na^+ and K^+ in these samples were well below the WHO recommended limit of 200 mg/l for both. These values obtained agree with those reported by Bolaji and Tse (2009), and has no health implication to the public consuming the water in terms of Na^+ and K^+ anions.

Fe^{2+} concentration values in the sampled river water samples ranged from 0.8 to 1.9 mg/l with a mean value of 1.3±0.2 mg/l. This shows that Fe^{2+} concentration in all the five location sampled exceeded the WHO limit of 0.3 mg/l. The relatively high concentration of metallic iron may be attributed to the observed turbid nature of the water and the brownish colouration when allowed to settle. This is due to the oxidation of Fe^{2+} to Fe^{3+} which causes nuisance to laundry and sanitary wares (Aiyesanmi et al., 2004).

Lead ion (Pb) was detected at location B, C and E, though it's mean value of 0.06±0.004 mg/l is bellow the WHO 0.3 mg/l limit for drinking water, the level obtained portends some future health hazard as accumulative effect of these levels may possibly lead to Pb poison (Ademoroti,1996). The Cadmium mean value of 0.07±0.003 mg/l obtained in this study is far above the 0.01mg/l WHO recommended limit. This confirmed the reported value in soil and show that oil spillage has impacted both the soil and water negatively. The mean concentration of other heavy metals in these river water samples (mercury) is 0.05±0.006 mg/l while As is below detection limits. The presence of Hg in some location water sample is another indication of the contamination of the water samples.

This may also be attributed to oil spillage and other anthropogenic activities in the area. The result of the water physico-chemical analysis shows that all the heavy metals analysis where present in the samples except Arsenal. This generally indicated that the water is not total safe for drinking, their present in and consumption may leads to toxic waste in the body and a high accumulative level may lead to water poison.

Table 3 present the result for gross alpha and beta activity concentration in both soil and water. Alpha activity in soil ranged from 0.049 Bq/g in location E to 0.96 Bq/g in location C with a mean value of 0.526±0.02 Bq/g. Beta activity concentration in the soil ranged from 1.52 to 119.91 Bq/g with a mean value of 29.29 Bq/g. These values obtained for activities in the soil are higher than other values reported elsewhere (Arogunjo et al., 2004) in similar environment. The high beta activity value obtained especially at location C is an indication of anthropogenic activity in the area. This confirmed that the present of oil spillage in this environment has elevated or enhances the level of radionuclides in the area.

Alpha activity in water ranged from 0.021 Bq/l in location E to16.950 Bq/l in location A with a mean activity of 4.02±0.09 Bq/l. Beta activity ranged from 5.84 Bq/l in location E to 135.88 Bq/l in location C with a mean

Table 3. Mean gross alpha and beta activity concentration in soil and water samples (±SD).

S/n	Location code	Soil activity conc.(Bg/g)		Water activity conc. (Bg/l)	
		α-activity	β-activity	α-activity	β-activity
1.	A	0.610±0.040	3.992±0.012	16.950±0.032	59.54±0.52
2.	B	0.620±0.015	17.430±0.096	1.240V0.042	15.220±0.25
3.	C	0.960±0.039	119.910±0.469	0.510±0.009	54.680±1.314
4.	D	0.390±0.003	3.600±0.149	1.380±0.124	135.88±6.713
5.	E	0.049±0.003	1.520±0.121	0.021±0.002	5.840±0.03
		0.526±0.020	29.29±0.169	4.02±0.09	54.232±1.763

activity value of 54.232±1.763 Bq/l. The generally low values of activity observe in location E is an indication of unpolluted soil and suitable water. The mean alpha emitters (4.02±0.009 Bq/l) and mean beta emitters (54.232±1.763 Bq/l) are far above the 0.1 Bq/l for alpha and 1.0 Bq/l for beta WHO recommended practical screening levels of radioactivity in drinking water (WHO, 2003). This shows that the activities of oil spillage and gas flare in the environment may have increased the radionuclide levels and concentration of these river water and are therefore not suitable radiologically for drinking as portable water.

A regression analysis on heavy meters and gross activity concentration show a strong correlation between the heavy metals examined and the concentration of alpha and beta activities in water and soil, with a value of 0.633 to 0.892 respectively. This shows that the presence of heavy metals has a direct bearing with the level of radionuclide in the environment.

Conclusion

The survey of radioactivity concentration and heavy metal assessment in soil and water in and around Imirigin oil field has been carried out. The level of the various heavy metals examined differs from location to locations. This is confirmed by the heterogeneity of radionuclide deposits in soil and water samples was mainly of geological origin. The present of some heavy metals especially cadmium in water is an indication of water pollution.

The measured alpha and beta activity in soil are above reported values in similar environment while in water, the values obtained are well above reported values and that of WHO recommended maximum permissible limit for drinking water. The research findings revealed that the water in the oil field is not radoiologically safe for drinking. The values obtained may pose some serious health hazards to the public users of these river waters.

REFERENCES

Abam TKS, Olu AW, Nwankwoala H (2007). Ground water monitoring for environmental liability assessment. J. Nig. Environ. Soc., 4(1): 42-49.
Adekola FA, Eleta OA, Atenda SA (2002). Determination of the level of some heavy metals in urban run-off sediments of Ilorin and Lagos, Nigeria. J. Appl. Sci. Environ. Mgt., 6(2):23-26.
Ademoroti CMA (1996). Standard methods for water and effluent analysis. Foludex Press Ltd. Ibadan, Pp. 145 – 151.
Aiyesanmi AF, Ipinmoroti KO, Oguntimehin II (2004). Impact of automobile workshop on groundwater quality in Akure Metropolis. J. Chem. Soc. Nig. (Supplement to 2004 Proceeding) 420. – 426.
Akpan EH, Offem JO (1993). Seasonal Variation in water quality of the Cross River, Nigeria. Rev. Hydrobiol. Trop., 26(2): 93 – 103.
American Public Health Association (APHA), (2000). American Water works Association and Water Pollution Control Federation. Standard Method for Examination of Water and Wastewater, APHA, AWWA and WPCF, New York.
Aregunjo AM, Farai IP, Fuwape IA (2004). Impact of oil and gas industry on the natural radioactivity distribution in the Delta region of Nigeria. Nig. J. Phys., 16:131-136.
Avwiri GO, Chad-Umoren YE, Enyima PI, Agbalagba EO (2009). Occupational radiation profile of oil and gas facilities during production periods in ughelli, Nigeria. Working and living. Environ. Hot., 6(1):11-19.
Avwiri GO, Agbalagba OE, Enyinna PI (2007). Assessment of natural radioactivity concentration and distribution in river Forcados, Delta State, Nigeria. Sci. Afr., 7(1): 1281-135.
Avwiri GO, Agbalagba EO (2007). Survey of gross alpha and gross beta radionuclide activity in Okpare-Creek Deltal State Nigeria. Asian Network for Scientific information J. Appl. SC., 7(22): 3542-3546.
Bolaji TA, Tse CA (2009). Spatial variation in ground water geochemistry and water quality index in Port Harcourt Scientia Africana, 8(1):134-155.
Chevron Nigeria limited (CNL), (2001). Draft report of the environmental impact assessment of Tubu Field Development. Chevron Nig. Ltd., Lagos Nigeria.
Egila JN, Terhemen A (2004). A preliminary investigation into the Quality surface d water in the environment of Benue cement company Plc Gboko Benue state, Nigeria Int. J. Sci. Tech., 3(1): 12-17.
Ekpete OA (2002). Determination of physico-chemical parameters in borehole water in Odihologboj, Community in Rivers State. Afr. J. interdisciplinary stud., 3(1): 23-27.
Etu- Efeotor JO, Akpokodje EG (1990). Aquifer systems of the Niger Delta . J. min. Geol., 26(2): 279-294.
FUGRO (2004). Environmental Impact assessment of East Area Projects- Additional oil Recovery Project. Procedures FUGRO Quality Assurance. Report No. S-0310.
ISO 9696 (1992). Water Quality-Measurement of gross alpha activity in non saline water (International Organization for Standardization, London), first edition. Pg. 12-14.
ISO 9697 (1992). Water Quality-Measurement of gross beta activity in non saline water (International Organization for Standardization, London), first edition, pg. 12-14.
Jones LW (1998). Corrosion and water technology: Oil and Gas Consultant International Inc., Tulsa Pp. 47 – 51.
Juliet NM (2006). Gross alpha and beta radioactivity in Kaduna River . M.sc. thesis. ABU Zaria.
Niger Delta Environmental Survey (NDES), 2000. Niger Delta Development Priorities and Action Plan. Phase II Report, 2: 156-176.

Nwala CO, Akaninwor JO, Abbey BW (2007). Physico-chemical parameters of mono pumps and well waters in Igbo Etche. J. Nig. Environ. Soc., 4(1): 78-87.

Odukoya OO, Arowolo TA, Bamgbose (2002). Effect of solid waste land fill under-ground and surface water quality at Ring-Road, Ibadan. Global J. Environ. Sci., (22): 235-242.

Oguzie EE, Ayochukwu IB, Offem JO (2007). Groundwater contamination: A simulation study of buried waste. Metallic contaminant penetration through the aquifers. J. chem. Soc. Nig., 27(1): 82-84.

Onoja RA, Akpan TC, Mallam SP, Ibeanu IGE (2004). Characterization of the gross alpha/beta counter in the center for energy research and training; ABU, Zaria . Nig. J. phys., 16: 13-18.

Onwuka OA, Uzoukwu BA (2008). Studies on the Physicochemical properties of soil from the Botanica Garden, University of Port Harcourt, Port Harcourt, Nigeria. Scientia Africana, 7(1): 157-164.

Oyinloye AO, Jegede GO (2004). Geographysical survey, Geo chemical and microbiology investigation of ground well water in Ado-Ekiti North, South western Nigeria. Global J. Geological Sci., 2(2):235-242.

Udom GJ, Etu - Efeotor JO, Esu EO (1999). Hydro-chemical evaluation of ground water in part of Port Harcourt and Tai-Eleme local Government Area, Rivers state. Global J. pure Appl. Sc., 5(5): 545-552.

US-EPA (United States Environmental Protection Agency), 2002. Current drinking water Standard. Office of groundwater and drinking water quality. Government printing office, Washington DC, USA.

Uzoukwu BA, Onomake OR (2003). Physico-chemical studies of soil along the bank of Milli Shallow Stream in Nnewi, Anambra state, Nigria for evidence of industrial pollution. Global J. Pure Appl. Sci., 10(1);111-119.

Uzoukwu BA, Onomake OR (2005). Physicochemical Studies along the bank of Ubu River in Ekwusigo and Nnewi Local government Areas of Anambra State, Nigeria. Annali di Chimca, 95(1&2):95-104.

6

Preconcentration and separation of some heavy metal ions by solid-phase extraction using silica modified with zirconium phosphate

M. A. Hamed[1]*, Kh. S. Abou El-Sherbini[2], Y. A. Soliman[1], M. S. El-Deek[1], M. M. Emara[3] and M. A. El-Sawy[1]

[1]National Institute of Oceanography and Fisheries, Red Sea branch, Egypt.
[2]Department of Inorganic Chemistry, National Research Center, Dokki, Giza, Egypt.
[3]Department of Chemistry, Faculty of Science, Al-Azhar University Egypt.

An analytical method was developed for the separation and preconcentration of Cu^{2+}, Zn^{2+}, Pb^{2+}, Cd^{2+}, Ni^{2+} and Fe^{3+} from surface water samples collected from ten locations at Suez Gulf, using silica gel chemically modified with zirconium (IV) phosphate. The effects of pH value, time of stirring, concentration of eluting acids and some common ionic species on the separation and preconcentration of the investigated metal ions in synthetic aqueous solutions were firstly studied. The results indicate that the optimum conditions for separation are pH = 5.0- 5.5 and time of stirring = 30 min. HNO_3 is better eluent for the investigated metal ions than HCl hence 2M HNO_3 was used as an eluent. Citrate and EDTA severely affect on the recovery of the metal ions therefore the water samples were previously oxidized to digest organic matter prior to the application process.

Key words: Separation, preconcentration, modified silica, metals, water.

INTRODUCTION

Metals differ from other toxic substances in that they are neither created nor destroyed by human. The pollution levels of the aquatic environmental by heavy metals can be estimated by analyzing water, sediment and marine organisms after a prerequisite separation and/or preconcentration step (Hamed et al., 2006). In recent years, solvent extraction and coprecipitation have been increasingly replaced by solid phase extraction (SPE). It has been extensively studied for different analytical application (Hamed, 2006). Advantages of silica as a base for chelating agents to be used as an ion sorbent, over the organic polymers, are its good mechanical and thermal stability, and it is less susceptible to swelling, shrinking and microbial and radiation decay (Lashein, 2005). The use of chelating agents, immobilized on silica, is a promising route for its efficiency and selectivity of extraction. 8-Quiolinol immobilized on silica gel (Obata et al., 1993). 3-(trimethoxysilyl)-1-ropanethiol modified silica gel (Köklü et al., 1995) and diethylenetriamine with salicylaldehyde and naphthaldehyde immobilized on silica gel (Soliman et al., 2001) were used to separate trace amounts of heavy metal ions from their parent solutions. It was found that the efficiency of the separation depends on the concentration of the metal ion, the nature of the extractant and its surface area, temperature and the stability of the formed chelate. Synthesis of chemically-modified porous silica with Zirconium phosphate as chelating agent (denoted SZP) was described in material and methods. The nature of bonding of IE with some metal ions was investigated by elemental analysis, infrared spectra and thermal analysis (Lopez et al., 1996). According to the importance of metals to the aquatic environs, it may be divided into three groups i) light metals such as sodium, potassium and calcium which are normally mobile cations in aqueous solution; ii) transitional metals such as iron, copper, cobalt and manganese which may be toxic in high concentrations and iii) heavy metals and metalloids such as mercury, lead, cadmium, tin, selenium and arsenic which are generally not required for metabolic activities and are toxic to the cell at quite low concentrations (Abou-El-Sherbini et al., 2003b).

*Corresponding author. E-mail: drhamed64@yahoo.com.

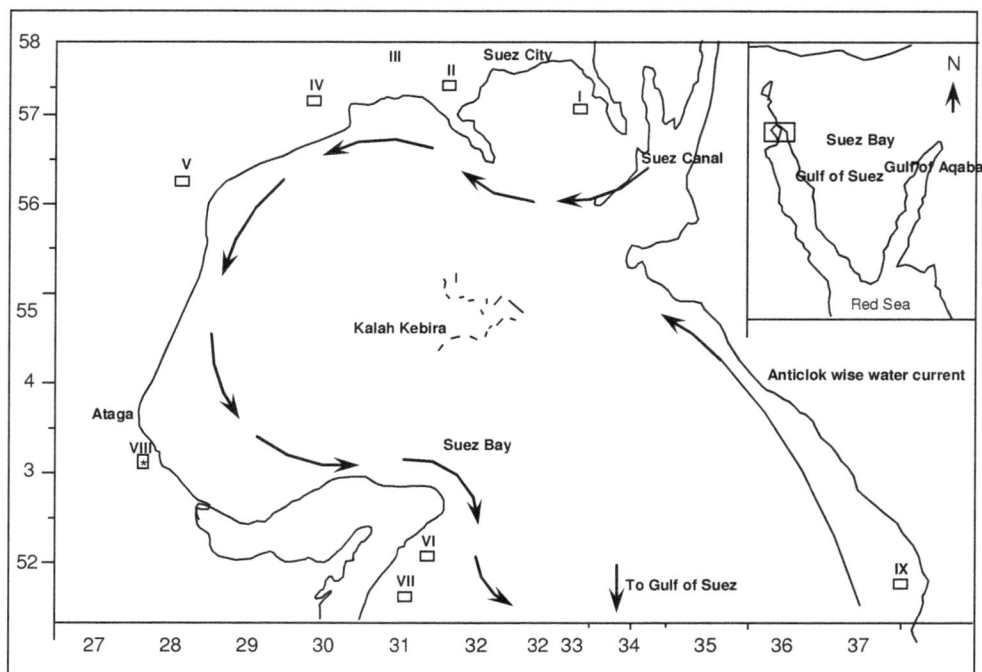

Figure 1. I) Summer Balas Hotel beach. II) Zaitiya. III) El-Kabanon beach. IV) Power Planned Station. V) National Institute of Oceanography and Fisheries (NIOF). VI) Ain Sokhna. VII) Sand Beach.   VIII) North of the Adabyia Harbour. IX) Uyun Mousa. X) Ras Sudr

Due to the interest of water quality, many studies had been performed on the water sources to estimate the levels of heavy metals periodically (Fifield and Haines 2000). The factors controlling and influencing the metal uptake capacity such as pH of solution and stirring time were determined to discuss the performance towards extraction of Cu^{2+}, Zn^{2+}, Pb^{2+}, Cd^{2+}, Ni^{2+} and Fe^{3+} (Hamed, 2004). As continuation, the present paper is aimed to study the uptake behavior of SZP towards Cu^{2+}, Zn^{2+}, Pb^{2+}, Cd^{2+}, Ni^{2+} and Fe^{3+} to suggest the optimum conditions for the separation, preconcentration and determination of these metal ions in aquatic environs.

MATERIALS AND METHODS

Area of study

The Suez Gulf extends for about 250 km South-Southeast from the Suez port in the north (lat. 29° 56') to Shadwan island in South (latutude 27° 36') The width of the Gulf varies between 20 and 40 km, and its depth, throughout its axis is fairly constant with a mean of 45 m (El-Sabh and Beltagy, 1983) Depth increases abruptly to about 250 m, at its mouth (Meshal, 1970).

Sampling stations

Samples were collected from ten stations along the northern part of the Gulf of Suez (Suez Bay) (Figure 1). Station (1): Summer Palace Hotel beach (west of Port Tawific Harbour), Station (2): El-Zeitiya, Station (3): El-Kabanon beach. Station (4): Suez Thermal Power

Station beach. Station (5): Beach of National Institute of Oceanography and Fisheries. Station (6): El-Sukhna. Station (7): Sand Beach. Station (8): North of the Adabyia Harbour, Station (9): Aeon Musa and Station (10): Ras Sudr.

Experimental

Synthesis

The ion sorbent (SZP) was prepared as follow: 56.4 cm^3 of water glass (sodium silicate, 36%) was diluted to one liter with double distilled water (DDW), and neutralized by adding HCl acid drop with drop with vigorous stirring to pH=1. The formed gel was crushed and aged for 1 day. Then, the gel was washed with DDW to remove the formed NaCl and excess acid by decantation till pH > 5 then 0.09, 1.0, 2.0 or 4.0 g $ZrOCl_2.8H_2O$ (Merck) dissolved in minimum amount of DDW was thoroughly mixed. Then the mixture was dried in a water bath to get xero gel which was further dried at 200°C for 2 h then, the sample was cooled and 100 ml H_2O was added. Then 0.03, 0.37, 0.74 or 1.49 cm^3 H_3PO_4 (98%) was added, respectively. The samples were stirred for 60 min, dried at 120°C and annealed at 150, 300 or 400°C for 5 h.

Optimum conditions of the ion sorbent (SZP)

A 0.02 g of SZP was added to 25 cm^3 of 10 $\mu g/cm^3$ of Cu^{2+}, Cd^{2+}, Pb^{2+}, Zn^{2+}, Ni^{2+} and Fe^{3+} (as nitrate) and the pH values of the solutions were adjusted in the range 5-5.5 using 2M NaOH and 0.1M HNO_3 was used instead. Then the solutions were stirred at constant rate for 30 min., filtered and the concentrations of the investigated ions in the filtrates were determined by AAS. After adjusting the pH of the solution at 5.0-5.5 to give maximum K_d

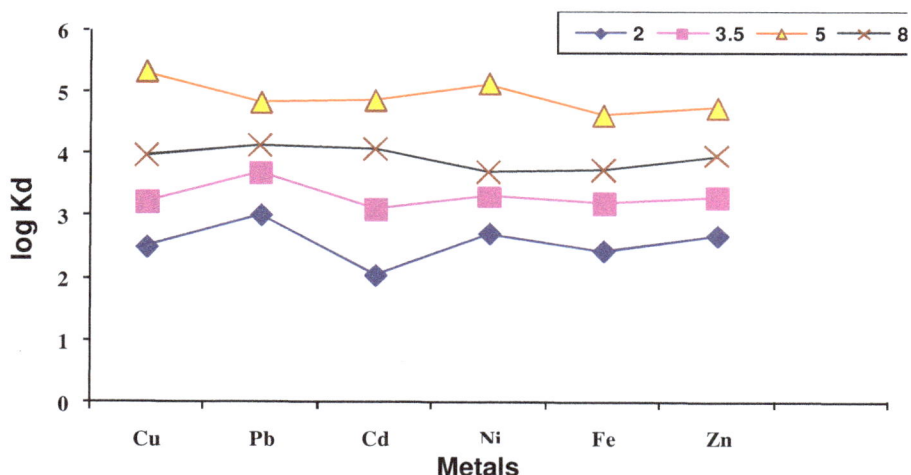

Figure 2. Effect of pH on the logarithmic values of distribution coefficient; log K_d of the preconcentration of heavy metal ions on the modified silica gel as ion sorbent.

stirring time and effect of weight of ion sorbent was studied to obtain optimum time of stirring and weight of ion sorbent for preconcentration and separation of the investigated metal ions. Then these optimum conditions (pH= 5.0-5.5, time of stirring= 30 min., and weight of SZP= 100mg) were applied during the study of the interfering effects of the different foreign ions on the efficiency of separation and the effect of concentration of the eluent acids (10 cm^3 of HCl or HNO_3) on the recovery. The distribution coefficient (K_d) is determined using the equation: (Helfferich, 1962; Korkisch, 1969).

$$K_d = C_{iex\,(mg/g)}/C_{sol\,(mg/cm3)} \qquad cm^3/g \qquad (1)$$

Where C_{iex} is the metal concentration in the ion sorbent (solid phase) and C_{sol} is the metal ion concentration in the solution phase. For the capacity determination, 100mg of the ion sorbent was added to 100 cm^3 of 100 $\mu g/cm^3$ of Cu and stirred for 24 h, at pH 5.0-5.5. The decrease in the metal concentration was determined by AAS and accordingly the capacity is calculated by the equation: 63.55

$$Capacity= \frac{(100 - C_{sol}(\frac{\mu g}{cm^3}))/63.55}{} \qquad (2)$$

Where C_{sol} is the concentration of the solution after extraction.

Application

100 mg of ion sorbent was added to 10 dm^3 of the sample and the pH value was adjusted to 5.0 - 5.5 and stirred for 30 min then the sample was filtered. To the filtrate another 50 mg of the ion sorbent was added and pH value was again controlled. The sample was stirred again for 30 min. and filtered. Both residue were gathered and the collected metal ions were released by 10 cm^3 2 M HNO_3 to give a concentration factor of 100 fold.

Equipment

The absorption measurements were made with a Perkin Elmer

model A-Analyst 100, equipped with hollow cathode lamps for copper, zinc, lead, cadmium, nickel and iron, portable digital Orion pH meter, model 230A, was used and also IR spectrum model (Perkin Elmer, 1430) was used for the identification of prepared compound.

RESULTS AND DISCUSSION

Optimization of separation conditions of some metal ions, Cu^{2+}, Zn^{2+}, Pb^{2+}, Cd^{2+}, Ni^{2+} and Fe^{3+}

Best stability against hydrolysis and capacity towards Cu^{2+} (1.09 mmole/g) was found for ion sorbent SZP functionalized with 1.0 g $ZrOCl_2.8H_2O$, 0.37 cm^3 H_3PO_4 and annealed at 400°C. Hence, the optimum conditions of separation of Cu^{2+}, Zn^{2+}, Pb^{2+}, Cd^{2+}, Ni^{2+} and Fe^{3+} were studied on SZP synthesized at these conditions.

Effect of pH

Figure 2 represents the effect of pH on the uptake behavior of Cu^{2+}, Zn^{2+}, Pb^{2+}, Cd^{2+}, Ni^{2+} and Fe^{3+} on SZP. The distribution coefficient of these metals on SZP increases with pH and reaches a maximum value at pH = 5.0 - 5.5. The decrease in the distribution coefficient at pH>7 is attributed to the hydrolysis of the silica based ion sorbents (Hamed et al., 2006). The order of decreasing the efficiency of separation at the pH 5.0 - 5.5 is arranged as follows: $Cu^{2+} > Ni^{2+} > Cd^{2+} > Pb^{2+} > Zn^{2+} > Fe^{3+}$.

Effect of stirring time

Effect of stirring time 1 to 60 min on the separation efficiency of the investigated heavy metal ions was studied at pH 5.0 - 5.5. It was found that the recovery

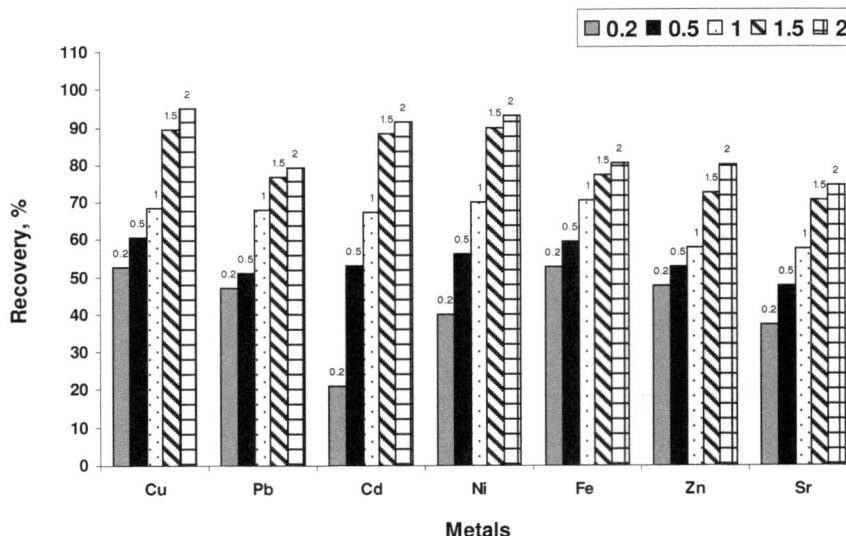

Figure 3. Effect of time of stirring on the separation of Cu^{2+}, Cd^{2+}, Pb^{2+}, Zn^{2+}, Ni^{2+} and Fe^{3+}.

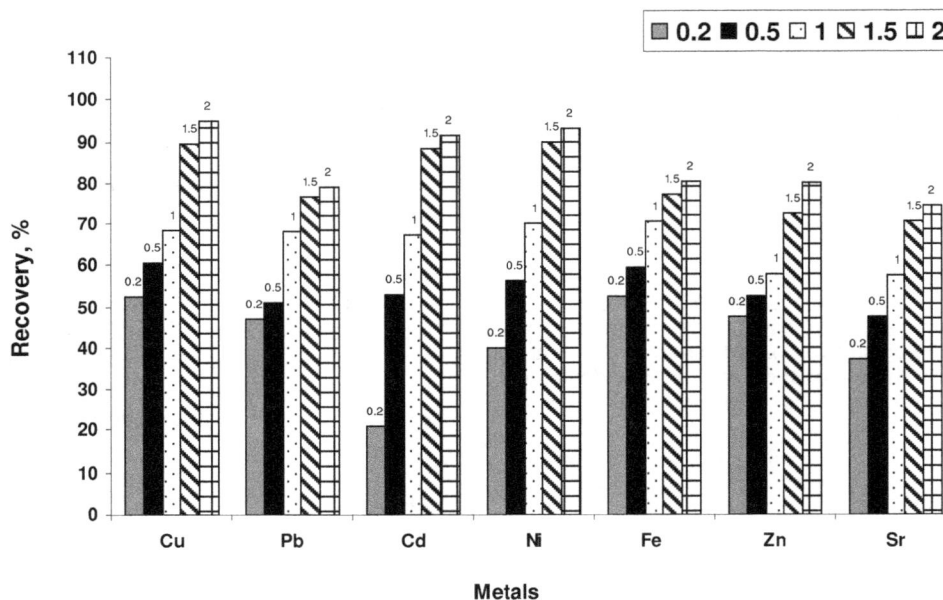

Figure 4. Effect of concentration of HCl used for elution on the recovery of Cu^{2+}, Cd^{2+}, Pb^{2+}, Zn^{2+}, Ni^{2+} and Fe^{3+}.

(%) increases with time of stirring up to 30 min, then becomes independent on the time of stirring at pH = 5.0-5.5 using SZP (Figures. 3).

Effect of eluting acid concentration on metal recovery

The optimum conditions of separation of studied heavy metal ions are pH= 5.0 -5.5, time of stirring= 30 min, and mass of modified silica gel ion sorbent≥20 mg. These conditions were applied to collect studied metal ions as well as to study the effect of eluting acid (10 ml of HCl and HNO_3). Figures 4 and 5) show the effect of concentration of eluting acids on the recovery of the investigated metal ions. The results showed that HNO_3 (Figure 5) releases the investigated metal ions from the ion sorbent more efficiently than HCl (Figure 4).

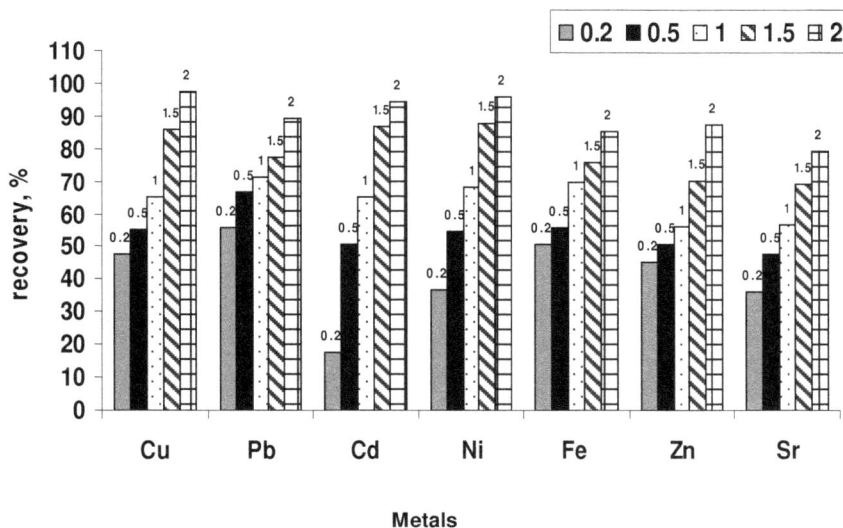

Figure 5. Effect of concentration of HNO_3 used for elution on the recovery of Cu^{2+}, Cd^{2+}, Pb^{2+}, Zn^{2+}, Ni^{2+} and Fe^{3+}.

Table 1. Effect of some common

Ionic Species	Recovery %					
	Cu	Pb	Zn	Cd	Ni	Fe
None	92.2	90.7	86.6	92.8	89.7	88.7
EDTA	60.6	77.8	43.1	61.5	58.7	57.3
Acetate	92.3	90.7	86.5	92.3	89.2	88.5
Sulphate	80	91.7	85.4	93.4	89.8	90.1
Citrate	90.7	61.7	81.9	83.2	80.1	82.4
Oxalate	93.2	92.7	90.7	93.5	90.7	87.9
Nitrate	91.8	90.2	85.1	90.3	87.4	87.3
Phosphate	97.4	93.5	94.9	95.6	92.9	89.2

Ionic species on the separation of investigated metal ions on SZP.

Effect of interfering ions

The analytical preconcentration procedure for trace metal ions can strongly interfere with matrix constituents, and then the tested preconcentration procedure must be examined in the presence of the possible matrix elements. The effect of EDTA, acetate, sulphate, citrate, oxalate, nitrate and phosphate which represent the common ionic species in natural water samples were studied applying the optimum conditions detected above. As shown in Table 1, EDTA caused serious interference with the detected metal ions; this interference was attributed to competition between the complexation of the metal ions with the phosphate substrate on silica and with the interfering species in solution. Therefore, EDTA could be used successfully with HNO_3 acid as a good eluent for metal ions from the loaded SZP. On the other hand, oxalate, sulphate and citrate showed slight interference for all metal ions, while the investigated metal ions using silica gel-based zirconyl phosphate ion sorbent, showed a good resistance to most interfering ions such as acetate and nitrate.

Application

The SPE method was applied for the preconcentration and separation of Cu^{2+}, Zn^{2+}, Pb^{2+}, Cd^{2+}, Ni^{2+} and Fe^{3+} in water samples where the procedure was applied to one litre of sample. The analytical results of different samples of water collected from the northern part of Suez Gulf starting from summer 2004 until spring 2005 using SZP are shown in Tables 2 to 7. A comparison was performed with the standard solvent extraction method (APDC/MIBK).

In the case of using APDC/MIBK for the

Table 2. Copper concentration (µg/L) in seawater of Suez Gulf during period of study (2004 - 2005).

Station	Summer SZP	Summer APDC/MIBK	Winter SZP	Winter APDC/MIBK	Autumn SZP	Autumn APDC/MIBK	Spring SZP	Spring APDC/MIBK	Annual Mean SZP	Annual Mean APDC/MIBK
1	3.99±0.04	3.73±0.03	4.59±0.02	4.42±0.02	4.48±0.03	4.37±0.03	4.27±0.04	3.98±0.04	4.33±0.03	4.13±0.03
2	2.34±0.01	2.04±0.01	3.60±0.01	3.26±0.01	3.72±0.03	3.35±0.03	3.75±0.04	2.59±0.02	3.35±0.02	2.81±0.02
3	2.44±0.03	2.10±0.02	2.95±0.02	1.94±0.01	3.97±0.01	3.69±0.01	2.76±0.02	2.18±0.01	3.03±0.02	2.48±0.01
4	3.78±0.02	2.37±0.01	3.92±0.02	2.84±0.01	4.14±0.03	3.91±0.04	3.61±0.01	2.93±0.01	3.86±0.02	3.01±0.02
5	3.76±0.04	3.37±0.03	4.03±0.02	3.80±0.02	4.27±0.03	3.73±0.02	3.77±0.04	3.54±0.03	3.95±0.03	3.61±0.03
6	1.97±0.01	1.26±0.01	3.11±0.01	2.31±0.02	3.54±0.01	2.48±0.02	2.07±0.02	1.82±0.01	2.67±0.01	1.97±0.01
7	2.36±0.02	1.57±0.01	3.16±0.02	2.88±0.01	3.59±0.01	3.45±0.01	2.19±0.01	1.91±0.01	2.82±0.01	2.45±0.01
8	3.26±0.02	3.18±0.02	3.75±0.02	3.60±0.01	4.12±0.02	3.85±0.01	4.12±0.05	3.77±0.04	3.81±0.03	3.60±0.02
9	3.50±0.03	2.39±0.01	3.74±0.01	2.94±0.01	4.28±0.02	3.78±0.01	3.29±0.03	2.85±0.02	3.70±0.02	2.99±0.01
10	2.78±0.01	2.44±0.01	3.35±0.01	3.05±0.01	4.16±0.02	3.88±0.01	3.06±0.03	2.98±0.02	3.33±0.02	3.09±0.01
Mean	3.02±0.02	2.45±0.01	3.62±0.01	3.10±0.01	4.03±0.02	3.65±0.02	3.29±0.03	2.86±0.02	3.49±0.02	3.02±0.02

Table 3. Zinc concentration (µg/L) in seawater of Suez Gulf during period of study (2004 - 2005).

Station	Summer SZP	Summer APDC/MIBK	Winter SZP	Winter APDC/MIBK	Autumn SZP	Autumn APDC/MIBK	Spring SZP	Spring APDC/MIBK	Annual Mean SZP	Annual Mean APDC/MIBK
1	6.63±0.05	6.16±0.04	11.92±0.07	11.72±0.07	10.71±0.06	10.61±0.06	9.11±0.06	8.97±0.04	9.59±0.02	9.37±0.05
2	5.30±0.03	3.92±0.02	10.17±0.05	9.11±0.06	8.68±0.03	8.29±0.03	7.70±0.02	7.48±0.03	7.96±0.01	7.20±0.04
3	3.89±0.02	3.26±0.02	10.35±0.04	9.98 ±0.01	8.10±0.02	7.93±0.02	6.09±0.01	5.96±0.01	7.11±0.01	6.78±0.02
4	5.10±0.03	4.13±0.02	10.53±0.05	10.39±0.04	8.32±0.03	8.17±0.06	7.12±0.03	6.92±0.03	7.76±0.01	7.40±0.03
5	8.16±0.04	7.19±0.04	15.33±0.09	13.69±0.08	11.21±0.09	11.13±0.07	10.71±0.04	10.46±0.05	11.35±0.03	10.62±0.06
6	2.09±0.01	1.76±0.01	5.12±0.03	4.83±0.03	5.09±0.06	4.65±0.05	4.09±0.01	4.15±0.01	4.09±0.03	3.85±0.03
7	3.49±0.02	2.71±0.01	7.23±0.04	6.54±0.01	5.10±0.05	4.75±0.01	4.29±0.01	4.41±0.02	5.03±0.03	4.60±0.01
8	4.74±0.02	3.77±0.01	8.38±0.04	8.16±0.04	6.92±0.03	6.81±0.03	6.01±0.02	5.73±0.02	6.51±0.03	6.12±0.02
9	4.41±0.02	4.33±0.02	6.90±0.03	5.61 ±0.02	6.29±0.02	4.81±0.04	4.92±0.03	4.48±0.04	5.63±0.02	4.81±0.03
10	4.34±0.02	4.27±0.02	5.32±0.08	4.93±0.07	6.50±0.03	5.96±0.01	5.01±0.05	4.84±0.04	5.29±0.05	5.00±0.04
Mean	4.82±0.02	4.15±0.02	9.13±0.05	8.49±0.04	7.69±0.04	7.31±0.04	6.51±0.03	6.34±0.03	7.04±0.04	6.57±0.03

preconcentration and separation of Cu^{2+}, Zn^{2+}, Pb^{2+}, Cd^{2+}, Ni^{2+} and Fe^{3+} in water, the metal ion concentrations were in the range of [(3.02-4.03), (4.82–9.13), (0.59-1.41), (0.39-0.69), (0.89-1.57) and (7.82-12.08) µg/L], respectively. On a seasonal scale, trace metal concentrations increased gradually from their minimum values during spring and summer to their maximum values during winter and autumn except for Pb, which recorded its highest concentration during spring. This decrease in concentration during spring may be due to their consumption by

Table 4. Lead concentration (µg/L) in seawater of Suez Gulf during period of study (2004-2005).

Season Station	Summer		Winter		Autumn		Spring		Annual Mean	
	SZP	APDC/MIBK	SZP	APDC/MIBK	SZP	APDC/MIBK	SZP	APDC/MIBK	SZP	APDC/MIBK
1	0.61±0.02	0.51±0.01	0.97±0.06	0.85±0.06	0.88±0.03	0.57±0.03	1.77±0.04	1.59±0.03	1.06±0.04	0.88±0.03
2	0.37±0.01	0.33±0.01	0.92±0.01	0.75±0.01	0.27±0.01	0.23±0.01	1.58±0.03	1.48±0.03	0.79±0.02	0.69±0.02
3	0.49±0.01	0.42±0.01	1.03±0.02	0.93±0.02	0.63±0.01	0.49±0.01	0.91±0.02	0.77±0.01	0.77±0.02	0.65±0.01
4	0.66±0.03	0.45±0.02	1.09±0.02	0.97±0.02	0.72±0.02	0.69±0.01	1.07±0.02	0.91±0.02	0.89±0.02	0.75±0.02
5	1.26±0.02	1.13±0.01	1.52±0.02	1.37±0.02	0.95±0.03	0.77±0.02	1.55±0.03	0.98±0.02	1.32±0.03	1.06±0.02
6	0.26±0.02	0.23±0.01	0.45±0.03	0.39±0.03	0.47±0.03	0.39±0.02	0.48±0.02	0.40±0.02	0.42±0.03	0.35±0.02
7	0.29±0.02	0.26±0.03	0.65±0.03	0.57±0.02	0.58±0.02	0.42±0.02	0.60±0.03	0.49±0.02	0.53±0.03	0.43±0.02
8	1.27±0.03	1.22±0.02	2.03±0.03	1.90±0.01	1.30±0.04	1.22±0.03	2.12±0.06	1.95±0.04	1.68±0.04	1.57±0.03
9	0.31±0.05	0.15±0.03	0.61±0.03	0.54±0.01	0.90±0.04	0.75±0.04	1.95±0.04	1.62±0.03	0.94±0.04	0.76±0.03
10	0.34±0.02	0.19±0.03	0.51±0.04	0.49±0.02	0.94±0.03	0.87±0.03	2.04±0.05	1.73±0.04	0.96±0.03	0.82±0.03
Mean	0.59±0.02	0.49±0.01	0.98±0.03	0.88±0.02	0.76±0.02	0.64±0.02	1.41±0.03	1.19±0.02	0.94±0.02	0.80±0.02

Table 5. Cadmium concentration (µg/L) in seawater of Suez Gulf during period of study (2004-2005).

Season Station	Summer		Winter		Autumn		Spring		Annual Mean	
	SZP	APDC/MIBK	SZP	APDC/MIBK	SZP	APDC/MIBK	SZP	APDC/MIBK	SZP	APDC/MIBK
1	0.29±0.01	0.24±0.01	0.72±0.02	0.45±0.01	0.38±0.01	0.29±0.01	0.36±0.03	0.29±0.01	0.44±0.02	0.31±0.01
2	0.61±0.02	0.40±0.01	1.08 ±0.02	0.94 ±0.01	0.72±0.02	0.50±0.01	0.64±0.01	0.48±0.03	0.76±0.02	0.58±0.01
3	0.29±0.03	0.13±0.01	0.73±0.02	0.65±0.01	0.49±0.02	0.25±0.01	0.36±0.01	0.29±0.01	0.47±0.02	0.33±0.01
4	0.36±0.03	0.15±0.02	0.74±0.02	0.66±0.01	0.47±0.03	0.30±0.01	0.33±0.02	0.24±0.01	0.47±0.03	0.33±0.01
5	0.41±0.05	0.28±0.02	0.80±0.03	0.83±0.02	0.56±0.03	0.49±0.02	0.59±0.02	0.45±0.03	0.59±0.03	0.51±0.02
6	0.19±0.01	0.10±0.01	0.42±0.01	0.29±0.02	0.30±0.01	0.13±0.01	0.25±0.02	0.12±0.01	0.29±0.01	0.16±0.01
7	0.22±0.01	0.11±0.01	0.45±0.01	0.36±0.01	0.30±0.01	0.15±0.01	0.39±0.01	0.25±0.02	0.34±0.01	0.21±0.01
8	0.58 ±0.03	0.37±0.03	0.84 ±0.01	0.72±0.01	0.53±0.02	0.39±0.01	0.50±0.01	0.38±0.03	0.61±0.02	0.46±0.02
9	0.47±0.06	0.22±0.04	0.55±0.02	0.37±0.01	0.35±0.03	0.28±0.02	0.43±0.02	0.32±0.04	0.45±0.03	0.29±0.03
10	0.44±0.04	0.23±0.03	0.61±0.02	0.44±0.01	0.40±0.02	0.31±0.02	0.49±0.04	0.33±0.03	0.48±0.03	0.32±0.02
Mean	0.39±0.03	0.22±0.01	0.69±0.02	0.57±0.01	0.45±0.02	0.31±0.01	0.43±0.02	0.32±0.02	0.49±0.02	0.35±0.01

phytoplankton. During the period of study, it can be seen that, spring showed the highest concentration of Pb (2.12 µg/dm³) at El-Adabyiastation and this may be due to that, this station considered the most active and largest fishing harbour in the Gulf of Suez and the loaded gasoline spelt in the water surface during the fishing ships loaded. The high concentration of trace metals in winter as compared with other seasons may be attributed to terrestrial inputs of dust particles containing metal ions and strong winds throughout the water column including

Table 6. Nickel concentration (µg/L) in seawater of Suez Gulf during period of study (2004-2005).

Season / Station	Summer		Winter		Autumn		Spring		Annual Mean	
	SZP	APDC/MIBK	SZP	APDC/MIBK	SZP	APDC/MIBK	SZP	APDC/MIBK	SZP	APDC/MIBK
1	0.88±0.01	0.78±0.01	2.07±0.05	0.95 ±0.04	1.90±0.05	1.10±0.02	1.89±0.03	1.10±0.02	1.68±0.03	0.98±0.02
2	1.32±0.02	1.21±0.02	2.21±0.04	1.99 ±0.03	2.20±0.04	1.80±0.03	2.34±0.02	1.82±0.03	2.01±0.03	1.70±0.03
3	1.14±0.01	1.03±0.01	1.55±0.03	1.39±0.01	1.10±0.02	0.87±0.01	1.24±0.02	0.79±0.01	1.25±0.02	1.02±0.01
4	0.92±0.01	0.89±0.01	1.78±0.04	1.26±0.03	1.50±0.02	0.98±0.01	1.35±0.03	0.83±0.01	1.38±0.03	0.99±0.01
5	0.97±0.03	0.91±0.02	1.50±0.06	1.18±0.04	2.10±0.06	1.80±0.04	2.11 ±0.03	1.62±0.03	1.67±0.05	1.37±0.03
6	0.67±0.01	0.54±0.01	1.03±0.02	0.83 ±0.01	0.90±0.02	0.74±0.01	0.72±0.02	0.65±0.01	0.83±0.02	0.69±0.01
7	0.69±0.01	0.56±0.01	1.14 0.02	0.98 ±0.01	0.95±0.01	0.82±0.01	0.84±0.01	0.68±0.01	0.90±0.01	0.76±0.01
8	0.81±0.02	0.78±0.02	1.68±0.02	1.13 ±0.02	1.07±0.02	0.92±0.01	1.36±0.03	0.86±0.01	1.23±0.02	0.92±0.01
9	0.75±0.02	0.60±0.02	1.40±0.02	1.05±0.02	1.10±0.03	0.97±0.02	1.48±0.03	1.01±0.03	1.18±0.03	0.90±0.02
10	0.77±0.02	0.67±0.02	1.30±0.03	1.02±0.02	1.08±0.03	0.91±0.02	1.32±0.02	0.96±0.01	1.11±0.03	0.89±0.02
Mean	0.89±0.01	0.79±0.01	1.57±0.03	1.18±0.02	1.38±0.03	1.09±0.02	1.47±0.02	1.03±0.01	1.32±0.03	1.02±0.01

Table 7. Iron concentration (µg/L) in seawater of Suez Gulf during period of study (2004-2005).

Season / Station	Summer		Winter		Autumn		Spring		Annual Mean	
	SZP	APDC/MIBK	SZP	APDC/MIBK	SZP	APDC/MIBK	SZP	APDC/MIBK	SZP	APDC/MIBK
1	7.29 ±0.01	6.71±0.03	17.12±0.09	9.58±0.04	12.10 ±0.05	7.80±0.03	9.50±0.04	7.95±0.04	11.50±0.05	8.01±0.03
2	7.39±0.01	6.89±0.05	13.48±0.06	11.49 ±0.05	12.40±0.06	8.90±0.04	9.57±0.01	7.76±0.04	10.71±0.03	8.76±0.04
3	8.67±0.01	5.97±0.02	11.84±0.05	10.95±0.05	10.40±0.04	8.50±0.04	9.05±0.01	7.33±0.04	9.99±0.03	8.18±0.03
4	9.84 ±0.03	7.02±0.03	11.68±0.03	10.87±0.04	10.50±0.05	9.70±0.04	10.14±0.05	8.63±0.01	10.54±0.04	9.05±0.03
5	13.34±0.04	10.44±0.04	18.42±0.07	15.93±0.07	15.10±0.02	13.4 ±0.06	14.42±0.06	12.31±0.02	15.32±0.05	13.02±0.05
6	5.33 ±0.02	5.05 ±0.02	7.95±0.04	6.76 ±0.01	7.09±0.04	6.45±0.01	7.01±0.01	5.45±0.02	6.84±0.03	5.92±0.01
7	4.59 ±0.02	4.19 ±0.01	8.03±0.09	6.95 ±0.03	7.24±0.04	6.63±0.01	7.14±0.03	6.27±0.02	6.75±0.04	6.01±0.02
8	7.45 ±0.01	5.92 ±0.02	11.51±0.05	8.66±0.01	10.70±0.05	7.50±0.02	8.34±0.01	6.82±0.03	9.50±0.03	7.22±0.02
9	7.29±0.03	6.39±0.04	10.08±0.02	7.88±0.02	9.40±0.05	8.85±0.04	8.05 ±0.06	7.49±0.05	8.70±0.04	7.65±0.04
10	7.05±0.04	6.03±0.03	10.70±0.02	7.26±0.06	9.70±0.07	8.92±0.05	8.32±0.05	8.17±0.04	8.94±0.04	7.59±0.03
Mean	7.82±0.02	6.46±0.03	12.08±0.05	9.63±0.04	10.46±0.05	8.67±0.03	9.15±0.03	7.82±0.03	9.87±0.04	8.14±0.03

favourable conditions for metal ions to be transferred from solid suspended. The results indicated that, the northern part of Suez Gulf was more polluted than the southern part may be due to presence of different sources of pollutants such as the corrosion of ships hull coating with antifouling paints and the presence of many oil piers in this area. On the other hand, the western part of the gulf was highly polluted than the

Table 8. Efficiency for removal of metal ions (%) at ambient conditions.

Metal \ Station	Efficiency of separation, %				
	Zaitiya	NIOF	Adabyia	El-Kahraba	Uyun Mousa
Cu	34.39	9.16	26.43	6.20	17.99
Cd	34.43	32.14	70.27	42.5	18.42
Ni	43.55	8.38	45.38	59.09	20.69
Pb	21.65	23.85	47.57	5.08	26.15
Zn	3.39	4.79	32.52	5.02	9.98
Fe	3.08	3.33	29.06	3.06	7.01
Sr	59.02	26.08	11.32	26.00	27.14
Mg	38.52	42.26	12.13	15.61	62.00
Ca	9.56	7.68	14.23	10.62	7.41

eastern part due to the presence of much sources of pollution in this area except for Ni and Cd; their high concentration in the eastern side may be due to the presence of power thermal station at Aeon Mousa. Tables 2 to 7 showed that, Ain Sukhna station (St. VI) has the lowest concentrations of heavy metal ions, these values for Cu^{2+}, Zn^{2+}, Pb^{2+}, Cd^{2+}, Ni^{2+} and Fe^{3+} were (1.97, 2.09, 0.26, 0.19, 0.67 and 5.33 µg/L, respectively). According to the data about metal level in water, it can be observed that, Cd and Ni give their maximum values at El-Zeitiya station 1.08 and 2.34 µg/L, respectively. These results may be due to this station characterized by the presence of refinery factories and power plant discharge. While the lowest value of Cd and Ni (0.19 and 0.67) µg/L, respectively were recorded at Ain Sukhna station. Regionally the lowest value of copper (1.97 µg/L) occurred at Ain Sukhna station and this may be due to remoteness of this station from any sources of pollution. On the other hand the highest concentration of copper was found at 4.59 µg/L at Summer Balas. This is probably due to the emotion of this station by small fishing boats, touristic impact and corrosion of ships hull coating with antifouling paints. The presence of the highest concentration of Zn and Fe at NIOF may be due to the leaching of metals from the ships passing through the Suez Canal. While the lowest concentrations of these metals were recorded at Sand Beach station and this may be due to remoteness of this station from any sources of pollution. Ion exchange method (SZP) showed the same results but with values lower than that of SZP method. However it was found that there is no significant difference was found between the values obtained by the ion exchange and that of solvent extraction methods; the reliability of the two methods will be examined statistically. Also the results of APDC/MIBK system showed that, Ain Sukhna station has the lowest concentrations of heavy metal ions (Tables 2 to 7). Comparing the results obtained from the present study with those reported in the early studies (Hamed, 1996) (Abdel-Azim, 2002), there is an increasing in the heavy metal concentrations can be observed. This was explained by the increasing of factories along the coast of northern part of the Gulf of Suez and presence of more sources of pollution along this area.

Utilization of silica for the removal of metal ions at ambient conditions

Marine pollution may be due different sources; almost 80% of this pollution and coastal deterioration in Egypt coasts are attributed to different land activities whether industrial, agricultural, urban or physical, in particular those produced wastes and emissions that are not treated in a sound environmental manner (GESAMP, 1993). The remaining 20% is due to other sources ahead of which are marine sources i.e. different offshore activities such as oil, mineral and natural gas exploration and drilling operations, besides fishes, shipping and unloading and marine transportation (EL-Shenawy et al., 2006). Most toxic heavy metals have been discharged into the environment as industrial wastes, causing serious soil and water pollution. (Lin and Juang 2002) Pb^{2+}, Cu^{2+}, Fe^{3+}, and Cr^{3+} are especially common metals that tend to accumulate in organisms, causing numerous diseases and disorders (Erdem et al., 2004). Various treatment processes are available; among which ion exchange is considered cost- effective if low-cost ion sorbents such as zeolites are used (Bailey et al., 1999).

Table 8 shows the results of treatment of trace metals for the highest polluted locations at ambient conditions (natural pH of the environment of studied area).It was found that, the highest percentage of the efficiency for copper (II) (34.39%) was recorded at Zaitiya location at which strontium also was co-sorbed with 59.02%. Adabyia location represented the maximum percent for the removal of cadmium (II) (70.27%). Power Station (St. IV) represented the highest percent for Ni^{2+} efficiency (59.09%) while; Adabyia station has the maximum percentage for Pb^{2+}, Zn^{2+}, Fe^{3+} and Ca^{2+} ions (47.57, 32.52, 29.06 and 14.23%, respectively).

Table 9. Statistical evaluation for Cu (II) analysis in natural water samples after preconcentration by solvent extraction (method 1) and Ion sorbent (method 2), n =5.

Location	Method 1		Method 2		S_P	t- Test	Two- tailed F- test
	$\overline{X_1}$ ng/ml	S_1	$\overline{X_2}$ ng/ml	S_2			
1	4.59	0.069	4.55	0.067	0.068	1.31	1.06
2	3.60	0.062	3.59	0.058	0.060	0.37	1.14
3	2.95	0.042	2.92	0.039	0.040	1.65	1.15
4	3.92	0.052	3.89	0.049	0.050	1.32	1.12
5	4.03	0.079	3.99	0.065	0.072	1.23	1.47
6	3.11	0.039	3.09	0.034	0.036	1.22	1.31
7	3.16	0.051	3.13	0.042	0.046	1.43	1.47
8	3.75	0.070	3.72	0.062	0.066	1.01	1.27
9	3.74	0.081	3.70	0.063	0.072	1.23	1.65
10	3.35	0.053	3.31	0.045	0.049	1.81	1.38

Table 10. Statistical evaluation for Zn (II) analysis in natural water samples after preconcentration by solvent extraction (method 1) and Ion sorbent (method 2), n =5.

Location	Method 1		Method 2		S_P	t- Test	Two- tailed F- test
	$\overline{X_1}$ ng/ml	S_1	$\overline{X_2}$ ng/ml	S_2			
1	11.92	0.179	11.89	0.129	0.156	0.43	1.92
2	10.16	0.168	10.11	0.162	0.165	0.68	1.07
3	10.35	0.169	10.30	0.135	0.152	0.73	1.56
4	10.53	0.149	10.50	0.139	0.144	0.47	1.15
5	15.33	0.301	15.31	0.229	0.267	0.17	1.72
6	5.12	0.070	5.09	0.067	0.068	0.98	1.09
7	7.23	0.121	7.19	0.113	0.117	0.76	1.14
8	8.37	0.150	8.34	0.124	0.137	0.48	1.46
9	6.91	0.129	6.87	0.121	0.125	0.71	1.14
10	5.32	0.088	5.28	0.08	0.084	1.06	1.21

Statistical evaluation of the ion exchange method compared with the standard solvent extraction method

The statistical analysis of the results was performed on commercial software. The agreement between a series of results measured by computing their mean deviation. This was evaluated by determining the arithmetical mean of the results (X), then calculating the deviation of each individual measurement from the mean and finally dividing the sum of the deviations, regardless of sign, by the number of measurements (n).

$$\text{Mean} = \overline{X} = X_1 + X_2 + \ldots\ldots X_n \qquad \frac{X_1 + X_1 + \ldots\ldots X_n{}^2}{n} \qquad (3)$$

In analytical chemistry, one of the most employed common statistical terms is the standard deviation (s) of a population of observations. This is also called the root mean square deviation, as it is the square root of the mean of the sum of the squares of the differences between the values and the mean of those values (El-Moursi, 2001).

$$S = \sqrt{\frac{\Sigma(X_i - \overline{X})^2}{n-1}} \qquad (4)$$

The comparison between the means of the two analytical methods can be performed by $|t_2|$ – test (Miller and Miller, 1986) at P= 0.05. F-test was also used for comparison of standard deviation, that is, the random error for two sets of data given by two analytical methods (EL-Shenawy et al., 2006). Tables 9-14 show the statistical analysis (mean values, standard deviation and " $|t_2|$ " values) for heavy metal samples collected during

Table 11. Statistical evaluation for Pb (II) analysis in natural water samples after preconcentration by solvent extraction (method 1) and Ion sorbent (method 2), n =5.

Location	Method 1		Method 2		S_P	t- Test	Two- tailed F- test
	$\overline{X_1}$ ng/ml	S_1	$\overline{X_2}$ ng/ml	S_2			
1	0.97	0.080	0.92	0.075	0.077	1.44	1.13
2	0.92	0.071	0.89	0.069	0.070	0.95	1.05
3	1.03	0.070	0.99	0.059	0.064	1.38	1.40
4	1.09	0.050	1.06	0.044	0.047	1.42	1.29
5	1.52	0.029	1.50	0.023	0.026	1.70	1.58
6	0.45	0.072	0.41	0.066	0.069	1.29	1.19
7	0.66	0.022	0.65	0.017	0.019	1.13	1.67
8	2.03	0.061	1.99	0.052	0.056	1.57	1.37
9	0.61	0.030	0.59	0.021	0.025	1.72	2.04
10	0.49	0.019	0.48	0.016	0.017	1.27	1.41

Table 12. Statistical evaluation for Cd (II) analysis in natural water samples after preconcentration by solvent extraction (method 1) and Ion sorbent (method 2), n =5.

Location	Method 1		Method 2		S_P	t- Test	Two- tailed F- test
	$\overline{X_1}$ ng/ml	S_1	$\overline{X_2}$ ng/ml	S_2			
1	0.72	0.061	0.68	0.059	0.060	1.49	1.06
2	1.08	0.069	1.03	0.057	0.063	1.76	1.46
3	0.75	0.052	0.72	0.048	0.050	1.34	1.17
4	0.74	0.071	0.71	0.061	0.066	1.01	1.35
5	0.84	0.061	0.83	0.052	0.056	0.39	1.37
6	0.42	0.050	0.40	0.047	0.048	0.92	1.13
7	0.45	0.070	0.44	0.060	0.065	0.34	1.36
8	0.84	0.051	0.81	0.041	0.046	1.44	1.54
9	0.55	0.059	0.51	0.054	0.056	1.58	1.19
10	0.61	0.059	0.59	0.050	0.054	0.81	1.39

Table 13. Statistical evaluation for Ni (II) analysis in natural water samples after preconcentration by solvent extraction (method 1) and Ion sorbent (method 2), n =5.

Location	Method 1		Method 2		S_P	t- Test	Two- tailed F- test
	$\overline{X_1}$ ng/ml	S_1	$\overline{X_2}$ ng/ml	S_2			
1	2.07	0.032	2.05	0.022	0.027	1.60	1.95
2	2.21	0.039	2.19	0.036	0.037	1.17	1.13
3	1.55	0.026	1.54	0.019	0.022	0.97	1.85
4	1.78	0.092	1.75	0.089	0.090	0.73	1.06
5	1.50	0.099	1.46	0.086	0.093	0.96	1.32
6	1.03	0.072	0.99	0.064	0.068	1.31	1.26
7	1.14	0.090	1.11	0.050	0.072	0.92	3.23
8	1.68	0.073	1.65	0.069	0.071	0.94	1.11
9	1.40	0.028	1.39	0.025	0.026	0.83	1.19
10	1.32	0.072	1.30	0.062	0.067	0.66	1.35

Table 14. Statistical evaluation for Fe (III) analysis in natural water samples after preconcentration by solvent extraction (method 1) and Ion sorbent (method 2), n =5.

Location	Method 1		Method 2		S_P	t- Test	Two- tailed F- test
	$\overline{X_1}$ ng/ml	S_1	$\overline{X_2}$ ng/ml	S_2			
1	17.12	0.265	17.09	0.191	0.231	0.29	1.91
2	13.48	0.238	13.44	0.225	0.232	0.38	1.11
3	11.85	0.199	11.81	0.146	0.174	0.51	1.84
4	11.68	0.175	11.66	0.166	0.170	0.26	1.10
5	18.42	0.368	18.40	0.261	0.319	0.14	1.98
6	7.95	0.112	7.93	0.102	0.107	0.41	1.19
7	8.03	0.134	7.99	0.110	0.122	0.72	1.47
8	11.51	0.217	11.49	0.209	0.213	0.20	1.07
9	10.07	0.201	10.02	0.184	0.192	0.57	1.18
10	10.70	0.184	10.65	0.172	0.178	0.62	1.13

Table 15. Correlation between parameters using Solvent extraction (SE) method during period of study (2004-2005) at P <0.05.

Metals	Cu (SE)	Zn (SE)	Pb (SE)	Cd (SE)	Ni (SE)	Fe (SE)	Temp.	pH	S‰	D O	NO$_2$	NO$_3$	NH$_3$	PO$_4$
Cu (SE)	1.00													
Zn (SE)	0.73	1.00												
Pb (SE)	0.73	0.49	1.00											
Cd (SE)	0.40	0.52	0.59	1.00										
Ni (SE)	0.61	0.81	0.37	0.79	1.00									
Fe (SE)	0.70	0.97	0.57	0.58	0.77	1.00								
Temp.	0.25	-0.04	-0.06	-0.02	0.03	-0.03	1.00							
pH	0.16	0.47	0.29	0.28	0.24	0.57	-0.05	1.00						
S‰	0.44	0.17	0.20	0.27	0.32	0.21	0.66	0.25	1.00					
D O	0.11	-0.17	0.28	-0.22	-0.26	-0.26	-0.35	-0.61	-0.45	1.00				
NO$_2$	0.51	0.65	0.81	0.65	0.50	0.68	-0.19	0.29	0.00	0.25	1.00			
NO$_3$	0.59	0.76	0.79	0.74	0.67	0.79	-0.18	0.45	0.12	0.12	0.94	1.00		
NH$_3$	0.61	0.81	0.73	0.69	0.69	0.80	-0.16	0.28	0.00	0.19	0.95	0.95	1.00	
PO$_4$	0.44	0.86	0.52	0.52	0.63	0.85	-0.23	0.54	-0.09	-0.03	0.83	0.87	0.90	1.00

the period of study (2004-2005). $|t_2|$ and F values were less than the tabulated ones indicating that the SPE method is precise and accurate.

Correlation coefficients were carried out using statistical computer program. They were calculated between all pairs of measured variables

for the two methods; solvent extraction with APDC/MIBK (Table 15) and SPE using SZP (Table 16). The results showed that:

Table 16. Correlation between parameters using ion exchange (SZP) method during period of study (2004-2005) at P <0.05.

Metals	Cu (SZP)	Zn (SZP)	Pb (SZP)	Cd (SZP)	Ni (SZP)	Fe (SZP)	Temp.	pH	S‰	D O	NO2	NO3	NH3	PO4
Cu (SZP)	1.00													
Zn (SZP)	0.71	1.00												
Pb (SZP)	0.75	0.42	1.00											
Cd (SZP)	0.47	0.61	0.60	1.00										
Ni (SZP)	0.26	0.61	0.20	0.89	1.00									
Fe (SZP)	0.48	0.84	0.36	0.67	0.67	1.00								
Temp.	-0.07	-0.09	-0.09	-0.07	-0.01	0.09	1.00							
pH	0.17	0.42	0.22	0.42	0.39	0.69	-0.05	1.00						
S‰	0.23	0.09	0.18	0.23	0.23	0.23	0.66	0.25	1.00					
D O	0.24	-0.12	0.35	-0.24	-0.47	-0.47	-0.34	-0.62	-0.45	1.00				
NO2	0.55	0.64	0.82	0.73	0.46	0.57	-0.19	0.29	0.00	0.25	1.00			
NO3	0.62	0.74	0.77	0.83	0.63	0.69	-0.18	0.45	0.11	0.11	0.94	1.00		
NH3	0.61	0.80	0.71	0.78	0.60	0.67	-0.16	0.28	0.00	0.18	0.95	0.95	1.00	
PO4	0.46	0.85	0.47	0.67	0.61	0.78	-0.23	0.54	-0.09	-0.03	0.83	0.87	0.90	1.00

1. Temperature is correlated with salinity, where positive correlation (r= 0.66). This due to salinity increases during hot seasons and vice versa.

2. The high significant positive correlation coefficient was found between nutrient salts and trace metals (Tables 15 and 16) indicate that the factors responsible for their distribution are similar.

3. There is high significant positive correlation between Cu and Fe (r =0.70) in the case of using APDC/MIBK method for preconcentration and separation of metal ions and the moderate correlation for SZP (r =0.48). This positive correlation indicates that Cu is mostly adsorbed by amorphous iron (Johnson, 1986).

4. Zn is highly correlated with Fe and Cu, where high significant positive correlations for APDC/MIBK method (r= 0.97 and 0.73, respectively) while, for SZP (r = 0.84 and 0.71, respectively). These correlations are due to adsorbing Zn by hydrous iron oxide (Kester, 1974). However, the obtained correlation between Zn and Cu is due to that both metals are insoluble in the oxidized states and form sulphides in reducing conditions.

Conclusion

The optimum conditions for the concentration of Cu^{2+}, Zn^{2+}, Pb^{2+}, Cd^{2+}, Ni^{2+} and Fe^{3+} using chemically-modified silica with zirconium phosphate are pH= 5.0-5.5, time of stirring=30 min and eluent acid 2M HNO_3. Citrate and EDTA cause strong interference therefore digestion of organic matter should be performed prior to the separation process with the ion sorbent.

The SPE method was applied for preconcentration and separation of Cu^{2+}, Zn^{2+}, Pb^{2+}, Cd^{2+}, Ni^{2+} and Fe^{3+} in the water samples in comparison with the standard solvent extraction method. On a seasonal scale, trace metal concentrations increased gradually from their minimum during spring and summer to their maximum during winter and autumn except for Pb, which recorded its high concentration during spring. This decrease in concentration during spring may be due to their consumption by phytoplankton. During the period of study, it can be seen that, spring showed the highest concentration of Pb (2.12 µg/L) at El-Adabyia station and this may be due to that, this station considered the most active and largest fishing harbour in the Gulf of Suez and the loaded gasoline spelt in the water surface during the fishing ships loaded. The ion sorbent (SZP) can be used in the separation and preconcentration of Cu^{2+}, Zn^{2+}, Pb^{2+}, Cd^{2+}, Ni^{2+} and Fe^{3+} with high distribution coefficient at optimum conditions with no interference from common ionic species.

REFERENCES

Abdel-Azim H (2002). Heavy metals in Suez Canal relevant to

the impacts of landbased sources, Ph. D. Thesis, Faculty of Science, Mansoura University.

Abou-El-Sherbini KhS, Khalil MSh, El-Ayaan U (2003b). Study on the complex formation between N-prpylsalicyldimine based on silica as ion exchanger and some heavy metals. J. Mat. Sci. Techn., pp. 10, 12, 23.

El-Moursi RM (2001). Preconcentration, Separation and Determination of some metal ions using silica-based ion exchanger, M. Sc. Thesis, Faculty of Sience, Tanta University.

El-Sabh MI, Beltagy AI (1983). Hydrography and chemistry of the Gulf of Suez during September 1966. Bull. Inst. Oceanogr. Fish. 9: 78.

EL-Shenawy M, Farag A, Zaky M (2006).: Sanitary and aesthetic quality of Egyptian coastal waters of Aqaba Gulf, Suez Gulf and Red Sea. Egyptian J. Aquatic Res., 32(1): 220-234.

Erdem E, Karapinar N, Donat R (2004). The removal of heavy metal cations by natural Zeolite, J. Colloid and Interface sci., 280: 309 – 314.

Fifield FW, Haines PJ (2000). Environmental Analytical Chemistry, 2nd edn., Black Wall Science Ltd., Cambridge, (2000) pp. 363.

GESAMP, joint group of experts on the scientific aspects of marine pollution (1993). Impacts of oil and released chemicals and wastes on the marine environment". International Maritime Organization, London. Reports and studies No. 50

Hamed MA (1996). Determination of some micro-elements in aquatic ecosystems and their relation to the efficiency of aquatic life, Ph. D. Thesis, Faculty of Science, Mansoura University.

Hamed MA (2004). Determination of some metal ions in aquatic environments by atomic absorption spectrometry after concentration with modified silica. Bull. Nat. Inst. Oceanogra. Fish., A.R.E., Vol. 30.

Hamed MA, Abou-El-Sherbini KhS, Lotfy HR (2006). Determination of some metal ions in aquatic environs by atomic absorption spectrometry after concentration with modified silica.

Hamed MA, Emara AM (2006). Marine molluscs as biomonitors for heavy metal levels in the Gulf of Suez, Red Sea.

Helffferich F (1962). Ion exchange. McGraw-Hill, London.

Johnson CA (1986). The regulation of the trace element concentrations in river and estuarine waters contaminated with acid mine drainage, the adsorption of Cu and Zn on amorphous Fe oxyhydroxides, Geochemica et. Cosmochimica Acta, 50: 2443-8.

Kester DR (1974). Chemical behaviour of trace metals in the ocean. Maritimes, 11: 14-18.

Köklü Ü, Akman S, Göcer Ö, Döner G (1995). Separation and preconcentration of cobalt and nickel with 3-(trimethoxysiliyl)-1-propanethiol loaded on silica gel. Anal. Lett., 28: 357.

Korkisch J (1969). Modern methods for the separation of rarer metal ions. Pergamon Press, Oxford. pp. 10.

Lashein RR (2005). Preconcentration and separation of some heavy metal ions in different environmental samples on a modified chelating silica gel and their analytical determination. PhD thesis, El-Mansoura University, Egypt.

Lin SH, Juang RS (2002). Heavy Metal Removal from Water by Sorption using Surfactant-Modified Montmorillonite. J. Hazard. Mater., B 92: 315.

Lopez AM, Camean A, Repetto M (1996). Preconcentration of heavy metals in urine using chelating ion exchange resin and quantification by ICP AES. At. Spectrosc. 17: 83.

Meshal (1970). Water pollution in Suez Bay. Bull. Inst. Oceanogr. Fish., ARE. 1: 463-473.

Miller JC, Miller JN (1986): Statistics for analytical chemistry. Ellis Horwood Limited, England, 1st Ed. pp. 43, 53, 59, 189, 192.

Obata H, Karatani H, Nakayama E (1993). Automated determination of iron in seawater by chelating resin concentration and chemiluminescence detection. Anal. Chem., 65: 1524.

Soliman EM, Mahmoud ME, Ahmed SA (2001). Synthesis, characterization and structure effects on selectivity properties of silica gel covalently bonded diethylenetriamine mono- and bis-salicyaldehyde and naphthaldehyde schiff's bases towards some heavy metal ions. Talanta, 54: 243.

7

Analysis of heavy metals found in vegetables from some cultivated irrigated gardens in the Kano metropolis, Nigeria

Abdulmojeed O. Lawal[1]* and Abdulrahman A. Audu[2]

[1]Nuclear Technology Centre, Nigeria Atomic Energy Commission, Sheda, P. M. B. 007, Gwagwalada, Abuja, FCT, Nigeria.
[2]Department of Chemistry, Bayero University, Kano. P. M. B. 3011, Kano, Nigeria.

The use of industrial and domestic wastewaters for irrigation on vegetable gardens is a public health concern. Using atomic absorption spectrophotometry (AAS), concentrations of Cr, Co, Cu, Ni, Pb and Zn were determined in four different vegetables, including, spinach, okra, onions and tomatoes, grown in effluent irrigated fields at Sharada, Kwakwachi and Jakara in the Kano metropolis, Nigeria. Similar vegetable samples from another irrigated field at Thomas Dam, in the remote part of Kano, were used as control. Samples were collected during dry and rainy seasons. The mean level of metals obtained ranged widely from 0.28 mg/Kg Cr to 18.89 mg/Kg Zn. Samples from Jakara garden indicated highest mean levels of Co (1.14 ± 0.17 mg/Kg), Cu (7.50 ± 1.08 mg/Kg), Zn (18.89 ± 1.93 mg/Kg) and Cr (0.85 ± 0.10 mg/Kg) while those from Sharada indicated highest levels of Ni (2.02 ± 0.35 mg/Kg) and Pb (1.60 ± 0.53 mg/Kg). Comparison of results with the control showed significant levels ($p<0.05$) of all the metals analyzed in the vegetable samples obtained from the effluent irrigated gardens. However, the levels were within the National Agency for Food and Drug Administration and Control (NAFDAC) tolerable limits for metals in fresh vegetables.

Key words: Industrial effluents, irrigation, vegetables, metals, atomic absorption spectrophotometry (AAS), Kano.

INTRODUCTION

Irrigation is the artificial addition of water to soils in order to meet plants' needs to overcome drought limitations and improve the crops' yields. However, other factors such as soil and water quality and management practices are also important. Wastewater irrigation is known to contribute significantly to the heavy metal contents of soils (Mapanda et al., 2005; Devkota and Schmidt, 2000).

In Zimbabwe, Nyamangara and Mzezewa (1999) implicated land disposal of sewage and industrial effluents as the chief source of heavy metal enrichment of pasturelands and agricultural fields. Barrow and Webber (1972), Pike et al. (1975) pointed out the dangers

of repeatedly treating soils with metallurgical slag because of the possible build up of elements to toxic concentrations. Juste (1974) observed that the spreading of some organic wastes (town refuse, domestic and industrial effluents etc) might contribute to increased levels of nonessential metals in soil, which could cause poor plant growth. Studies conducted by Kisku et al. (2000) in Kalipur, Bangladesh, on the uptake of Cu, Pb, Ni and Cd by *Brassica oleracea* from fields irrigated with industrial effluent indicated widespread contamination from heavy metals despite showing a healthy and gigantic external morphology. High levels of accumulation of heavy metals from soil by common garden vegetables have been reported by many environmental researchers (Boon and Soltanpour, 1992; De Pieri et al., 1997; Xiong, 1998). Therefore, heavy metal contamination of vegetables cannot be underestimated as these foodstuffs

*Corresponding author. E-mail:abdulmojeedlawal@yahoo.com.

are important components of human diet.

Vegetables are rich sources of vitamins, minerals, and fibers, and also have beneficial anti-oxidative effects. However, intake of heavy metal-contaminated vegetables may pose a risk to the human health. This is because, heavy metals have the ability to accumulate in living organisms and at elevated levels they can be toxic. It has been reported that prolonged consumption of unsafe concentrations of heavy metals through foodstuffs may lead to the chronic accumulation of the metals in the kidney and liver of humans causing disruption of numerous biochemical processes, leading to cardiovascular, nervous, kidney and bone diseases (Trichopoulos,1997; Jarup, 2003).

Determination of the chemical composition of plants is one of the most frequently used methods of monitoring environmental pollution. Various plants have been used as bioindicators (Kasanen and Venetvaara, 1991). Several studies have been reported on the accumulation of environmental pollutants in plants. In Israel, for example lichen and higher plant species were exposed near industrial areas in order to detect the accumulation of heavy metals in these plants (Naveh et al., 1979). Tree barks and their leaves remain in the environment for a long period and are sensitive indicators of the environmental contamination with heavy metals, sulphur and fluorine (Ayodele and Ahmed, 2001). Batagarawa (2000), analyzed moss plant in Kano metropolis for heavy metals and reported high levels of lead, zinc and cadmium from industrial areas of Sharada, Bompai and Challawa. Nuhu (2000) also reported high levels of cadmium, manganese and lead in mango leaves obtained from industrial areas of Bompai, Challawa and Sharada in Kano metropolis.Kano is one of the highly populated cities in Nigeria. It lies within longitude $8°32'E$ and latitude $11°58'N$, within a topographical drainage of River Jakara flowing north east. The vegetation of the area is the savannah type, with more grasses than hard wood trees. The average annual rainfall of the area is 817 mm and the temperature varies between 27 to 35°C with a moderate relative humidity.

Study area

Jakara (JKR) and Kwakwachi (KKC) gardens are irrigation sites alongside Jakara river valley at Ahmaddiya and Sabon-gari areas respectively, while Sharada (SRD) garden is located in the middle of industries at Sharada industrial estate all in the Kano metropolis. In these three sites, farming activities are carried out throughout the year but with domestic and industrial wastewaters being used to treat the soils during dry seasons. Thomas (TMS) Dam is another irrigation site outside Kano metropolis where fresh water from the dam is being used to treat the soils during dry seasons.

The objectives of this study were to analyze the vegetable samples from the irrigation sites for heavy metals and to compare results obtained with one another and with those of National Agency for Food and Drugs Administration and Control (NAFDAC) safe limits, while using vegetable (spinach, okra, onions and tomatoes) samples from Thomas Dam as control. The metals of interest include cobalt (Co), chromium (Cr), copper (Cu), nickel (Ni), lead (Pb) and zinc (Zn). The results obtained from this study will be useful for assessing the metals contamination and as well as determining the need for remediation. The results would also provide information for background levels of metals in the vegetables in the study area.

MATERIALS AND METHODS

Analytical reagent (AnalaR) grade chemicals and distilled water were used throughout the study. All glassware and plastic containers used in this work were washed with detergent solution followed by 20%(v/v) nitric acid and then rinsed with tap water and finally with distilled water.

Sampling and sample treatment

The vegetables analyzed include spinach, okra, tomatoes and onions. Samples were collected twice in the year 2002 from three different farms in each site. The first round of sampling was carried out in May towards the end of the dry season while the second round was in September at the peak of the rainy season. Each sample was randomly handpicked, wrapped in a big brown envelope and labeled.

In the laboratory, each sample was washed with tap water and thereafter with distilled water and then dried in an oven at 80°C (Larry and Morgan, 1986). At the end of the drying, the oven was turned off and left overnight to enable the sample cool to room temperature. Each sample was grounded into a fine powder, sieved and finally stored in a 250 cm^3 screw capped plastic jar appropriately labeled.

Digestion procedure

A 2.0 g of the sample was weighed out into a Kjaedahl flask mixed with 20 cm^3 of concentrated sulphuric acid, concentrated perchloric acid and concentrated nitric acid in the ratio 1: 4: 40 by volume respectively and left to stand overnight. Thereafter, the flask was heated at 70°C for about 40 min and then, the heat was increased to 120°C. The mixture turned black after a while (Erwin and Ivo, 1992). The digestion was complete when the solution became clear and white fumes appeared. The digest was diluted with 20 cm^3 of distilled water and boiled for 15 min. This was then allowed to cool, transferred into 100 cm^3 volumetric flasks and diluted to the mark with distilled water. The sample solution was then filtered through a filter paper into a screw capped polyethylene bottle.

Instrumental analysis

An Alpha 4 model atomic absorption spectrophotometer (Chemtec Analytical, UK) equipped with photomultiplier tube detector and hollow cathode lamps was used for the determination of metal concentrations. Working standards were also prepared by further dilution of 1000 ppm stock solution of each of the metals and a calibration curve was constructed by plotting absorbance versus concentration. By interpolation, the concentrations of the metals in sample digests were determined.

Table 1A. Average levels of heavy metals (mg/Kg) in vegetables in the dry season.

Sampling site	Metals					
	Co	Cr	Cu	Ni	Pb	Zn
Spinach						
SRD	0.68±0.29[a]	0.60±0.12[b]	0.69±0.12[b]	2.02±0.35[a]	1.60±0.53[a]	6.63±0.64[c]
KKC	0.69±0.06[a]	0.53±0.09[b]	0.75±0.11[b]	1.39±0.10[b]	1.47±0.10[a]	8.37±0.67[b]
JKR	0.73±0.26[a]	0.85±0.15[a]	1.02±0.25[a]	1.68±0.13[b]	0.97±0.05[b]	10.81±1.49[a]
TMS	0.24±0.09[b]	0.19±0.03[c]	0.33±0.15[c]	B.D.L.	B.D.L.	1.69±0.49[d]
Okra						
SRD	0.94±0.21[a]	0.64±0.08[a]	4.24±0.86[b]	1.13±0.08[a]	1.10±0.04[a]	3.26±0.38[c]
KKC	0.82±0.12[b]	0.69±0.08[a]	6.11±0.67[a]	0.88±0.06[b]	1.00±0.03[a]	4.23±0.56[b]
JKR	0.97±0.18[a]	0.74±0.07[a]	6.74±0.70[a]	1.10±0.23[a]	0.71±0.04[b]	5.55±0.64[a]
TMS	0.24±0.16[c]	0.18±0.03[b]	2.42±0.31[c]	B.D.L.	B.D.L.	1.32±0.21[d]
Onion						
SRD	0.77±0.25[b]	0.64±0.07[a]	4.77±1.00[b]	0.99±0.02[b]	0.95±0.05[a]	11.78±1.00[c]
KKC	0.86±0.21[b]	0.51±0.12[b]	6.81±0.85[a]	0.54±0.04[c]	0.78±0.03[b]	15.48±0.83[b]
JKR	1.14±0.17[a]	0.59±0.06[a]	7.50±1.08[a]	1.70±0.01[a]	0.64±0.04[b]	18.89±1.93[a]
TMS	0.17±0.05[c]	0.33±0.07[c]	3.16±0.42[c]	B.D.L	B.D.L	4.79±0.49[d]
Tomato						
SRD	0.55±0.17[b]	0.49±0.11[b]	0.74±0.21[b]	1.20±0.11[a]	1.56±0.18[b]	1.55±0.19[b]
KKC	0.64±0.13[b]	0.45±0.04[b]	1.05±0.21[a]	1.01±0.06[a]	1.44±0.11[b]	1.79±0.36[b]
JKR	0.75±0.16[a]	0.58±0.03[a]	1.11±0.20[a]	1.05±0.07[a]	1.79±0.12[a]	2.54±0.22[a]
TMS	0.13±0.09[d]	0.16±0.05[c]	0.51±0.15[c]	B.D.L	B.D.L	0.67±0.22[c]

Values are mean ± SD of three samples of each vegetable, analyzed individually in triplicate. Mean values in the same column followed by the same superscript letters are not significantly different ($p<0.05$). B.D.L: Below detection limit.

Statistical analysis

All analysis was performed in triplicates. Results were expressed by means of ±SD. Statistical significance was established using one way analysis of variance (ANOVA). Means were separated according to Duncan's multiple range analysis (p < 0.05) using software SPSS 16.0.

RESULTS AND DISCUSSION

The mean concentrations of Co, Cr, Cu, Ni, Pb and Zn in different vegetable samples from the three effluent irrigated sites and the control are listed in Tables 1A and B. The results generally show significant levels (p < 0.05) of metals in vegetable samples obtained from the effluent irrigated gardens (JKR, KKC and SRD) compared with those obtained from the control (TMS) garden. The high concentrations of heavy metals observed in the vegetable samples from the effluent irrigated gardens might be related to the concentrations of the metals in the soil (Al Jassir et al., 2005; Akinola and Ekiyoyo, 2006). Also from the results, general reductions in metal levels were observed in vegetables sampled during the rainy season when compared with those sampled during dry season.

This may be due to the fact that during the rainy season, the gardens were not irrigated with the wastewater. There is also the possibility of rainwater leaching away parts of the metals that have accumulated in the soil, thus reducing the quantity of these metals available to plants in the soil.

However, there are a few cases in the control site where negative values were recorded for percentage loss of metal in rainy season samples over those of the dry season, thus indicating an increase in metal levels in the rainy season samples over those of the dry season (Table 2). This may be attributed to the possibility of the runoffs from the surrounding land containing metal salts being washed into the control site.

Generally, the mean concentration range of Cu in all vegetables analyzed was 0.30 to 7.50 mg/Kg, with the highest concentration recorded for Jakara onions and the lowest for spinach from the control site. The maximum value recorded is below the National Agency for Food and Drug Administration and Control's (NAFDAC) maximum tolerable Cu concentration of 40 mg/Kg in fresh vegetables (Figure 1). Ni was below detectable level in the control samples while the highest level of 2.02 mg/Kg was obtained in SRD spinach. Also Pb was below the

Table 1B. Average levels of heavy metals (mg/Kg) in vegetables in the rainy season.

Sampling site	Metal					
	Co	Cr	Cu	Ni	Pb	Zn
Spinach						
SRD	0.60±0.04[a]	0.45±0.06[b]	0.53±0.16[b]	1.39±0.06[a]	1.03±0.02[a]	5.29±0.94[c]
KKC	0.54±0.19[a]	0.34±0.03[c]	0.54±0.15[b]	0.99±0.16[b]	0.91±0.11[a]	6.86±0.94[b]
JKR	0.61±0.13[a]	0.60±0.11[a]	0.66±0.21[a]	1.07±0.20[b]	0.65±0.12[b]	8.18±0.47[a]
TMS	0.22±0.08[c]	0.18±0.03[d]	0.30±0.07[c]	B.D.L.	B.D.L.	1.59□0.72[d]
Okra						
SRD	0.82±0.13[a]	0.38±0.11[a]	3.54±0.58[b]	0.85±0.13[a]	0.66±0.02[a]	2.47±0.57[c]
KKC	0.72±0.15[b]	0.40±0.09[a]	4.97±0.60[a]	0.67±0.6[b]	0.67±0.02[a]	3.29±0.36[b]
JKR	0.76±0.21[b]	0.43±0.07[a]	5.45±0.32[a]	0.57±0.08[c]	0.53±0.02[b]	4.51±0.53[a]
TMS	0.18±0.09[c]	0.16±0.06[b]	2.44±0.40[c]	B.D.L.	B.D.L.	1.22±0.47[d]
Onion						
SRD	0.64±0.31[b]	0.51±0.09[a]	4.20±0.65[b]	0.73±0.06[a]	0.74±0.02[a]	10.33±0.95[b]
KKC	0.72±0.08b	0.43±0.05[b]	5.83±0.61[a]	0.41±0.10[b]	0.59±0.02[b]	13.79±1.16[a]
JKR	0.93±0.08[a]	0.44±0.08[b]	6.12±0.51[a]	0.45±0.04[b]	0.53±0.02[b]	15.75±0.65[a]
TMS	0.18±0.09[c]	0.29±0.06[c]	3.23±0.06[c]	B.D.L.	B.D.L.	4.60±0.57[c]
Tomato						
SRD	0.46±0.05[b]	0.34±0.04[a]	0.63±0.10[b]	0.85±0.15[a]	1.05±0.13[a]	1.30±0.36[b]
KKC	0.54±0.24[b]	0.34±0.09[a]	0.84±0.19[a]	0.78±0.16[b]	0.95±0.11[a]	1.55±0.49[b]
JKR	0.63±0.05[a]	0.35±0.07[a]	0.87±0.19[a]	0.75±0.06[b]	0.48±0.12[b]	1.84±0.71a
TMS	0.12±0.17[c]	0.18±0.02[b]	0.45±0.11[c]	B.D.L.	B.D.L.	0.71±0.31[c]

Values are mean ± SD of three samples of each vegetable, analyzed individually in triplicate. Mean values in the same column followed by the same superscript letters are not significantly different ($p>0.05$) B.D.L.: Below detection limit.

detectable level in control samples while SRD spinach recorded the highest level of 1.60 mg/Kg. The highest values obtained for Pb and Ni are below the NAFDAC safe limits for these metals (2.00 and 2.70 mg/Kg, respectively) in fresh vegetables.

The mean concentration range for Co was found to be 0.12 to 1.14 mg/Kg with the highest concentration recorded in JKR onions and the lowest in tomato from control site. The mean concentration range for Cr was found to be 0.16 to 0.85 mg/Kg with the highest concentration recorded in JKR spinach and the lowest in the okra from control site. The results indicated the mean concentration range of Zn to be 0.67 to 18.89 mg/Kg with the highest concentration found in JKR onions and the lowest in tomatoes from the control site. However, the highest value obtained is still below the NAFDAC safe limit of Zn (50 mg/Kg) in fresh vegetables. The results obtained in this study are comparable with some literature values of similar studies reported previously (Onianwa et al., 2001; Erwin and Ivo, 1992; Pennington et al., 1995).

Consequently, from the results, the general trend for the mean levels of metals analyzed in all vegetables sampled from the three effluent irrigated sites as well as the control for both dry and rainy seasons showed that for the concentrations of Cu and Zn, JKR > KKC > SRD > Control; for Co and Cr concentrations, JKR> SRD> KKC > Control; for Ni concentration, SRD> JKR > KKC > Control, and for Pb concentrations, SRD > KKC > JKR > Control (Figure 1).These sequences indicated that the metal contents of the vegetables are higher in areas being treated with wastewater. The observation is in good agreement with other studies elsewhere (Sharma et al., 2006; Sawidis et al., 2001) which suggested that uptake of metals by plants is proportional to their concentrations and availabilities in soils. Dasuki (2000) had earlier reported high levels of Cr (1.5 to 3.8 mg/Kg) in effluents from Sharada and Challawa industrial estates while Batagarawa (2000) had also reported high levels of Cu (1.74 to 11.54 mg/Kg), Pb (10.38 to 154.64 mg/Kg), and Zn (11.40 to 87.34 mg/Kg) in the samples of moss plant from Bompai and Sharada industrial estates in Kano metropolis.

The trend also shows that JKR garden recorded highest mean concentrations in four out of six metals analyzed (Co, Cu, Zn and Cr), while SRD garden recorded highest concentrations in two metals (Ni and Pb). Hence, the trend for the level of contamination by

Table 2. Percentage loss of heavy metals in vegetables in rainy season compared with levels in dry season.

Metals	SRD	KKC	JKR	Control	Mean ± % loss
			Vegetables		
			Spinach		
Co	11.77	19.40	16.44	8.33	13.99±4.91
Cr	25.00	35.84	29.41	5.26	23.88±13.19
Cu	23.19	28.00	35.29	9.09	23.89±11.05
Ni	31.19	28.78	36.31	B.D.L.	32.09±3.84
Pb	35.63	38.10	32.10	B.D.L.	35.28±3.02
Zn	24.23	18.04	24.33	5.92	18.13±8.66
			Okra		
Co	12.77	12.20	21.65	4.17	12.70±7.14
Cr	40.63	42.03	41.89	11.11	33.92±15.22
Cu	16.51	18.65	19.14	-0.83	13.37±9.53
Ni	24.78	28.41	21.24	B.D.L.	24.81±3.59
Pb	40.00	33.00	45.19	B.D.L.	39.40±6.12
Zn	21.27	22.22	18.74	7.58	17.45±6.74
			Tomato		
Co	16.36	15.63	16.00	7.69	13.92±4.16
Cr	30.61	24.44	39.65	-12.50	20.55±22.90
Cu	14.86	20.00	21.62	11.76	17.06±4.56
Ni	29.17	22.77	28.57	B.D.L	26.84±3.54
Pb	33.97	35.42	39.24	B.D.L	36.21±2.72
Zn	16.12	13.41	27.56	-5.97	10.96±5.25
			Onion		
Co	16.88	16.28	18.42	-5.88	11.43±11.57
Cr	20.31	15.69	25.42	12.12	18.39±5.77
Cu	11.77	14.39	18.40	-2.22	10.59±8.96
Ni	26.26	24.07	35.71	B.D.L	28.68±6.19
Pb	22.10	24.36	17.19	B.D.L	21.22±3.67
Zn	12.31	10.92	16.62	3.97	10.96±5.25

metals in the irrigation gardens is JKR > SRD > KKC > control (Figure 1). The high mean levels of Pb and Ni in SRD samples could be attributed to industrial emissions (Yilmaz and Zengin, 2004) while the high level of Pb in KKC could be attributed to automobile emissions as a result of its proximity to the road side in addition to the possible high levels of metal in contaminated wastewater being used for irrigation. The close relationship between lead concentrations and traffic intensity has been demonstrated in detail by many authors (Li et al., 2001; Viard et al.,2004). Furthermore, the relative high levels of Zn, Cu, Co and Cr in JKR and KKC samples may be attributed to the contaminated Jakara stream (Ogbalor, 1991; Dasuki, 2000) used for treating soils at the two sites, as many industrial and domestic waste waters are discharged into it.

Conclusion

This study further confirms the increased danger of growing vegetables on soils irrigated with contaminated industrial and domestic wastewaters. However, the levels of the metals are currently within the NAFDAC safe limits guidelines. But, if the practice of treating the soils in the irrigation gardens with contaminated waters is not controlled, it may lead to health hazard on the part of consumers of the vegetables on the long term. Therefore, there is the need to continually monitor, control and take necessary policy decisions so as to limit and ultimately prevent these avoidable problems. However, in the mean time, farmers from the study areas are hereby encouraged to use well water for irrigation in their gardens instead of contaminated streams.

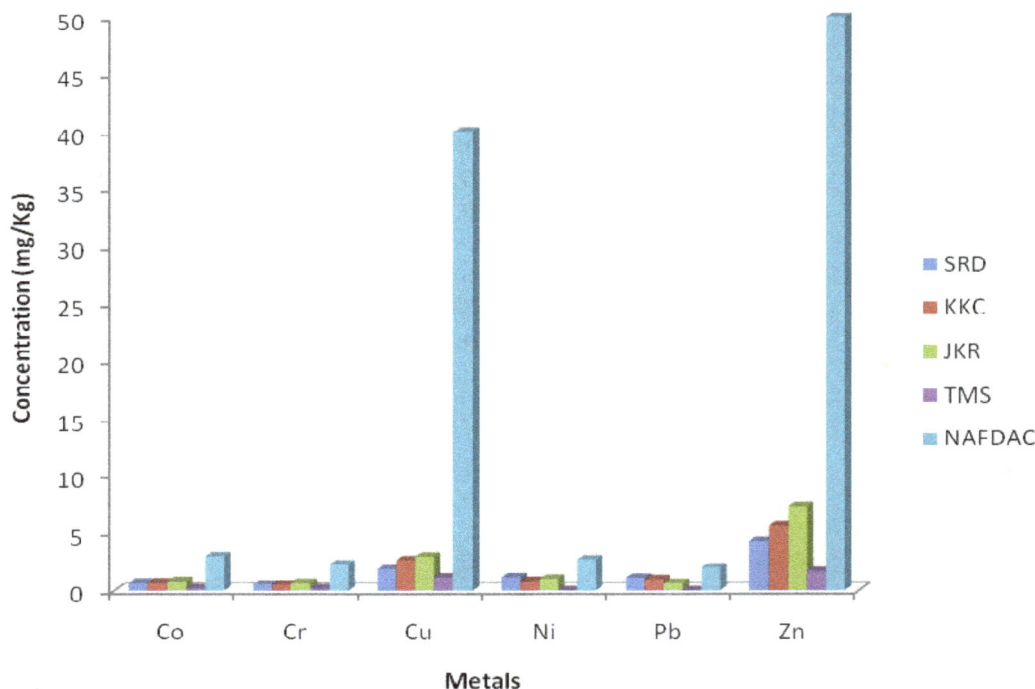

Figure 1. Variation of total mean levels of metals in relation to the garden sites.

REFERENCES

Akinola MO, Ekiyoyo TA (2006). Accumulation of lead, cadmium and chromium in some plants cultivated along the bank of river Ribila at Odo-nla area of Ikorodu, Lagos State, Nigeria. J. Environ. Biol., 27(3): 597-599.

Al Jassir MS, Shaker A, Khaliq MA (2005). Deposition of heavy metals on green leafy vegetables sold on roadsides of Riyadh City, Saudi Arabia. Bull. Environ. Contam. Toxicol., 75: 1020-1027.

Ayodele JT, Ahmed A (2001). Monitoring air pollution in Kano by chemical analysis of Scots pine (*Pinus sylvestris* L) needle for sulphur. Environmentalist, 21: 145-151.

Barrow ML, Webber J (1972). Trace elements in Sewage Sludge. J. Sci Food Agric., 23: 93-100.

Batagarawa SM (2000). *Funaria hygrometrica* as bio-indicator of heavy metals in Kano metropolis. Bayero University, Kano-Nigeria, M. Sc. Thesis.

Boon DY, Soltanpour PN (1992). Lead, cadmium, and zinc contamination of aspen garden soils and vegetation. J. Environ. Qual., 21: 82-86.

Dasuki IA (2000). Analysis of trace metals and some selected chemical pollutants in effluents from Challawa, Bompai and Sharada industrial areas of Kano. Bayero University Kano, Nigeria, M. Sc. Thesis.

De Pieri LA, Buckley WT, Kowalenko CG (1997). Cadmium and lead concentrations of commercially grown vegetables and of soils in the lower Fraser valley of British Columbia. Can. J. Soil Sci., 77: 51-57.

Devkota B, Schmidt GH (2000). Accumulation of heavy metals in food plants and grasshoppers from the Taigetos Mountains, Greece. Agric. Ecosyst. Environ., 78: 85-91.

Erwin JM, Ivo N (1992). Determination of Lead in tissues: A pitfall due to wet digestion procedures in the presence of sulphuric acid. Analyst. 17: 23-26.

Jarup L (2003). Hazards of heavy metal contamination. Br. Med. Bull., 68: 167–182.

Juste C (1974). Long-term trend analysis of Heavy metal content and translocation in soils. C.R. Acad. Agric., 60: 975-982.

Kasanen P, VenetVaara FT (1991). Comparison of biological collection

of airborne heavy metals near ferrochrome and steel work. Water Air Soil Pollut., 60: 337-359.

Kisku GC, Barman SC, Bhargava SK (2000). Contamination of soil and plants with potentially toxic elements irrigated with mixed industrial effluent and its impact on the environment. Water Air Soil Pollut., 120: 121-137.

Larry RW, Morgan JT (1986). Determination of Plant Iron, Manganese and Zinc by wet digestion procedures. J. Food Agric., 37: 839-844.

Li X, Poon C, Liu PS (2001). Heavy metal contamination of urban soils and street dusts in Hong Kong. Appl. Geochem., 16(11-12): 1361-1368.

Mapanda F, Mangwayana EN, Nyamangara J, Giller KE (2005). The effect of long term irrigation using wastewater on heavy metal contents of soils under vegetables in Harare, Zimbabwe. Agric. Ecosyst. Environ., 107: 151-165.

Naveh Z, Steinberg EH, Chaim S (1979). The use of bio-indicators for monitoring of air pollution by fluorine, ozone and sulphur dioxide: In Environmental Bio-monitoring, Assessment, Prediction and Management of certain case studies and related Quantitative issues. Ed. Cains, G.I.P. and Waters, W.E. International Cooperative Publishing House, Fatiland, USA, pp. 21-47.

Nuhu A (2000). Heavy metal determination in mango leaves (*Magnifera indica*) from Kano metropolis as bio-indicators of environmental pollution. Bayero University Kano, Nigeria. M. Sc. Thesis.

Nyamangara J, Mzezewa J (1999). The effects of long-term sludge application on Zn, Cu, Ni and Pb levels in clay loam soil under pasture grass in Zimbabwe. Agric. Ecosyst. Environ., 73: 199-204.

Ogbalor B (1991). An evaluation of domestic waste water quality for irrigation of vegetable along Jakara stream. Bayero University Kano, Nigeria. M. Sc. Thesis.

Onianwa PC, Adaeyemo AO, Odowu EO, Ogabiela EE (2001). Copper and Zinc contents of Nigerian foods and estimates of the adult dietary intakes. J. Chem., 72: 89-95.

Pennington JAT, Schoen SA, Salmon GD, Young B, John RD, Matrt RW (1995). Composition of core foods of the USA foods of the USA food supply 1982-1991. J. Food Compos. Anal., 8(2): 129-169.

Pike ER, Graham LC, Fogden MW (1975). Redevelopment of

contaminated land tentative guidelines for acceptable levels of selected elements in soil. J. Ass. Pub. Anal., 13: 19-48.

Sawidis T, Chettri MK, Papaionnou A, Zachariadis G, Stratis J (2001). A study of metal distribution from lignite fuels using trees as biological monitors. Ecotoxicol. Environ. Saf., 48: 27-35.

Sharma RK, Agrawal M, Marshall F (2006). Heavy Metals Contamination in Vegetables grown in wastewater irrigated areas of Varanasi, India. Bull. Environ. Contam. Toxicol., 77: 312-318.

Viard B, Pihan F, Promeyrat S, Pihan JC (2004). Integrated assessment of heavy metal (Pb, Zn, Cd) highway pollution: Bioaccumulation in soil, Graminaceae and land snails. Chemosphere, 55(10): 1349-1359.

Trichopoulos D (1997). Epidemiology of cancer. In: DeVita, V.T. (Ed.),. Cancer, Principles and Practice of Oncology. Lippincott Company, Philadelphia, WHO (1992). Cadmium. Environmental Health Criteria, Geneva 134: 231–258.

Xiong ZT (1998). Lead uptake and effects on seed germination and plant growth in a Pb hyperaccumulator Brassica pekinensis Rupr. Bull. Environ. Contam. Toxicol., 60: 285-291.

Yilmaz S, Zengin M (2004). Monitoring environmental pollution in Erzurum by chemical analysis of Scot pine (Pinus sylvestris L.) needles. Environ. Int., 29(8): 1041-1047.

8

Heavy metal speciation trends in mine slime dams: A case study of slime dams at a goldmine in Zimbabwe

Mark Fungayi Zaranyika* and Tsitsi Chirinda

Chemistry Department, University of Zimbabwe, P. O. Box MP 167, Mount Pleasant, Harare, Zimbabwe.

Heavy metal speciation trends in the slime dams at a typical gold mine in Zimbabwe were studied using a six-step sequential extraction technique to give six fractions, namely, the exchangeable, acid soluble, easily reducible, moderately reducible, oxidizable and residual fractions. The elements studied include Cd, Co, Ni, Cu, Zn, Cr, Fe and Mn. Cadmium was found mainly in the acid soluble fraction, and to a lesser extent in the exchangeable and easily reducible fractions. For the other elements, speciation trends within the slime dams appear to depend, in the main, on the stability of metal cyano complexes formed during the cyanide extraction process, and the ease with which the metals co-precipitate with Mn oxides and/or CaCO$_3$. For Fe speciation trends depend mainly on the pH within the slime dam.

Key words: Heavy metal speciation, heavy metal mobility, slime dam, mine tailings, sequential extraction.

INTRODUCTION

Mining is a major activity in Southern Africa. The result is that the region is home to thousands of mine solid waste dumps. Release of toxic metals from such solid waste dumps depends on the mobility of the elements within the dumps. Transformations in the chemical form of toxic metals and other toxic substances in these mine solid waste dumps will affect the mobility and bioavailability of such toxic substances (Tessier et al., 1989; Sager, 1992). There is therefore need to study the speciation trends of toxic metals in mine solid waste dumps, with the view of understanding how the release of such toxic metals from mine solid waste dumps can be controlled. Although several workers have studied the impact of mine solid waste dumps in the region on the quality of surrounding surface and ground water (Rosner and van Schalkwyk, 2000; Winde and Sandham, 2004; Zaranyika et al., 1994, 1997; Rafiu, 2007; Nsimba, 2009), there are no reports on speciation studies of trace metals within the mine solid waste dumps themselves.

Speciation of metals in soils, sediments and solid wastes is often studied using sequential extraction techniques whereby the target metals are fractionated into several fractions using extractant solutions of increasing strength (Tessier et al., 1979) Several such sequential extraction schemes have been described (Chester and Hughes, 1967; Tessier et al., 1979; Gibson and Farmer, 1983; Brown et al., 1984; Gibson and Farmer, 1986). The technique has been used to study the speciation of heavy metals in soils, street sweepings and urban aquatic sediments, lake sediments, pelagic sediments, semiarid soils, dredged sediment derived surface soils, and solid waste materials (Gibson and Farmer, 1986; Kabala and Singh, 2000; Lu et al., 2003; Mehra et al., 1999; Wilbur and Hunter, 1979; Banerjee, 2003; Tessier et al., 1989; Vuorinon and Carlson, 1985; Chester and Hughes, 1967; Navas and Lindhorfer, 2003; Singh et al., 1998; Flyhammar, 1997). The technique has also been used to study the speciation, mobility and bioavailability of radionuclides (Riise et al., 1990; Salbu et al., 1994; Blanco et al., 2004).

The aim of the present work was to study the speciation trends of selected heavy metals in the slime dams of a typical gold mine in Zimbabwe, with the view of

---

*Corresponding author. E-mail: zaranyika@science.uz.ac.zw.

Table 1. Sequential extraction scheme used*.

Fraction	Extractant	Shaking Time (h)	Temp(°C)	Designation
FI	1 M NH$_4$AC, pH 7	8	RT**	Easily exchangeable
F2	1 M CH$_3$COONa, pH 5	15	RT	Acid soluble
F3	0.1 M NH$_2$(OH)Cl/0.2M HNO$_3$	1	RT	Easily reducible
F4	1 M NH$_2$(OH)Cl/25% CH$_3$COOH	15	RT	Reducible
F5	30% H$_2$O$_2$/0.02 M HNO$_3$	6	85	Oxidizable
F6	HClO$_4$/HF			Residual fraction

*Source: Gibson and Farmer (1986), **RT = room temperature.

understanding how such speciation trends affect the mobility of the metals within the slime dams. The elements studied include Cr, Co, Ni, Cu, Zn, Cd, Fe, and Mn. Major trends studied are speciation as a function of time, speciation as a function of depth in the slime dam, inter-conversion between speciation fractions, and inter-element speciation relationships.

MATERIALS AND METHODS

The gold mine selected for the study is situated about 50 km north of Harare. The ore mined consists of pyrite (FeS$_2$), pyrrhotite (Fe$_{1-0.8}$S), and arsenopyrite (FeAsS) as the major minerals, and chalcopyrite (CuFeS$_2$), bornite (Cu$_6$FeS$_4$) and pyrolucite (MnO$_2$) as minors. The ore is crushed and milled to fine particles, then treated with alkali cyanide and compressed air to dissolve the gold (Puddephatt, 1978). The solution is filtered, and the tailings are treated with lime and ferrous sulphate to complex any excess cyanide ion in the slurry before being discharged onto slime dams.

Slime dam soil samples were collected from two slime dams, the current slime dam (or CSD) still receiving tailings, and an old slime dam (OSD) no longer receiving tailings, using a soil corer 15 cm in diameter. In the current slime dam (CSD), samples were collected from three points approximately 30 m apart. At each sampling point, samples were collected at three depths of 15, 50 and 100 cm from the surface. The Old Slime Dam (OSD), about 10 m high, was in the process of being reworked, and samples were collected at the surface, about 100 cm, and at the bottom. The samples were collected into clean plastic bags previously soaked in ultra-pure dilute nitric acid (Merck, Germany) overnight and then rinsed with distilled de-ionized water. All soil samples were tightly sealed and immediately taken to the laboratory where they were stored in a refrigerator prior to analysis.

For analysis soil samples were thawed then partially dried in an oven at 80°C in order to evaporate off most of the entrapped water. Analysis of the soil was done after sequential extraction into 6 fractions after Gibson and Farmer (1986). An aliquot of 1 g of the ground soil was placed in an Erlenmeyer flask and 10 ml of 1-M ammonium chloride solution added. The flask was stoppered and shaken on a mechanical shaker for 1 h, after which the sample was centrifuged at 4000 rpm for 10 min. The supernatant solution was transferred to a plastic vial for storage (Fraction I (FI), Table 1). To the residue in the centrifuge tube, 10 ml of de-ionized water were added, the mixture centrifuged again and the supernatant combined with the first extract.

For subsequent fractions the residue after extraction of the preceding fraction was washed into a clean Erlenmeyer flask with a solution of the appropriate reagent (Table 1), and the extraction carried out as given in Table 1. Analytical grade reagents were used to prepare standard solutions for all the analyses carried out. The results obtained, expressed as percentages of the sum of all the six fractions, are shown in Figure 1 in bar graph form for samples from the current slime dam, while Figure 2 shows the percentages for each fraction at each depth sampled in the old slime dam. The use of percentages eliminates variations in the quantities of the metals measured caused by variations in the composition of the tailings deposited on the slime dam at different times. Figure 3 shows bar graphs and standard error bars, for the mean total element concentrations in the 1 m surface layers of the two slime dams. Table 2 shows the mean total element concentrations at the three depth levels sampled in the old slime dam. Statistical calculations were done using Minitab for Windows Release 10 statistical package (Minitab, 1994).

The efficiency of the sequential extraction technique was tested by performing total element metal determination using the acid digestion method, and comparing the result to the total element obtained by the sequential extraction method. The results obtained for the efficiency of the sequential extraction technique, expressed as %recovery, are shown in Figure 4. The pH of the slime dam soil was measured using a pH meter after shaking a 2 g aliquot of the ground sample with 20 ml of deionised distilled water (McLean, 1982). The pH values obtained for the various samples are shown in Table 3.

RESULTS AND DISCUSSION

pH within the slime dams

Tailings are deposited on the slime dams at a minimum pH of 10.5. Table 3 shows that pH in the surface layers of the current slime dam is very variable, ranging from an average of 6.28 at Sampling point 2 to 8.14 at Sampling point 3. This is probably related to the time elapsed since the slime effluent discharge was directed at each sampling point. The tailings discharge is rotated so that there is an even build up of the slime dam. It is also apparent from Table 3 that there is a well defined pH increase with depth at Point 2 from 4.95 at the surface to 7.94 at 1-m depth. On the other hand the pH at Points 1 and 3 appears to be fairly uniform.

Another cause of the variations in pH is probably variations in the composition of the tailings discharged, especially its sulphide content and lime content. The ore being processed at any one time determines the sulphide

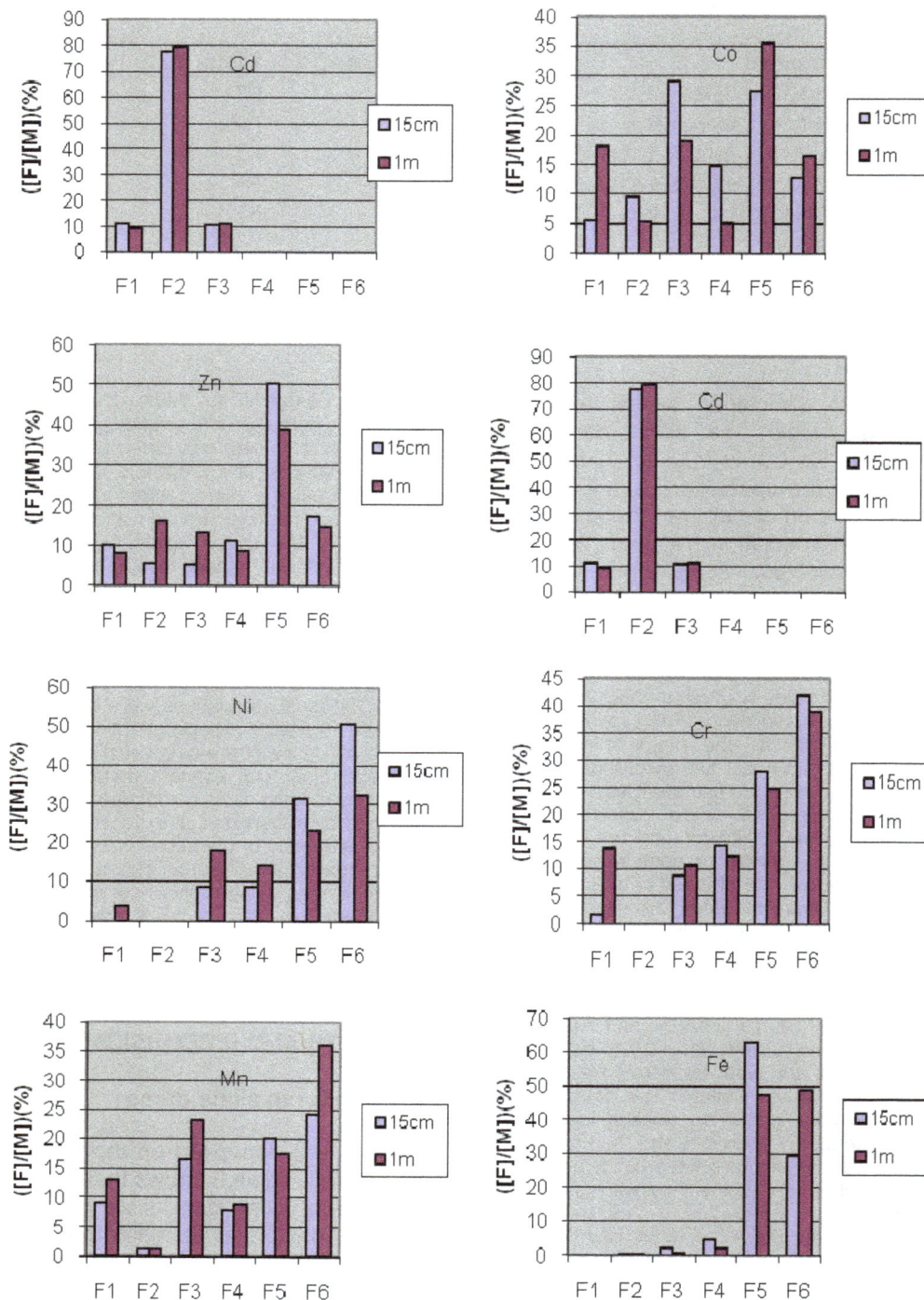

Figure 1. Speciation of Cd, Cu, Zn, Co, Ni, Cr, Mn and Fe in the current slime dam: [F] = concentration of speciation fraction; [M] = total element concentration.

content. The old slime dam shows a drastic pH drop from 6.56 at the surface to about 3.16 at approximately 1 m depth, and 2.21 at its base. This is attributed to the accumulation of acidic conditions at the base of the dump as a result of sulphide oxidation in the surface layers.

Total element concentrations

From Figure 3, it is apparent that for Ni, Cu, Zn, Cd and Co, total element concentrations in the 1 m surface layer of the current slime dam are much lower than those in the

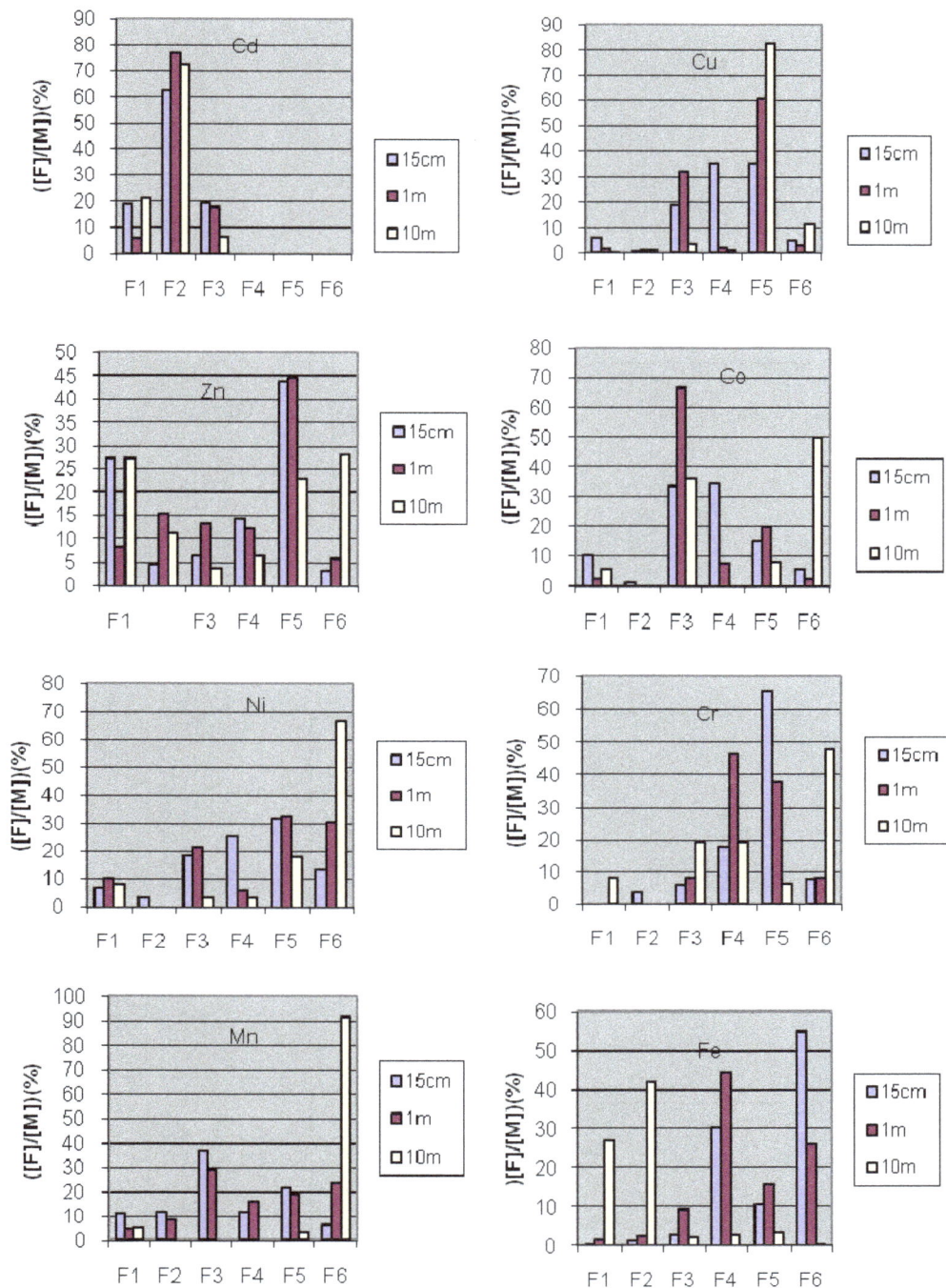

Figure 2. Speciation of Cd, Cu, Zn, Co, Ni, Cr, Mn and Fe in the old slime dam: [F] = concentration of speciation fraction; [M] = total element concentration.

old slime dam, while the reverse is true for Cr, Mn and Fe. The low concentration of Fe, Mn and Cr in the surface 1 m layer of the old slime dam is attributed to leaching that has taken place since the time that deposition of tailings was discontinued. A possible explanation for the fact that the concentrations of Ni, Cu, Zn, Cd and Zn are higher in the old slime dam is that the current ore being processed in the plant is low in these elements compared to the ore that was being processed when tailings were being deposited on the old slime dam. Total element data for the old slime dam in Table 2 shows that for Cd and Cr, concentrations drop steadily with depth, while for Mn, Co, Cu, and Ni concentrations drop in going from the surface to the 1 m level, then

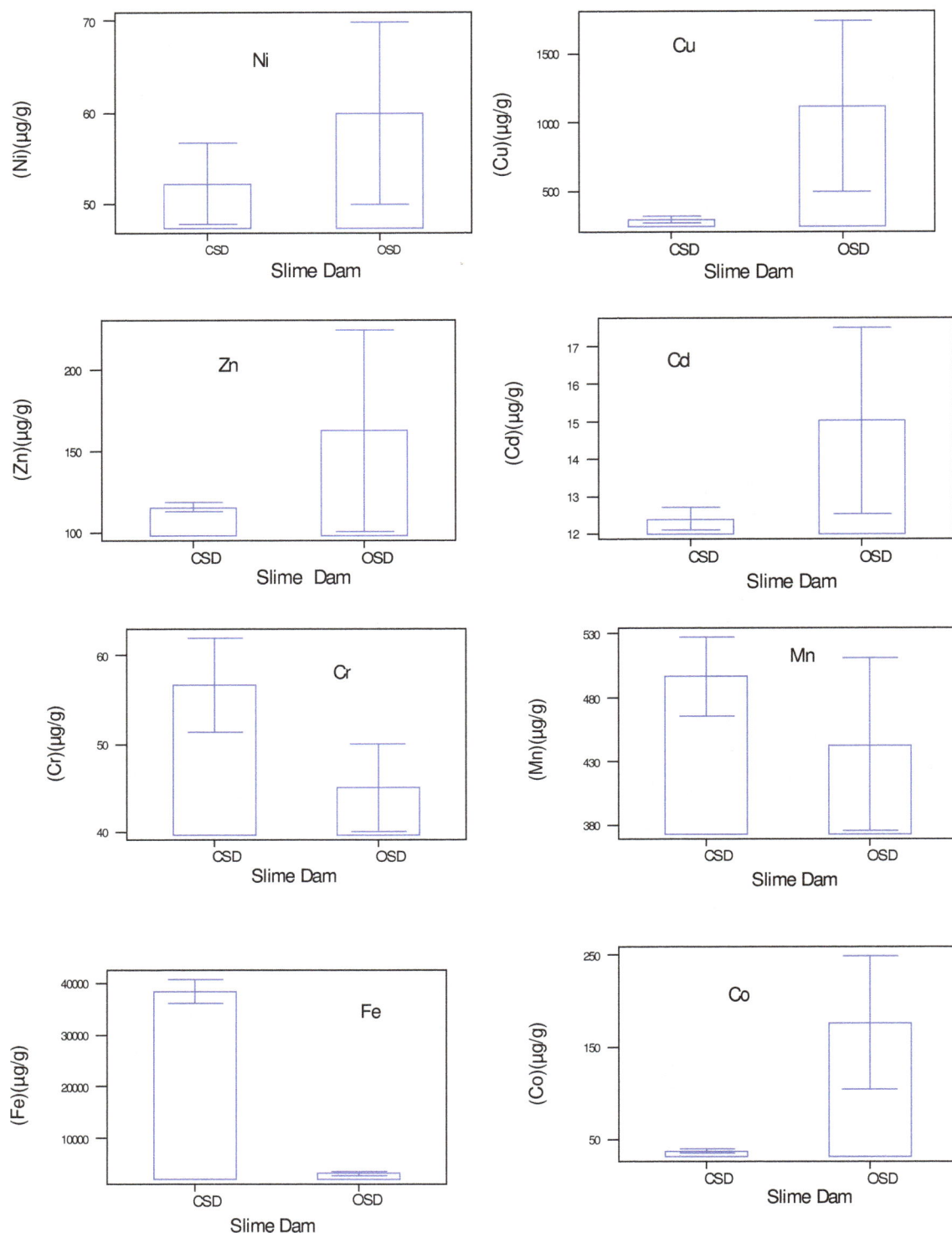

Figure 3. Mean total element concentrations in the 1m surface layer of the current and old slime dams: (a) Cr, Mn, Fe and Co; (b) Ni, Cu, Zn and Cd.

increase at the base of the slime dam. Zinc concentration drops in going from the surface to the 1 m level, then

levels off.

On the other hand, Fe shows a steady increase with

Table 2. Mean total element concentrations (µg/g) in the old slime dam (n = 3).

Element	Surface	1 m depth	Base (10 m)
Cd	17.6±0.1	12.3±0.3	13.5±0.4
Mn	512±11	375±20	650±10
Ni	75±1	50.0±0.8	70±5
Cu	1750±100	500±148	525±6
Fe	1975±100	2500±300	3500±400
Co	250±5	105±2	135±4
Cr	95±10	50±6	40±4
Zn	225±20	100±6	100±4

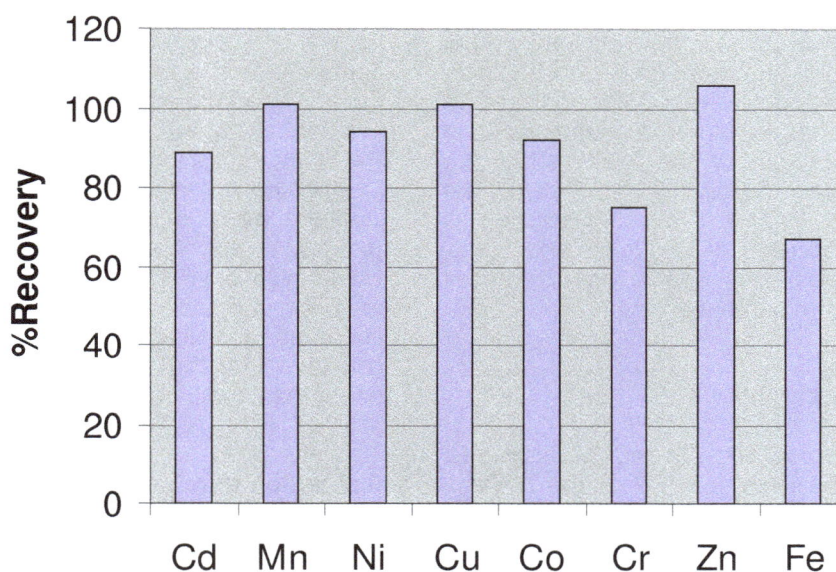

Figure 4. Sequential extraction recovery efficiency.

depth, probably due to the conversion of Fe(II) cyano complex to insoluble Fe(III) oxides. Table 2 further shows that at the surface of the old slime dam, the concentrations of the elements studied decrease in the order: Fe > Cu > Mn > Co > zn > Cr > Ni > Cd. At the 1 and 10 m levels, the order remains the same, except for Cr and Ni which are approximately equal at the 1 m level, while Ni > Cr at the base of the slime dam.

Extraction efficiency

Sequential extraction percentage recoveries of 89±3, 101±1, 94±8, 101.7±0.2, 67±44, 92±6, 75±5 and 106±2 were obtained for Cd, Mn, Ni, Cu, Fe, Co, Cr and Zn respectively, (Figure 4). The poor recovery for Fe is attributed to filtration loss of colloidal Fe hydroxides during the sequential extraction.

Speciation trends in the current slime dam

Figure 1 shows the percentage of the various fractions for the different elements in the form of bar graphs. Examination of this figure shows that the following groups of elements show roughly similar trends in speciation: (a) Ni, Cr, Mn and (b) Cu, Zn, Co. Cadmium and Fe show unique speciation patterns.

Cadmium

Cd in the surface layers of current slime dam is available mainly in the acid soluble fraction, F2, at 80%, and the balance as the exchangeable fraction, F1, and the fraction associated with Mn oxides, F3. It is interesting to note that, although other elements are available in the residual fraction, Cd is not found in the residual fraction. It

Table 3. pH in the current slime dam and the old slime dam.

Slime dam	Depth (m)	SP1	SP2	SP3
Current	0	7.25	4.95	8.07
	0.5	6.88	5.96	8.20
	1.0	6.75	7.94	8.16
	Mean	6.96	6.28	8.14
Old	1	6.56		
	4	3.16		
	10	2.21		

SP = sampling point.

would appear from this that Cd is completely leached out of the mineral fraction during the ore processing and gold extraction process. An alternative explanation is that Cd is introduced into the slimes through the chemicals used in the gold extraction process or the treatment of the tailings prior to their discharge on the slime dam. The predominance of Cd in the acid soluble fraction is attributed to co-precipitation by $CaCO_3$ (Jones and Jarvis, 1981). There are no significant speciation differences between the 15 cm and 1 m samples.

Manganese, nickel and chromium

This group comprises Group 6B, 7B and 8B elements. For this group the residual fraction, F6, is the highest, while the acid soluble fraction, F2, is the lowest at 1% for Mn and Ni, and zero for Cr. For all three elements, the oxidizable fraction, F5, is the next highest fraction, followed by the moderately reducible fraction, F4, for Cr, and the easily reducible, F3, for Mn and Ni. The exchangeable fractions for the three elements are 11, 5 and 10% respectively for Mn, Ni and Cr. The very low levels of the acid soluble fraction suggest that these elements are not co-precipitated by $CaCO_3$.

Copper, cobalt and zinc

The oxidizable fraction, F5, is the most prevalent for Cu, Co, and Zn. This group differs from the Mn, Ni and Cr group in the levels of the acid soluble fraction which are higher at 3, 10 and 13% respectively for Cu, Co, and Zn, suggesting co-precipitation by $CaCO_3$ to a limited extent. Results from soil leaching experiments with slightly acidic buffers show that, in most cases, carbonates can take up only 1 to 6% Cu (Sager, 1992).

Iron

For Fe only the oxidizable fraction, F5, and the residual fraction, F6, are found at 52 and 47% respectively. The content of iron sulphide minerals in the ore is high. Since $FeSO_4$ is added to the tailings to complex any excess cyanide ion, the absence of fractions 1 to 4 (only 1% of F4 is found), suggests that the $Fe_2[Fe(CN)_6]$ complex formed shows in fraction 5.

All the metals studied form complexes with the CN^- ion (Moeller, 1963; Cotton and Wilkinson, 1972a), and we would also expect these to appear in Fraction 5. This may explain the high levels of this fraction for the elements Mn, Ni, Cr, Cu, Co and Zn. On this basis we conclude that the cyanide complex of Cd is not stable under the conditions found in the slime dam.

Speciation trends in the old slime dam

In discussing speciation trends in the old slime dam attention is focussed on (a) comparison of speciation trends in the surface layers of the current slime dam and the old slime dam in order to highlight the effect of time on speciation trends in the surface layers of the slime dams, (b) speciation trends as a function of depth in order to highlight the effects of leaching and change in pH, (c) inter-conversion between speciation fractions, and (d) inter-element speciation relationships.

Cadmium

For Cd, comparison of Figures 1 and 2 shows very little difference in the speciation pattern in the surface layers of the slime dams. The acid soluble fraction, F2, is still the highest in the old slime dam, although there is a drop from 80 to 62%. This is accompanied by increases in the exchangeable and moderately reducible fractions, F1 and F3 respectively, to 19% from 9 to 10%. The increase in F1 and F3 in going from the current slime dam to the old dump suggests slow release of Cd from $CaCO_3$ precipitates, possibly via replacement by Ca.

Figure 5 shows that the profiles of the F2 and F3 fractions are parallel, suggesting very little or no

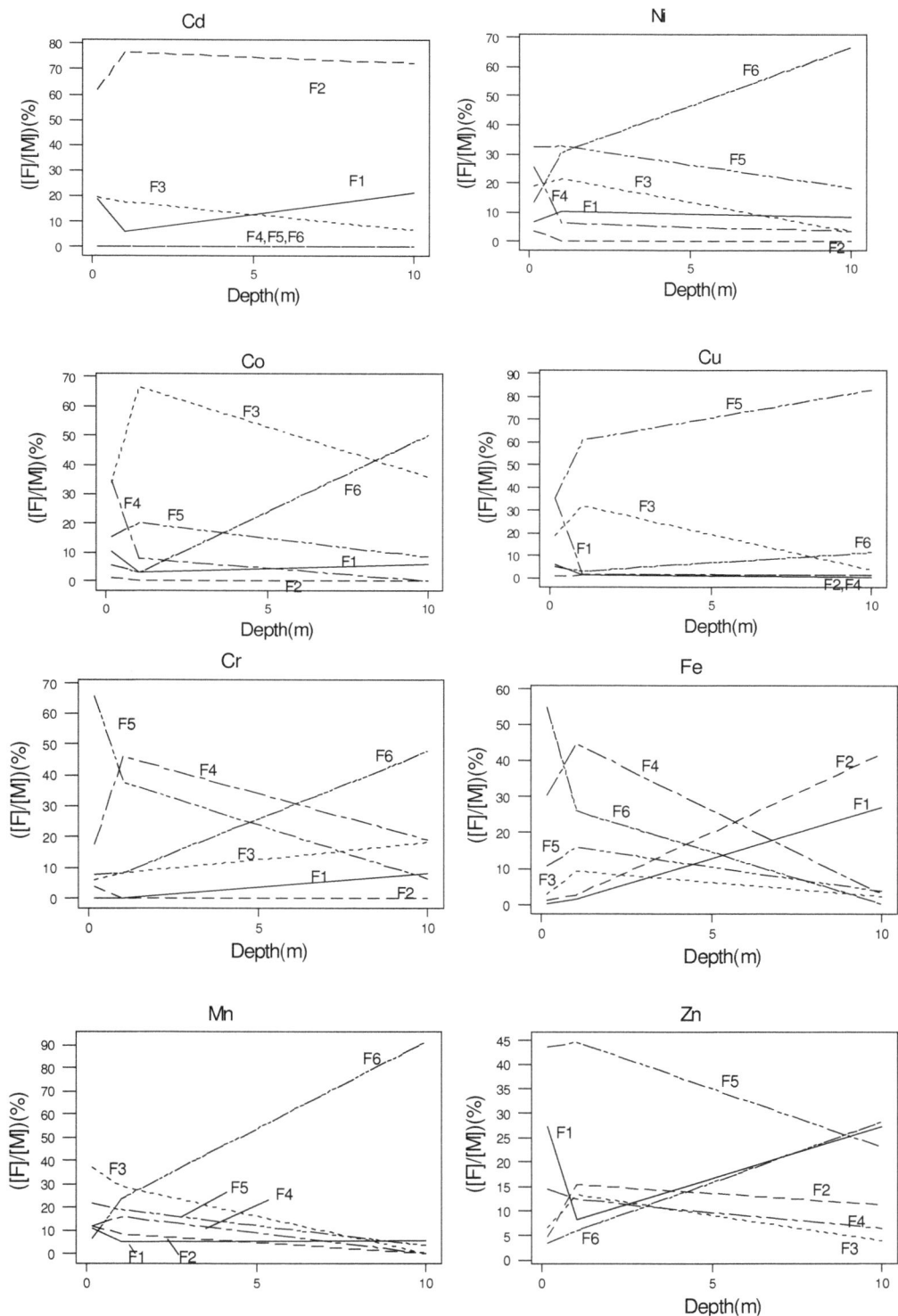

Figure 5. Speciation as a function of depth in the old slime dam: [F] = concentration of speciation fraction; [M] = total element concentration.

interconversion between these two fractions. The profiles of the F1 and F2 factions are, on the hand, mirror images, as are the profiles of F1 and F3, suggesting interconversion between these pairs of speciation

Table 4. Pearson's correlation factor* for residual fractions.

	F6(Ni)	F6(Cu)	F6(Co)	F6(Zn)	F6(Cr)	F6(Mn)
F6(Cu)	0.861					
F6(Co)	0.932	0.987				
F6(Zn)	0.975	0.953	0.989			
F6(Cr)	0.951	0.976	0.998	0.996		
F6(Mn)	0.992	0.919	0.971	0.995	0.983	
F6(Fe)	-0.973	-0.720	-0.823	-0.897	-0.854	-0.935

* Minitab (1994).

fractions. Sorption of Cd by hydrated $CaCO_3$ has been shown to be reversible (Sager, 1992). At the surface (15 cm) the levels of F1 and F3 are approximately equal. At 1 m depth %F3>%F1, probably due to leaching of F1, or conversion of F1 to F3 via co-precipitation with Mn oxides. At 10 m %F1>%F3. This can be due to leaching of F1 to the base of the slime dam, coupled with conversion of F3 to F1 as a result of the increased acidic conditions at the base of the slime dam.

Iron

For Fe the residual fraction is the highest in the surface layers of the old slime dam, followed by F4, then F5, F3 and F2. The exchangeable fraction, F1, is negligible. In the middle of the slime dam, F4 is the highest, followed by F6, then F5, F3, F2 and lastly F1. In the bottom layers, F6 becomes negligible, while F2 is now the highest, followed by F1. Fractions F5, F4 and F3 are very low at 4, 3 and 2% respectively. From Figure 5 the major speciation change in the surface layers, 15 cm to 1 m, with time is the weathering of the residual fraction F6 to yield the oxidizable Fraction F5, and fractions associated with Mn and Fe oxides, that is, F3 and F4 respectively. Between 1 and 10 m the mineral bound fraction, F6, the oxidizable fraction F5, and the oxy-hydrates fractions F3 and F4, are all converted to the acid soluble fraction, F2, and the exchangeable fraction, F1, as a result of the increased acidic conditions at the base of the slime dam. The net result is the build-up of the acid soluble and the exchangeable fractions at the base of the slime dam.

Cobalt

From Figure 5 the major speciation change for Co with time in the surface layers, 15 cm to 1 m, is the conversion of the fraction associated with Fe oxy-hydrates, F4, to the fraction associated with Mn oxy-hydrates, F3. For the elements studied, Co is unique in this respect, that is, the increase in the fraction associated with Mn oxides. The association of Co with Mn oxides has been reported

previously, and it has been suggested that the adsorption of Co by Mn oxides is formed by oxidation of Co^{2+} to Co^{3+} followed by lattice replacement of Mn^{3+} by Co^{3+} (Sager, 1992; Jones and Jarvis, 1981).

A drop in the exchangeable fraction, F1, is also observed, and may be explained by conversion to the fraction associated with Mn oxides, F3. A slight increase in the oxidizable fraction, F5, is also observed. The major speciation change below 1 m is the conversion of the fraction associated with Mn oxides, F3, and oxidizable fraction, F5, to the residual fraction, F6.

Manganese

The easily reducible, or fraction associated with Mn oxides, is the highest in the surface layers of the old slime dam. This is followed by the oxidizable fraction, then the acid soluble, and the exchangeable fraction. The residual fraction is the lowest at 6%. From Figure 5 the behaviour of Mn appears to be characterised by the conversion of the easily reducible, oxidizable, acid soluble and exchangeable fractions to the reducible (F4) and residual (F6) fractions in the surface layers. With depth the reducible fraction F4 is further converted to the Mn oxides, with the consequent increase in the residual fraction, F6. The increase in the residual fraction, F6, for manganese is due to precipitation of hydrous Mn oxides.

Other elements which show an increase in the residual fraction are Cr, Zn, Co and Ni. Table 4 shows Pearson's correlation factors (Minitab, 1994) for the residual fractions for the elements studied (except Cd). The high positive correlation factors for Cu, Co, Zn and Cr with the residual fraction of Mn, all greater than 0.9, suggest that the increase in the residual fraction for these elements is due to co-precipitation with Mn oxides. Co-precipitation of Ni and Zn in Mn oxides was reported previously (Jones and Jarvis, 1981), while uptake of Pb, Cu and Cd by freshly prepared hydrous Mn(IV) oxide, from acetate buffer, was reported to be nearly complete over the pH range 3 to 9 (Sager, 1992; Aualiitia and Pickering, 1987). It is interesting to note from Table 4 that Ni, Cu, Co, Zn, Cr, and Mn all show negative correlation factors with the residual fraction of iron, Fe(F6), showing that these elements do not co-precipitate with hydrous Fe(III) oxide under the conditions prevailing in the slime dams.

Chromium

Figure 2 shows that the oxidizable fraction F5 is the highest fraction in the surface layers of the slime dam. From Figure 5 the behaviour of Cr in the slime dam is characterised by the oxidation of this fraction to the reducible fraction F4. Below the 1 m level, both the F4 and F5 fractions are converted to the residual fraction, F6. The concentration of Cr^{3+} in well aerated soils is controlled by the formation of chromic oxides or hydroxides which are stable and of very low solubility (Jones and Jarvis, 1981).

Copper

From Figure 5, the major speciation change for Cu in the surface layers with time appears to be conversion of the reducible fraction, F4, to the oxidizable fraction, F5, and the acid soluble fraction, F3. Below 1 m, the acid soluble fraction is also converted to the oxidizable fraction, F5. The net result is a continuous increase in the oxidizable fraction, F5, with depth. Cu is unique in this respect. Cu^{2+} is known to form very stable complexes with organic matter over a wide range of pH (Sager, 1992; Jones and Jarvis, 1981). It was pointed out above that Cu forms a complex anion with the CN^- ion. The complex ion, $Cu(CN)_2^-$, has a spiral polymeric structure in which each Cu(I) atom is bound to two CN-carbon atoms, and one CN-nitrogen atom in a nearly coplanar array (Cotton and Wilkinson, 1972b). The increase in F5 for Cu can only be explained by assuming that this polymeric ion is very stable, and is formed as the CN^- ion is released by the decomposition of the cyanide complexes of the other metals under the increasing acidic conditions.

Some conversion of the acid soluble fraction to the residual fraction also occurs, probably as a result of co-precipitation with Mn oxides. Co-precipitation of Cu in Fe and Mn oxides has been reported (Sager, 1992; Jones and Jarvis, 1981).

Zinc

Figure 2 shows that the oxidizable fraction, F5, is the highest in the surface layer of the old slime dam. This is followed by the exchangeable fraction, F1, then the reducible fraction, F4. The acid soluble (F2), easily reducible (F3) and residual (F6) fractions are very low in the surface layer. From Figure 5 the major speciation change in going from the surface to 1 m depth is the conversion of the exchangeable fraction, F1, and

reducible fraction, F4, to the acid soluble (F2) and easily reducible or fraction associated with Mn oxides (F3). Zinc carbonates and hydroxo carbonates are known (Cotton and Wilkinson, 1972c).

The major speciation change for Zn below the 1 m level is the conversion of the acid soluble (F2), the easily reducible (F3), the reducible (F4) and the oxidizable (F5) fractions to the exchangeable fraction (F1), and residual fraction (F6). The conversion to free Zn^{2+} ions results from the reduced pH at the base of the slime dam, while the increase in the residual fraction arises from co-precipitation with Mn oxides as discussed under Mn.

Nickel

Figure 5 shows that the major speciation changes for Ni above the 1 m level involve the conversion of (a) the acid soluble fraction, F2, to the exchangeable fraction F1 probably as a result of the dropping pH, and (b) conversion of the fraction associated with hydrous Fe oxides, F4, to the residual fraction F6, via co-precipitation with Mn oxides. Below the 1 m level, we see the conversion of the acid soluble fraction (F2), the fraction associated with hydrous Mn oxides (F3), the fraction associated with hydrous Fe oxides (F4), and oxidizable fraction (F5) to the residual fraction (F6) via co-precipitation with Mn oxides as discussed above.

Mobilities of the elements

Table 5 shows the overall mobilities of the elements in the two slime dams. Cd is 100% mobile in both slime dams. The order of mobilities in the surface layers of the current slime dam for the rest of elements studied increases as follows: Fe = Ni < Cr < Cu < Mn < Co < Zn. Fe and Ni have the lowest mobilities at 54%, while Zn has the highest at 83%.

Figure 6 shows the profiles of mobilities of the elements as a function of depth in the old slime dam. Cd is again 100% mobile throughout the slime dam. In general, mobilities are very high, >90%, in the surface layers of the slime dam for all the elements studied with the exception of Fe. For Fe, mobility increases with depth, which is evidence of acid mine drainage.

At the 1 m level the order of mobilities is Ni < Fe < Mn < Cr < Zn < Cu = Co < Cd. This order is maintained at the base of the slime dam, that is, 10 m level, except that (a) the order for Mn and Ni is reversed, and (b) mobilities drop further. Mobility at 10 m increases in the order Mn < Ni < Cr < Zn < Cu < Fe = Cd. Cadmium and Fe show unique mobility patterns, while the rest of the elements studied show a definite pattern of high mobilities in the surface layers which decreases with depth. The decrease in mobilities follows the order Mn > Ni > Cr > Zn > Cu > Co. This trend is attributed to co-precipitation with Mn

Table 5. Mobility of the elements in the current slime dam (CSD) and old slime dam (OSD).

Element	CSD: Surface	OSD: Surface	OSD:1 m	OSD:10 m
Cd	100	100	100	100
Mn	73	94	76	8
Ni	54	86	69	33
cu	68	95	97	89
Fe	54	45	74	100
Co	80	95	97	92
Cr	60	93	92	53
Zn	83	97	94	72

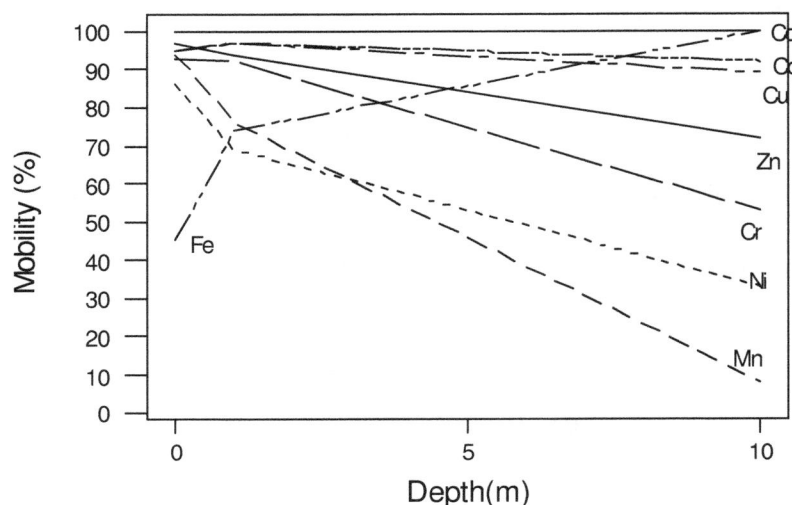

Figure 6. Potential mobility as a function of depth in the old slime dam.

oxides. Table 6 shows ionic radii for the metals studied (Evan, 1992), while Figure 7 shows a regression plot of percentage residual fraction, F6, for the elements as a function of ionic radii. It is apparent from Table 6 and Figure 7 that co-precipitation of the metals by Mn oxides or $CaCO_3$ is related to the size of the ion. Thus because of its large size, Cd(II) does not co-precipitate with Mn oxides, but co-precipitates with $CaCO_3$. On the other hand Fe does not co-precipitate with Mn because of its small size, while for Ni, Cr, Co, Cu and Zn co-precipitation with Mn decreases with increase in the size of the ion.

This trend suggests a tunnel or porous structure for the precipitated Mn oxides, such that the large Cd ion cannot enter the pores and therefore does not co-precipitate with the Mn oxides, while the small Fe(III) co-ordination 4 ion moves freely through the pores and is not retained. For Ni, Cr, Co, Cu and Zn the extent of co-precipitation decreases with increase in the size of the ion. Various forms of hydrous Mn(IV) oxides with tunnel structures have been reported (Burns and Burns, 1980;

Clearfield, 1988; Tsuji and Abe, 1984; Tsuji et al., 1992; Tsuji and Komameni, 1993; Tsuji et al, 1993). Figures 2 and 5 show that Fe exists mainly as the carbonate (F2) and the exchangeable fraction (F1) at the base of the slime dam. The low pH at the base of the slime dam results from the oxidation of sulphides; hence the major ligand in the exchangeable fraction must be the sulphate ion. Considering the large size of Ni(II) (0.69 Å), the high extent to which Ni co-precipitates with Mn oxides in Figure 7 suggests oxidation of Ni(II) to Ni(III) before co-precipitation as has been suggested by Sager (1992) and by Jones and Jarvis (1981).

Conclusions

From the foregoing discussion we conclude that speciation of Cd, Mn, Ni, Cr, Cu, Co and Zn within the slime dams depends on the stability of the metal cyano complexes formed during the cyanidation process, and on the ease with which the metals co-precipitate with Mn

Table 6. Ionic radii (R_i) of the elements studied.

Ion	Coordination number	R_i(Å)
Cd(+2)	4	0.78
	8	0.95
	8	1.10
Ca(+2)	6	1.00
	8	1.12
Cd(+2)	6	0.95
Co(+2)	6	0.65
Co(+3)	6	0.55
Cu(+2)	6	0.73
Zn(+2)	6	0.74
Cr(+3)	6	0.62
Ni(+2)	6	0.69
Ni(+3)	6	0.56
Mn(+4)	6	0.53
Fe(+3)	6	0.55
Fe(+3)	4	0.49

Source: Evan Jr. (1992).

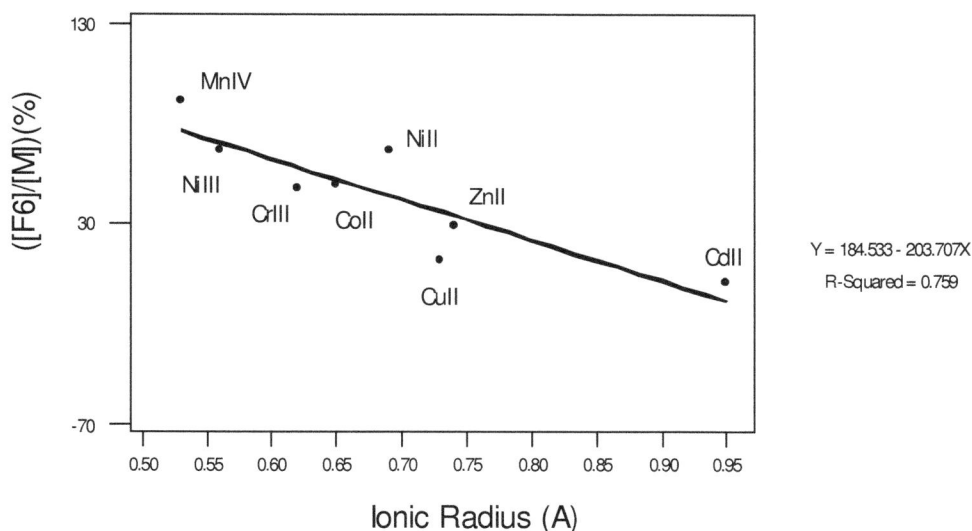

$Y = 184.533 - 203.707X$

R-Squared = 0.759

Figure 7. Percentage residual fraction (F6) at the base of the old slime dam as a function of ionic radius (co-ordination number (CN) 6): [F] = concentration of speciation fraction; [M] = total element concentration.

oxides and/or $CaCO_3$.

ACKNOWLEDGEMENTS

This work was supported by a grant from the research Board of the University of Zimbabwe.

REFERENCES

Aualiitia TU, Pickering WF (1987). Specific sorption of trace amounts of Cu, Pb and Cd by inorganic particles. Water Air Soil Pollut., 35(1-2): 171-185. http://www.springerlink.com/content/u931452297m23008k/

Banerjee ADK (2003). Heavy metal levels and solid phase speciation in street dusts of Delhi, India. Environ. Pollut., 123(1): 95-105.

Blanco P, Tome VF, Lozano JC (2004). Sequential extraction for radionuclide fractionation in soil samples: A comparative study. Appl.

Radiat. Iso., 61(2-3): 345-350.

Brown RM, Pickford CT, Davison WL (1984). Speciation of metals in soils. Int. J. Environ. Anal. Chem., 18: 135-141.

Burns RG, Burns MB (1980). Structures and reactivities of manganese (IV) oxides. In Proceedings of the Manganese Dioxide Symposium, Schumm B, Joseph HM, Kozawa A, Eds., Electrochemical Society, Cleveland, OH, 2: 97.

Chester R, Hughes MJ (1967). A chemical technique for the speciation of ferro-manganese minerals, carbonate minerals and adsorbed trace elements from pelagic sediments. Chem. Geol., 2: 249-263.

Clearfield A (1988). The role of ion exchange in solid-state chemistry. Chem. Rev., 88(1): 125-148.

Cotton FA, Wilkinson G (1972a). Advanced Inorganic Chemistry: A Comprehensive Text, 3rd Ed., Wiley Interscience, New York, pp. 721-723.

Cotton FA, Wilkinson G (1972b).Advanced Inorganic Chemistry: A Comprehensive Text, 3rd Ed., Wiley Interscience, New York, p. 907.

Cotton FA, Wilkinson G (1972c). Advanced Inorganic Chemistry: A Comprehensive Text, 3rd Ed., Wiley Interscience, New York, p. 513.

Evan HT, Jr (1992). Ionic radii in crystals. In Handbook of Chemistry and Physics, 73rd Ed., Lide DR, Ed., CRC Press, Boca Raton, Florida, pp. 12-9.

Gibson MJ, Farmer JG, Lovell MA (1983). Trace element speciation and partitioning in environmental geochemistry and health. Mineral Environ., 5(2-3): 57-66.

Flyhammar P (1997). Estimation of heavy metal transformation in municipal solid waste. Sci. Total Environ., 198(2): 123-133.

Gibson MJ, Farmer JG (1986). Multi-step sequential chemical extraction of heavy metals from urban soils. Environ. Pollut. Ser. B, 11(2): 117-135.

Jones LHP, Jarvis SC (1981). The fate of heavy metals. In The Chemistry of Soil Processes, J.D. Greenland and M.H.B. Hayes, Eds., Wiley & Sons, NY, pp. 593-620.

Kabala C, Singh BR (2000). Fractionation and mobility of copper lead and zinc in soil profiles in the vicinity of a copper smelter. J. Environ. Qual., 30(2): 485- 492.

Lu Y, Gong Z, Zhang G, Burghardt W (2003). Concentration and chemical speciation of Cu, Zn, Pb and Co in urban soils in Nanjing, China. Geoderma, 115(1-2): 101-111.

McLean EO (1982). Soil pH and lime requirement. In Methods of soil analysis, Part2, 2nd Ed., Page A.L., R.H. Millerand D.R. Keeney, Eds., American Society of Agronomy, Madison, pp. 199-224.

Mehra A, Cordes KB, Chopra S, Fountain D (1999). Distribution and bioavailability of metals in soils in the vicinity of a copper works in Staffordshire, UK. Chem. Speciation and Bioavalabilty, 11(2): 57-66.

Minitab (1994). MINITAB Release 10 for Windows Reference Manual, Minitab Inc., USA, pp. 8-19.

Moeller T (1963). Inorganic Chemistry: An Advanced Textbook. Wiley, NY, p. 719.

Navas A, Lindhorfer H (2003). Geochemical speciation of heavy metals in semiarid soils of central Ebro Valley (Spain). Environ. Int., 29(1): 61-68.

Nsimba EB (2009). Cyanide and cyanide complexes in the goldmine polluted land in the East and Central rand Goldfields, South Africa. M.Sc. Dissertation, University of Witwatersrand, S. Africa. (Electronic Theses and Dissertations: http://hdl.handle.net/10539/7052).

Puddephatt RJ (1978). The Chemistry of Gold, Elsevier Scientific, Amsterdam, p. 5.

Rafiu AO (2007). Influence of discharged effluent on the quality of surface water utilized for agricultural purposes. African J. Biotechnol., 6(19): 2251-2258.

Riise G, Bjornstad HE, Lien HN, Oughton DH, Salbu B (1990). A study on radionuclide association with soil components using a sequential extraction procedure. J. Radioanal. Nuclear Chem., 142(2): 531-538.

Rosner T, van Schalkwyk A (2000). The environmental impact of gold mine tailing footprints in the Johannesburg region of South Africa. Bull. Eng. Geol. Environ., 59: 137-148.

Sager M (1992). Chemical speciation and environmental mobility of heavy metals in sediments. In Hazardous Metals in the Environment, Stoeppler M, Ed., Elsevier Science Publ., Amsterdam, pp. 133-174.

Salbu, B, Oughton DH, Ratnikov AV, Zhigareva TL, Kruglov SV, Petrov KV, Grebenshakikova NV, Firsakova SK, Astasheva N P, Loshchilov NA, Hove K, Strand P (1994). The mobility of 137Csand 90Sr in agricultural soils in the Ukraine, Belarus, and Russia, 1991. Health Phys., 67(5): 518-528.

Singh SP, Tack FM, Verloo MG (1998). Heavy metal fractionation and extractability in dedges sediment derived surface soils. Water Air Soil Pollut., 102(3-4): 313-328.

Tessier A, Campbell PG, Bisson M (1979). Sequential extraction procedures for the specification of particulate trace metals. Anal. Chem., 5: 844-855.

Tessier A, Carignan R, Dubbreuil B, Rapin F (1989). Partitioning of zinc between the water column and the oxic sediments in lakes. Geochim. Cosmochim. Acta, 53(7): 1511-1522.

Tsuji M, Abe M (1984). Synthetic inorganic ion-exchange materials XXXVI. Synthesis of cryptomelane-type hydrous manganese dioxide as an ion-exchange materials and their ion-exchange selectivities towards alkali and alkaline earth metal ions. Solv. Extr. Ion Exchange, 2(2): 253-274.

Tsuji M, Komameni S (1993). Selective exchange of divalent transition metal ions in cryptomelane-type manganic acid with tunnel structure. J. Materials Res., 8(3): 611-616.

Tsuji M, Komameni S, Abe M (1993). Ion-exchange selectivity for alkali metal ions on a cryptomelane-type hydrous manganese dioxide. Solv. Extr. Ion Exchange, 11(1): 143-158.

Tsuji M, Komameni S, Tamawa Y, Abe M (1992). Cation exchange properties of layered manganic acid. Mater. Res. Bull., 27(6): 741-751.

Vuorinen A, Carlson L (1985). Speciation of heavy metals in Finnish lake Ores; Selective extraction analysis. Int. J. Environ, Anal. Chem., 20: 179-186.

Wilbur WG, Hunter JV (1979). Distribution of metals in street sweepings, stormwater solids, and urban aquatic sediments. J. (Water Pollution Control Federation), 51(12): 2810-2822.

Winde F, Sandham LA (2004). Uranium pollution of South African streams – An overview of the situation in the gold mining areas of the Witwatersrand. Geo. J., 61(2): 131-149.

Zaranyika MF, Mukono TT, Jayatissa N, Dube MT (1997). Effect of seepage from a gold mine slime dam on the trace heavy metal levels of a nearby receiving stream and dam in Zimbabwe. J. Environ. Sci. Health A, 32(8): 2155-2168.

Zaranyika MF, Madungwe L. Gurira RC (1994). Cyanide ion concentration in the effluent from two goldmines in Zimbabwe and in a stream receiving effluent from one of the goldmines. J. Environ. Sci. Health A, 29(7):1295-1303.

9

Heavy metals accumulation and distribution pattern in different vegetable crops

S. Singh*, M. Zacharias, S. Kalpana and S. Mishra

Division of Environmental Sciences, Insian Agricultural Research Institute, New Delhi 110012, India.

Different vegetable crops grown on heavy metal contaminated soil showed marked difference in metal accumulation, uptake and distribution pattern. Crop species also showed remarkable difference in metal concentration of various plant parts. Based on metal accumulation in edible parts and whole plants, root vegetables namely radish and carrot registered lower accumulation of almost all heavy metals except zinc (Zn) in radish root. However, leafy vegetables namely spinach, amaranthus, mustard and fenugreek recorded higher accumulation of both essential and non-essential heavy metals, except cadmium (Cd) and nickel (Ni) which showed less accumulation in fenugreek. Potato and onion showed lower accumulation of zinc, copper and higher accumulation of cadmium and nickel. Cauliflower and cabbage, however, showed greater accumulation of lead and nickel but less accumulation of copper and cadmium. Among fruit type vegetables, pea, soybean and cluster bean showed greater accumulation of Pb and Ni and very less accumulation of Cd. Among different vegetables cauliflower and cabbage recorded highest uptake of Zn, Pb and Ni, while mustard showed higher uptake of Zn and Cd. In general the uptake of Cd was lowest in almost all the crops except mustard. Generally the root and leafy vegetables namely radish, carrot, spinach, amaranthus, mustard, cauliflower and cabbage showed higher distribution of metals to the edible parts, whereas fruit types vegetables specially tomato and Brinjal exhibited least transport of metals to fruits except leguminous fruit vegetables pea and soybean. Leafy vegetables namely spinach, amaranthus and mustard seemed to be unsafe and not suitable for cultivation on heavy metal contaminated soil. Most of the fruit type vegetables could be suggested for cultivation on Cd contained soil but not for Ni and Pb contained soil.

Key words: Heavy metals, accumulation, distribution, uptake, contamination, cultivation.

INTRODUCTION

Due to rapid urbanization, the demand for food crops is rising day by day, and as vegetables can be grown in small fields with intensive use of inputs within shorter period, its cultivation is gaining popularity and fetching profitability in peri-urban areas of mega cities. This is a matter of serious concern since vegetables particularly leafy once, being prolific accumulators of heavy metals provide an easy entry into food chain to these dreaded metals. The excessive intake of these elements from the soil creates dual problems; first the harvested crops get contaminated, which serve as a source of heavy metal in

our diet, and secondly the crop yield decline due to the inhibition of metabolic processes (Sanders et al., 1987; Singh and Aggarwal, 2006). Increasingly higher quantities of heavy metals are being released into the environment through various anthropogenic activities such as smelting industries, sewage sludge, municipal solid wastes, burning of fossil fuel, pesticides etc (Sheila, 1994; Rattan et al., 2002). Zinc, copper, iron, manganese, lead and cadmium contents in soils receiving sewage sludge accumulated mostly in the top 0 to 15 cm layers. Heavy metals occur in the soil both in soluble and combined forms. However, only soluble exchangeable and chelated metal species in the soils are mobile and hence available to the plants (Mc Bride et al., 1981; Miller, 1986; Singh and Kumar, 2006). Vegetable crop plants have high ability to accumulate metals from the

*Corresponding author. E-mail: sdsingh14d@yahoo.co.uk.

environment, which may pose risks to human health when they are grown on or near contaminates lands and consumed. Metal accumulation in plant depends on plant species, growth stages, types of soil and metals, soil conditions, weather and environment (Asami, 1981; Chang et al., 1984; Khairiah et al., 2004). Thus accumulation of heavy metals in the edible parts of vegetables represents a direct pathway for their incorporation into the human food chain (Florijin, 1993). The health risk will depend upon the chemical composition of the waste material, its physical characteristics, types of vegetables cultivated and the consumption rate (Cobb et al., 2000).

Keeping in view the significance of metal contaminated fertile land in the peri-urban areas, and their judicious utilization for agricultural purposes, the present study was undertaken to examine the crop species differences in heavy metal accumulation and distribution in various edible and non-edible plant parts and to suggest the cultivation of different vegetable crops in metal contaminated soils based on their accumulation in edible plant part.

MATERIALS AND METHODS

Experimental set up and crop management

A field trial was conducted at the Research Farm, Indian Agricultural Research Institute during *Rabi* season of 2005 to 2006. 16 different vegetable crops comprising of root vegetables (radish and carrot), tuber and bulb vegetables (potato and onion), leafy vegetables (spinach, amaranthus, fenugreek, mustard and Cabbage), fruit vegetables (okra, brinjal, Tomato, Soyabean, Pea and Cluster bean) and others (cauliflower) were grown in the field pre-contaminated with copper (Cu), zinc (Zn), lead (Pb), cadmium (Cd) and nickel (Ni) at 20 kg/ha by incorporating their salts into the soils. The seeds and seedlings of different vegetable crops were procured from the division of vegetable science of the institute and cultivated in the field at recommended spacing of different crops with three replications each around twenty days after the incorporation of metals in soil. The experiment was conducted in simple Randomized Block Design (RBD).

Plant sampling and preparation of samples for metal analysis

The various plant parts (root, stem, leaf, fruit, curd, head) of fresh samples of vegetables (including edible and non-edible) were separated and dried properly in hot air oven at 70°C for 48 h and weighed on digital electronic balance to estimate heavy metal content in various edible and non-edible plant parts and also to measure their uptake from soil. After drying the samples, they were powdered to fine texture using grinder for heavy metal analysis of plant samples.

Heavy metal analysis

1 g ground sample from each plant part of different vegetables in triplicate were weighed and transferred to 150 ml conical flask followed by adding 15 ml di-acid mixture (Nitric acid and Perchloric acid) and thereafter kept overnight for partial digestion. The partially

digested samples in conical flasks were kept on hot plate for complete digestion. After digestion the samples were distilled water and filtered with Whatman No. 42 and finally the volume was made to 50 ml in volumetric flasks. After complete digestion, filtration and dilution, the samples were analyzed for heavy metals such as Cu, Zn, Pb, Cd and Ni by Atomic Absorption Spectrophotometer as per the procedure described by Singh et al. (1999). The statistical analysis was done to find out the critical difference (CD) at 0.05 probability level for measuring the significance difference in metal contents and uptake between different vegetable samples using SPSS (Statistical Package for Social Sciences).

RESULTS AND DISCUSSION

Heavy metal content in soil before and after their application

Soil samples were taken from the land used for experimental purpose before the application of heavy metal in the soil to know the initial level of metal in soil. Fifteen days after the application of heavy metal salts in soil, again soil samples were taken to know the heavy metal level in soil. The data presented in Table 1 show that the total as well as the available level of almost all the heavy metal in soil increased markedly following their application in soil. However, their availability percent was very low (0.84 to 17%) in the soil even after their application. Amongst the heavy metals the availability was recorded to be highest for cadmium (17%) followed by copper (11%) and the lowest was registered for nickel (0.84%).

Metal content and their distribution in different plant parts of vegetables

The accumulation of metals has been described as their content or concentration in different parts on the basis of their amount per unit dry weight of tissues (µg/g dry weight).

Metal accumulation in different plant parts of vegetable crops

The leaves contained maximum zinc followed by roots, stem and fruits. Average accumulation recorded in leaf, root, stem and fruit were 68, 62, 60 and 53 µg/g respectively. Among the crops and various plant parts, tomato stem showed the highest Zn accumulation (115 µg/g dry weight) followed by leaf and root of Brinjal (112 and 108 µg/g dry weight respectively) and potato leaves (106 µg/g dry weight). Roots of soybean, okra and carrot showed the lowest amount of zinc accumulation (21, 23 and 28 µg/g dry weight respectively) (Table 2).

Among the various plant parts, roots showed highest accumulation of copper followed by stem, leaf and fruits. When considered all parts of different crops, root of

Table 1. Heavy metal content of soil before and after their application.

Metal applied in soil	Total metal content (µg/g dry soil wt.)	Available metal content (µg /g dry soil wt.)	Metal availability (%)
Zinc (Zn) before application	114	2.39	2.10
Zinc (Zn) after application	146.39	3.99	2.73
Copper (Cu) before application	32.3	1.44	4.45
Copper (Cu) after application	60.34	7.17	11.53
Lead (Pb) before application	30.15	0.79	2.61
Lead (Pb) after application	52.69	3.91	7.17
Cadmium (Cd) before application	1.50	0.00	0.00
Cadmium (Cd) after application	25.91	9.31	17.08
Nickel (Ni) before application	34.60	0.00	0.00
Nickel (Ni) after application	61.05	0.56	0.84

Table 2. Heavy metal accumulation in various plant parts of different vegetable crops grown on metal contaminated soil.

Vegetable crop	Zn content (µg /g dry wt.)				Cu content (µg /g dry wt.)				Pb content (µg /g dry wt.)			
	Root	Stem	Leaf	Fruit	Root	Stem	Leaf	Fruit	Root	Stem	Leaf	Fruit
Radish	67	NA	59	NA	29	NA	60	NA	3	NA	10	NA
Carrot	28	NA	52	NA	13	NA	16	NA	15	NA	22	NA
Potato	90	96	106	NA	92	26	48	NA	46	43	74	NA
Onion	33	49	33	NA	23	17	18	NA	13	13	17	NA
Spinach	81	NA	86	NA	43	NA	29	NA	11	NA	23	NA
Amaranthus	58	54	87	NA	52	24	26	NA	17	24	35	NA
Fenugreek	58	51	70	NA	115	55	88	NA	21	33	36	NA
Mustard	37	48	64	NA	25	42	32	NA	14	15	27	NA
Cauliflower	104	33	60	53	90	10	14	13	40	26	38	29
Cabbage	92	37	32	NA	28	11	12	NA	35	30	34	NA
Soybean	21	31	30	71	29	101	64	51	10	13	34	18
Cluster bean	29	29	67	53	41	87	56	48	24	21	48	11
Tomato	90	115	73	29	19	26	32	24	10	14	25	6
Brinjal	108	87	112	29	45	39	58	43	30	46	87	2
Peas	79	87	80	71	33	13	24	17	22	23	36	19
Okra	23	66	72	65	21	18	34	71	26	19	48	33
Mean	63	60	68	53	44	29	38	38	21	25	37	17
CD at 5%	28	24	19	15	44	66	13	33	18	21	6	5

NA: Not Applicable in case where the particular plant part is not found in that vegetable.

fenugreek and stem of soybean showed its highest concentration that is 115 and 101 µg/g dry weight respectively. The lowest accumulations were recorded in cauliflower stem (10 µg/g dry weight), cabbage head (11 µg/g dry weight) and leaf (12 µg/g dry weight). Among the crops, potato, fenugreek, cauliflower and soybean recorded greater accumulation as compared to other crops (Table 2).

Regarding lead accumulation in different vegetables and their different plant parts, average content of lead was recorded highest in crop leaves followed by stem, roots and fruits. Leaves of Brinjal and potato showed highest content of lead accumulation (87 and 74 µg/g dry weight respectively), whereas fruit of Brinjal and root of radish recorded its lowest concentration (2 and 3 µg/g dry weight respectively) (Table 2).

Cadmium was found to be accumulated more in the roots of potato and onion (44 and 35 µg/g dry weight) followed by stem and leaves of potato, spinach roots and leaves, amaranthus leaves, fenugreek roots, and mustard leaves, while cluster bean, peas, carrot and soybean crops registered its lower accumulation. On an average Cd content was found highest in roots, followed by leaves, stem and fruits of vegetable crops (Table 3).

Table 3. Heavy metal accumulation in various plant parts of different vegetable crops grown on metal contaminated soil.

Vegetable crop	Cd content (µg /g dry wt.)				Ni content (µg /g dry wt.)				Metal content in edible parts (µg /g dry wt.)				
	Root	Stem	Leaf	Fruit	Root	Stem	Leaf	Fruit	Zn	Cu	Pb	Cd	Ni
Radish	6	NA	11	NA	10	NA	5	NA	67	29	3	6	10
Carrot	2	NA	3	NA	12	NA	41	NA	28	13	15	2	12
Potato	44	30	32	NA	45	27	31	NA	96	26	43	30	27
Onion	35	4	4	NA	26	28	28	NA	49	17	13	4	28
Spinach	28	NA	20	NA	16	NA	11	NA	86	29	23	20	11
Amaranthus	17	13	28	NA	22	10	16	NA	87	26	35	28	16
Fenugreek	25	8	4	NA	12	50	5	NA	64	88	36	4	5
Mustard	13	16	20	NA	20	16	26	NA	70	32	27	20	26
Cauliflower	10	4	7	3	84	32	38	40	53	13	29	3	40
Cabbage	18	4	8	NA	35	24	29	NA	32	12	34	8	29
Soybean	4	4	4	2	7	10	15	16	71	51	18	2	16
Cluster bean	3	1	6	2	14	12	23	17	53	48	11	2	17
Tomato	9	5	6	3	18	34	21	17	29	24	6	3	17
Brinjal	7	6	8	2	60	18	35	1	29	43	2	2	1
Peas	6	3	2	1	21	15	20	15	71	17	19	1	15
Okra	5	7	12	7	24	22	27	7	65	71	33	7	7
Mean	15	8	11	3	27	23	23	20	59	34	22	9	19
CD at 5%	10	7	5	2	19	8	10	11	22	11	9	7	8

In general, Ni content was registered highest in roots followed by stem, leaves and fruits with 27, 23, 22 and 20 µg/g dry weights respectively. Among the crops and their various parts, roots of cauliflower and Brinjal and stem of fenugreek were found to accumulate highest amount of nickel in the order of 84, 60 and 50 µg/g dry weight respectively (Table 3).

Metal accumulation in edible parts

Heavy metals showed differential level of their accumulation in different vegetable crops tested. In case of zinc, it varied from 29 to 87 µg/g dry weight of edible parts of different vegetables. Edible parts such as roots of carrot and fruits of tomato and Brinjal showed the minimum level of its accumulation in the order of 28, 29 and 29 µg/g dry weight respectively, whereas leaves of amaranthus, palak and fruits of pea recorded with its highest level of accumulation as 87, 86 and 71 µg/g dry weight respectively. However, copper was found to be highest in leaves of fenugreek (88 µg/g dry weight), followed by okra fruit (71 µg/g dry weight) and soybean fruit (51 µg/g dry weight). The lowest copper concentration was found in cauliflower curd (13 µg/g dry weight) followed by roots of carrot and cabbage head (13 and 17 µg/g dry weight respectively). In case of lead the range of accumulation varied from 3 to 43 µg/g dry weight, being lowest in radish and highest in potato tubers. Amaranthus leaves and potato tuber were found with higher concentration of cadmium (28 and 30 µg/g dry

weight respectively), while soybean fruit, carrot root, cluster bean and pea fruits registered its lowest concentration (2, 2, 2 and 1 µg/g dry weight respectively). For Nickel the highest accumulation was recorded in onion bulb (30 µg/g dry weight) followed by cabbage head and Brinjal fruit (29 µg/g dry weight) (Table 3).

Heavy metal uptake and their distribution in different plants parts

In general, the availability of heavy metals in soil to the plants (bio-availability) is meager and hardly a small fraction of total metal concentrations in soil are available to the plants. Despite the poor bioavailability of heavy metals in soil, the plants have high ability to accumulate the metals in their different parts from the environment, thus metals taken up by the crop plants may pose risks to human health when they are grown on or near metal contaminated areas through various food chains. The amount of heavy metals absorbed and the proportion of their translocation/ distribution to different edible and non-edible parts of tested vegetable crops plants are briefly discussed as below.

Total metal uptake by different crops

Irrespective of different crops and plant parts, the total uptake of metals was recorded in the order of Zn>Cu>Pb>Ni>Cd. Among the crops, zinc uptake ranged

Table 4. Heavy metal and their distribution in various plant parts of different vegetables.

Vegetable crop	Metal uptake (mg/m² crop area)					Zn distribution in different plant parts (%)				Cu distribution in different plant parts (%)			
	Zn	Cu	Pb	Cd	Ni	Root	Stem	Leaf	Fruit	Root	Stem	Leaf	Fruit
Radish	28	16	2	3	4	78	NA	22	NA	60	NA	40	NA
Carrot	39	15	18	2	25	40	NA	60	NA	51	NA	49	NA
Potato	20	11	8	3	4	5	75	20	NA	9	77	14	NA
Onion	7	3	3	1	6	4	66	30	NA	5	73	22	NA
Spinach	13	5	3	3	2	17	NA	83	NA	25	NA	75	NA
Amaranthus	35	14	14	10	7	10	36	54	NA	22	39	39	NA
Fenugreek	5	4	2	1	1	10	45	45	NA	16	38	46	NA
Mustard	67	45	25	22	25	7	42	51	NA	7	55	38	NA
Cauliflower	80	28	45	8	57	18	4	40	38	45	4	26	25
Cabbage	56	18	45	9	38	17	30	53	53	17	33	50	NA
Soybean	15	19	6	1	4	2	16	12	70	3	39	20	38
Cluster bean	24	36	15	2	9	9	28	41	22	9	55	23	13
Tomato	48	14	9	3	14	23	48	25	4	16	36	37	11
Brinjal	58	28	29	4	20	32	40	25	3	27	37	27	9
Peas	47	11	14	1	10	1	15	15	69	3	9	19	69
Okra	36	19	18	5	14	6	45	29	20	10	24	25	41
Mean	36	18	16	5	15	17	31	38	32	20	40	34	29
CD at 5%	21	12	10	3	5	12	9	8	9	8	12	6	8

between 5 to 80 mg/m² crop area, being the highest in cauliflower and lowest in fenugreek. Copper uptake ranged from 3 to 45 mg/m² crop area, which was found to be lowest in onion, spinach and fenugreek (3 to 5) and highest in mustard. Lead uptake was recorded to be lowest in onion, spinach, fenugreek and soybean (2 to 6 mg/m² crop area) and highest in cauliflower and cabbage (45 mg/m² crop area). Cadmium uptake was recorded lowest with a range of 1 to 5 mg/m² in spinach, onion, fenugreek, soybean carrot, radish, brinjal, okra, whereas mustard and amaranthus manifested maximum uptake of Cd (22 mg/m² crop area). Similarly nickel uptake varied from 1 mg/m² dry crop area in fenugreek to 57 mg/m² crop area in cauliflower. In general, fenugreek recorded lowest uptake of all the metals, while cauliflower recorded the highest uptake for almost all the metals except cadmium where mustard recorded the highest value of its uptake (Table 4).

Metal distribution in different plant parts

Among the different plant parts, spinach leaves recorded the highest proportion of zinc distribution (83%) followed by radish roots (78%), while lowest proportion was transported in roots of pea (1 %). In stem, the zinc distribution ranged between 4-75% in potato and cauliflower. Leaves of spinach manifested maximum proportion of Zn distribution (83%) and minimum in soybean 12%. Among the fruits of various crops under

study, the highest proportion of zinc was transported in soybean fruits (70%) and the lowest in brinjal (3%) Table 4. Among different types of vegetables, root types that are reddish showed greatest distribution of copper in their leaves (60%) and lowest was recorded in soybean and peas (3%). The lowest percentage of copper was recorded in cauliflower stem (4%) and highest in potato (77%), whereas in leaves the distribution pattern of Cu was maximum in spinach (75%) and lowest in potato (14%). Peas and Brinjal fruits manifested 69% and 9% respectively the distribution of copper (Table 4). Radish showed highest proportion of lead distribution in its roots (50%) and lowest in peas (2%). Stem of onion recorded the maximum distribution of lead (75%) and minimum in cauliflower with 6%. Almost all the plants showed highest proportion of lead distribution in their leaves. The leafy vegetables such as spinach, amaranthus and fenugreek recorded Pb distribution to the extent of 90, 53 and 41% respectively. Edible portion of peas (pods) recorded the highest distribution of lead (62%) followed by cabbage heads (54%) and the least was recorded in tomato fruits (5%) Table 5. Compared to the percent distribution of Cd in various crops, many of them showed higher distribution in their roots like radish (64%), carrot (44%) and 35% in tomato. However, cauliflower showed lowest proportion of Cd distribution in their stem (5%). Among the leaves of various crops, the highest proportion of Cd was recorded in spinach (76%) and the lowest was in fenugreek and peas (17%). In fruits the distribution was in the range of 3 to 58% being lowest in Brinjal and highest in peas (Table

Table 5. Heavy metal and their distribution in various plant parts of different vegetables.

Vegetable crop	Pb distribution in different plant parts (%)				Cd distribution in different plant parts (%)				Ni distribution in different plant parts (%)			
	Root	Stem	Leaf	Fruit	Root	Stem	Leaf	Fruit	Root	Stem	Leaf	Fruit
Radish	50	NA	50	NA	64	NA	36	NA	86	NA	14	NA
Carrot	46	NA	54	NA	44	NA	56	NA	27	NA	73	NA
Potato	6	68	26	NA	18	34	48	NA	12	55	29	NA
Onion	3	75	22	NA	24	52	24	NA	3	70	27	NA
Spinach	10	NA	90	NA	24	NA	76	NA	24	NA	76	NA
Amaranthus	7	40	53	NA	10	30	60	NA	18	33	49	NA
Fenugreek	7	52	41	NA	31	52	17	NA	4	90	6	NA
Mustard	7	34	59	NA	8	44	48	NA	10	36	54	NA
Cauliflower	12	6	44	38	20	5	52	23	21	6	35	38
Cabbage	8	38	54	NA	21	44	35	NA	10	39	51	NA
Soybean	4	17	32	47	10	32	26	32	4	18	22	56
Cluster bean	13	32	48	7	13	20	55	12	12	31	38	19
Tomato	14	33	48	5	35	28	32	5	17	50	25	8
Brinjal	17	43	40	NA	29	41	27	3	15	25	23	37
Peas	2	13	23	62	5	20	17	58	2	12	18	68
Okra	13	28	39	20	10	38	37	15	18	45	31	6
Mean	14	37	45	30	23	27	40	21	17	39	36	33
CD at 5%	12	15	9	11	12	10	8	6	13	12	11	9

5). The highest percent of Ni distribution was recorded in radish roots (86%) and lowest was in peas roots (2%). The magnitude of Ni distribution in stem was 90% in fenugreek and 6% in cauliflower. Spinach leaves recorded the highest percent of Ni distribution (76%), while lowest was found in fenugreek leaves (6%). Pea showed the highest degree of Ni distribution in pods (68%), while the lowest proportion of the same was recorded in okra fruits (6%) Table 5.

Crops suggested for heavy metal contaminated soils

Based on the pattern of metal accumulation and their distribution in edible plant part of different crop plants, it is concluded that carrot, tomato, Brinjal, clusterbean, cabbage, cauliflower, potato and onion could be safely grown on Zn and Cu contaminated soils. Contrary to this, several vegetables like spinach, fenugreek, mustard and soybean are not suitable for their cultivation on Cu and Zn contaminated soils. The study showed that mustard, amaranthus, spinach, cabbage, cauliflower, clusterbean, potato and onion should not be grown on lead, cadmium and nickel contaminated soils. However, some fruit type vegetables like tomato and Brinjal could be safely grown on almost all metal contaminated soils except tomato on Cd contaminated soil. Root type vegetable such as radish could be suggested for cultivation on Cu, Pb and Ni contaminated soil but not on Cd contaminated soil, while carrot could be grown safely on Cd and Ni contaminated

soil (Table 6).

Increased concentrations of heavy metals in different parts of vegetable crops, as recorded in the present investigation confirm the findings of other researchers (Allinson and Dzialo, 1981; Barman and Lal, 1994; Barman et al., 2000; Kim et al., 2002; Wang and Stuanes, 2003). Heterogeneous accumulation of heavy metals in different crop species, and different plant parts of same crop species under present investigation have also been reported by Barman et al. (2000) and Singh and Aggarwal (2006), which could be attributed to their diverse morphological characteristics and position of edible parts on the plants in respect of their distance from roots and selective uptake of metal by each crop (Mohamed and Rashed, 2003). Low metal accumulation in fruit type vegetables as compared to leafy vegetable crops, and in reproductive organs than in vegetative parts have also been observed by Allinson and Dzialo (1981), Iretskaya and Chien (1998), Kim et al. (2002) and Singh and Aggarwal (2006). This may possibly be due to poor metal mobility within the plants. In contrast, however, Barman and Lal (1994) reported higher accumulation of heavy metals (Cu, Zn, Pb, Cd) in edible parts than in non-edible plant parts. In general, lower levels of heavy metals particularly Pb and Cd in reproductive organs than in vegetative parts, may be due to their poor mobility in plants as compared to essential metal nutrients that is Cu and Zn. Very low concentration of Cd in fruits of fruit type vegetable and their poor uptake indicates the possibility of safe cultivation of such type of vegetables on cadmium

Table 6. Suggested vs not suggested vegetable crops in heavy metal contaminated soils.

Crop	Soil contaminated with heavy metal				
	Zinc	Copper	Lead	Cadmium	Nickel
Radish	X	√	√	X	√
Carrot	√	√	X	√	√
Spinach	X	X	X	X	√
Amaranthus	X	√	X	X	X
Fenugreek	X	X	X	√	√
Mustard	X	X	X	X	X
Soybean	X	X	X	√	√
Tomato	√	√	√	X	√
Pea	X	√	X	√	√
Okra	√	X	X	X	√
Brinjal	√	√	√	√	√
Clusterbean	√	√	X	X	X
Cabbage	√	√	X	X	X
Cauliflower	√	√	X	X	X
Potato	√	√	X	X	√
Onion	√	√	X	X	X

Where sign √ refers to crops suggested and X for crops not suggested for cultivation in different metal contaminated soils.

contaminated fields. Such variation in metals uptake and their distribution/compartmentalization between different parts of different crop plants may be useful for selecting crop species suitable for cultivation on metal contaminated soils to reduce the movement of metals into food chains.

It is clearly evident from the present findings that most of the leafy vegetables are hyper accumulators of most of the non-essential heavy metals such as lead and cadmium. The diverse vegetable crop species also showed marked differences in respect of metal uptake and their distribution to various plant parts especially to the edible part, which could be emphasized for selection of vegetable crops for cultivation on metals contaminated soils depending on their metal uptake potential and their transportation/distribution to edible part. In conclusion our results may be useful for selecting suitable crop species for different metal contaminated soils.

REFERENCES

Allinson DW, Dzialo C (1981). The influence of lead, cadmium, and nickel on the growth of ryegrass and oats. Plant Soil, 62: 81-89.

Asami T (1981). In: Heavy metal pollution in soils of Japan. Japan-Scientific Societies Press, Tokyo, pp. 257-274.

Barman SC, Lal MM (1994). Accumulation of heavy metals in (Zn, Cu, Pb and Cd) in soil and cultivated vegetables and weeds grown in industrially polluted fields. J. Environ. Biol., 15: 107-115.

Barman SC, Sahu RK, Bhargava SK, Chaterjee C (2000). Distribution of heavy metal in wheat, mustard and weed grown in field irrigated with industrial effluents. Bull. Environ. Contam. Toxicol., 64: 489-496.

Chang AC, Page AL, Warneke JE, Grgurevic E (1984). Sequential extraction of soil heavy metals following a sludge application. J. Environ. Qual., 1: 33-38.

Cobb GP, Sands K, Waters M, Wixon BG, Dorward-King E (2000). Accumulation of heavy metals by vegetables grown in mine wastes. Environ. Toxic. Chem., 19: 600-607.

Iretskaya SN, Chien SH (1998). Comparison of cadmium uptake by five different food grain crops grown on soils of varying pH. Comm. Soil Sci. Plant Anal., 30: 441-448.

Florijin PJ, Van Beusichem ML (1993). Uptake and distribution of Cd in Maize inbreed line. Plant Soil, 150: 25-193.

Kim JY, Kim K, Lee J, Lee JS, Cook J (2002). Assessment of As and heavy metal contamination in the vicinity of Duchum Au-Ag mine, Korea. Environ. Geochem. Health 24: 215-227.

Khairiah J, Zalifah MK, Yin YK, Aminah A (2004). The uptake of heavy metals by fruit type vegetables grown in selected agricultural areas. Pakistan j. Biol. Sci., 7:1438-1442

Mohamed AE, Rashed MN (2003). Assessment of essential and toxic elements in some kinds of vegetables. Ecotox. Environ. Safety, Environ. Res. Sect. B, 55: 251-260.

Mc Bride MB, Tyler LD, Hovde DA (1981). Cadmium adsorption by soils and uptake by plants as affected by soil chemical properties. Soil Sci. Soc. Am. J., 45: 739-744.

Miller KG (1986). Essential and toxic heavy metals in soils and their ecological relevance. Trans. XIII Congr. Intern. Soc. Soil Sci., 1: 29-44.

Rattan RK, Datta SP, Chandra S, Sharan N (2002). Heavy metals and Environnmental Quality. Fert. News, 47: 21-40.

Sanders JR, Mc Grath SP, Adams T (1987). Zn, Cu, and Ni concentration in soil extracts and crops grown on four soils treated with metal loaded sewage sludges. Environ. Pollut., 44: 193-210.

Sheila MR (1994). Source and forms of potentially toxic metals in soil – Plant systems. Wiley, New York, pp. 247-273.

Singh D, Chhnokar PK, Pandey RN (1999). Soil Plant Water Analysis: A methods Manual. IARI, New Delhi.

Singh S, Aggarwal PK (2006). Effect of heavy metals on biomass and yield of different crop species. Indian J. Agric. Sci., 76: 688-691.

Singh S, Kumar M (2006). Heavy metal load of soil, water and vegetables in peri-urban Delhi. Environ. Monit. Assess., 120: 79-90.

Wang HY, Stuanes AO (2003). Heavy metal pollution in air-water-soil-plant system of Zhuzhou City, Hunan Province, China. Water, Air Soil Pollut., 147: 1-4.

10

Heavy metals remediation from urban wastes using three species of earthworm (*Eudrilus eugeniae, Eisenia fetida* and *Perionyx excavatus*)

Swati Pattnaik[1] and M. Vikram Reddy[1]*

[1]Department of Ecology and Environmental Sciences, Pondicherry University, Puducherry – 605 014, India.

Accepted 27 September, 2011

Remediation of Cadmium (Cd), Lead (Pb), Zinc (Zn), Copper (Cu) and Manganese (Mn) by the three earthworm species - *Eudrilus eugeniae, Eisenia fetida* and *Perionyx excavatus* from three different urban wastes (Municipal Solid Waste (MSW), Market Waste (MW) and Flower Waste (FW)) through the vermicomposting process carried out for sixty days was investigated. The metals concentrations increased gradually from the initial stage till end (P<0.05). Metals concentrations in earthworm tissues was significantly higher in *E. eugeniae* than that of *E. fetida* and *P. excavatus* (P<0.05). The concentrtions of Cd and Cu were higher in earthworms sampled from vermicompost of MSW and of Pb, Zn and Mn were higher in that of MW. Bioaccumulation factor (BAF) of metals (Cd > Zn > Pb > Cu > Mn) implied that Cd accumulation in earthworm tissue was more than that of substrate whereas the reverse was true for other metals.

Key words: Bioremediation, urban waste, heavy metals.

INTRODUCTION

Urban waste is the used and left-over materials in urban systems comprising of household garbage included kitchen waste, green waste (the garden and lawn clippings), street sweepings, sanitation residues, etc. (CPCB, 2000) and is generated in huge quantities now-a-days (Kumar et al., 2009). The disposal of urban waste is of great concern as it poses serious management problems particularly in developing countries like India, where its management is mostly unsystematic and un-scientific (Chattopadhyay et al., 2009; Pattnaik and Reddy, 2010). It causes pollution of land, water and air, which threatens public health with the waste forming the breeding ground for various pathogen-carrying vectors such as mosquitoes, rodents, pigs and others (Forbes, 1996; Hardoy et al., 2001; Kansal, 2002). Moreover, the urban waste is known to contain hazardous persistent pollutants such as heavy metals, being detrimental to the ecosystem including human beings (Smith, 2009; Reddy and Pattnaik, 2009).

The organic component of the urban waste can be recycled and bio-converted into useful end-product called compost and vermicompost when processed by earthworms, which could be utilized as a supplement to chemical fertilizers on farm lands or in potting media. The earthworms not only convert the organic fractions of urban wastes into available nutrients (Pattnaik and Reddy, 2010) but also consequently remediate the per-sistent heavy metals from the wastes by bioaccumulation

---

*corresponding author. E-mail: venkateshsrinivas1@gmail.com.

in their bodies during vermicomposting process (Pare et al., 1998; Keener et al., 2001; Gupta et al., 2005; Suthar, 2008; Pattnaik and Reddy, 2010). Studies have shown that the earthworms accumulate heavy metals in their body tissue from contaminated substrates (Edwards and Bohlen, 1992; Gupta et al., 2005; Suthar, 2008). Various environmental factors contribute to the accumulation of As, Cd, Cu, Pb and Zn in earthworms (Vermeulen et al., 2009). Metal concentrations in earth-worm tissue were directly correlated to the level of their contamination in the substrates and soil (Rosciszewska et al., 2003; Hobbelen et al., 2006). Earthworms thus have been considered as important bio-accumulators and bio-indicators of environmental contamination of persistent pollutants like heavy metals (Gish and Christensen, 1973; Udovic et al., 2007; Suthar et al., 2008; Wang et al., 2009).

It is therefore important to know the metals that are accumulative in the earthworm tissue and their fate in the environment. Though several studies revealed the high concentrations of heavy metals pollutants in urban waste and sewage sludge (Smith, 2009; Reddy and Pattnaik, 2009), the problems associated with bioaccumulation and vermiremediation have received very little attention and is poorly understood (Fries, 1982), especially in India. Moreover, adequate information is not available on the vermiremediation as well as bioaccumulation of heavy metals in earthworm body tissue particularly in tropics (Shahmansouri et al., 2005; Suthar, 2008). The present study, therefore, attempted to assess the remediation of heavy metals - Cd, Pb, Zn, Cu and Mn from the organic components of three different urban wastes - the municipal solid waste (MSW), vegetable or market waste (MW) and floral waste (FW) through metal bioaccumulation in the body tissues of three earthworms species - two exotic species viz, *Eudrilus eugeniae* (Kinberg) and *Eisenia fetida* (Savigny) and a local species, *Perionyx excavatus* (Perrier) used for vermicomposting, heavy metal reduction in vermicompost, and also to find out the bio-accumulation factors (BAF). These heavy metals for the present study were chosen based on their toxicity - highly toxic cadmium (Cd), moderately toxic lead (Pb), and less toxic zinc (Zn), copper (Cu) and manganese (Mn) (Wizewardena and Gunaratne, 2004).

MATERIALS AND METHODS

Experimental set up and sampling

Three types of urban waste viz., MSW, MW and FW were chosen for the present study. Municipal solid waste was collected from one of the major garbage dumping site of Puducherry, a small town on the east coast of India and the erstwhile French colony, and the

vegetable MW was collected from its main vegetable market, which comprises of different left over vegetables such as cabbage, brinjal, tomato, potato, onion, carrot, turnip and leafy vegetables. The FW was obtained from the *Peltophorum pterocarpum,* belonging to the family Fabaceae and sub-family Caesalpinioideae - a widely-appreciated shade tree and a reclamation plant with dense spreading crown, planted along the road-sides of the Pondicherry Central University campus; it is usually planted on the road sides.

Five samples of each waste were collected randomly and then were mixed to form composite samples before taking smaller sub-samples for analysis. These wastes were characterized and segregated into biodegradable and non-biodegradable components. The MSW, MW and FW were separated and air-dried separately for 48 h and pre-composted for three weeks prior to vermicomposting and composting processes. During pre composting process, the temperature raised up to about 60°C. As such high temperature was lethal to earthworm survival; thermal stabilization was done prior to introducing earthworms into all the three substrates.

Earthen pots were used for vermicomposting and composting; in each pot, five kg of the substrate mixed with cow dung in 3:1 ratio was taken. In the present study, the cow-dung was used as inoculants to accelerate the vermicomposting process (Karthikeyan et al., 2007; Pramanik et al., 2007; Gupta and Garg, 2009), and when mixed with organic waste the mixture served as attractive feeding resource for earthworms (Suthar and Singh, 2009). A total of four sets of earthen pots each set containing six replicates was taken for each substrate material, of which three sets were used for vermicomposting each set using one species of earthworm and the fourth set was used for only composting without using any earthworm.

Similarly, all the three types of substrate materials were used for vermicomposting and composting. When the temperature of the pre composting process retrieved to 25°C, fifty adult earthworms of each species that is, *E. eugeniae, E. fetida,* and *P. excavatus* were introduced on the top of each of the pots of each of the substrates. All these earthworm species chosen for the experiment were collected from the vermiculture unit at Lake Estate (Aurobindo Ashram, Puducherry, India). All the pots were covered on the top by jute cloth cover and wire mesh to prevent and protect the earthworms from the predators - centipedes, moles and shrews. The bottom of each pot was filled with small stones up to a height of 5 cm for air circulation and small holes were drilled at bottom of each pot for good drainage of water. The processes of vermicomposting and composting were carried out for a period of 60 days. The sub-samples of compost and vermicompost from each replicate were collected un-destructively after each interval of 15, 30, 45 and 60 days.

Heavy metal analysis of earthworms' tissue

Earthworms were rinsed and placed in petri dishes on moist filter paper. They were kept in dark for 4 days at 20 to 22°C, during which time all organic material were eliminated from their gut (Ireland, 1975). The filter paper was changed daily. Worms were then frozen at − 10°C to prevent microbial decomposition between collection and analysis, and were oven dried at 80°C for 24 h (Kruse and Barrett, 1985). The earthworm's body tissues were digested using Katz and Jennis (1983) method and Shahmansouri et al. (2005) method for heavy metal analyses. The body tissue samples were individually dried, ground and finally burn to ash at

550°C. Afterwards the ash was placed in test tube and 10 ml of 55% nitric acid was added. It was left overnight at room temperature for digestion.

On the following day, the samples were heated at the temperature of 40 to 60° C for 2 h and then at a temperature of 120 to 130°C for one additional hour and then cooled to room temperature. One ml of 70% perchloric acid was added and the mixture was reheated to a temperature of 120 to 130° C for one hour. They were allowed to cool before adding 5 ml of distilled water. Samples were reheated to 120 to 130°C until white fumes emitted. The samples were allowed to cool finally before being filtered. The solutions were filtered through Whatman filter paper No. 41 to 100ml of flasks and heavy metals were estimated using ICP-AES.

Bioaccumulation factor (BAF)

The metal accumulation by earthworms in their body tissue is known as BAF. It is the ratio of the level of metals in earthworms to the substrate. BAF for earthworms were estimated for the metals in earthworm tissues and substrate materials using the method described by Mountouris et al. (2002). The BAF was calculated using the formula:

$$BAF = C_{biota}/C_{substrate,}$$

where C_{biota} and $C_{substrate}$ were the total concentrations of earthworm and substrate used for vermicomposting experiment, respectively. It is always compared to unity. If the metal concentration is high in earthworms' tissue than that of substrate, BAF would be more than a unity. If the concentration of metals in substratre is higher than that in earthworms' tissue, then BAF would be less than a unity.

Statistical analysis

The data in this study was analyzed using the XLStat computer software package (version 2009). ANOVA tests were used to analyze the significance difference between heavy metal contents of earthworm tissues. Relationship between heavy metal concentration in earthworm tissue and their respective vermicompost metal concentration were determined using regression equation.

RESULTS

Metal content of vermicompost

The heavy metal concentrations of the vermicompost produced by the three species of earthworms showed discernible difference and were significantly less compared to that of the respective waste substrates that is, MSW, MW and FW. The removal of Cd, Pb, Zn, Cu and Mn was 89.5, 94.4, 94.0, 89.2, 77.5% from MSW; 88.7, 96.1, 87.9, 95.1, 79.4% from MW and 84.1, 95.6, 94.7, 95.3 and 83.1% from FW through vermicompost processed by E. eugeniae whereas it was 85.9, 92.3, 91.0, 86.6, 70.9% from MSW; 87.8, 94.8, 85.2, 87.3, 76.1% from MW and 85.5, 90.4, 80.9, 87.5, 78.3% from FW through the vermicompost produced by E. fetida, and 81.8, 90.7, 84.4, 85.4, 68.6% from MSW; 81.5, 92.9, 82.7, 78.1, 75.7% from MW and 78.7, 85.9, 77.6, 80.9, 75.2% from FW through the vermicompost produced by P. excavatus, respectively.

The vermicompost produced by E. eugeniae possessed low concentration of metals as compared to other two earthworms, that is, E. fetida and P. excavatus; whereas it was just reverse for the accumulation of metal in earthworms' tissue that is , E. eugeniae showed higher accumulation of heavy metal in its tissue than that of E. fetida and P. excavatus, respectively. The tissue concentrations of the heavy metals – Cd, Pb, Zn, Cu and Mn in three earthworms grown in the waste (substrates) – MSW, MW and FW was negatively and significantly correlated with their respective vermicompost (P<0.0001). Thus, the concentrations of metals in earthworms' tissues increased gradually and that in vermicompost consequently decreased gradually as the vermicomposting process progressed from the initial stage of 15 days to final stage of 60 days.

Concentrations of Cd, Pb, Zn, Cu and Mn decreased from 15 to 60 days by 0.10 to 0.05, 0.61 to 0.31, 4.85 to 1.49, 0.74 to 0.47 and 7.98 to 5.16 ppm in vermicompost of E. eugeniae, by 0.12 to 0.07, 0.72 to 0.43, 5.85 to 2.24, 0.86 to 0.59, and 8.27 to 6.67 ppm in vermicompost of E. fetida, by 0.14 to 0.09, 0.83 to 0.52, 6.51 to 3.86, 0.97 to 0.64, and 9.32 to 7.20 ppm in vermicompost of P. excavatus prepared from MSW (Figure 1); the decrease was 0.09 to 0.02, 1.10 to 0.42, 3.89 to 2.19, 0.55 to 0.29, and 10.67 to 8.52 ppm in vermicompost of E. eugeniae, 0.11 to 0.04, 1.28 to 0.57, 4.64 to 2.69. 0.68 to 0.31, and 11.27 to 9.84 ppm in vermicompost of E. fetida, 0.14 to 0.07, 1.89 to 0.78, 5.23 to 3.15, 0.70 to 0.48, and 12.11 to 10.00 ppm in vermicompost of P.excavatus prepared from MW (Figure 2); and it was 0.04 to 0.006, 0.46 to 0.13, 1.60 to 1.04, 0.26 to 0.05, and 3.54 to 2.02 ppm in vermicompost of E. eugeniae, 0.05 to 0.02, 0.57 to 0.24, 1.84 to 1.25, 0.31 to 0.14, and 3.99 to 2.57 ppm in vermicompost of E. fetida; and 0.06 to 0.03, 0.68 to 0.35, 2.17 to 1.47, 0.40 to 0.20, and 4.53 to 2.95 ppm in vermicompost of P.excavatus prepared from FW (Figure 3), respectively.

Concentrations of Cd, Pb, Zn, Cu and Mn in earthworms' tissue were increasing from 15 to 60 days by 0.63 to 0.82, 4.11 to 5.11, 19.47 to 24.34, 2.01 to 3.53, and 15.09 to 20.50 ppm in E. eugeniae, 0.60 to 0.80, 3.52 to 4.84, 15.01 to 21.91, 1.61 to 3.13, and 13.77 to 18.46 ppm in E. fetida and 0.52 to 0.75, 3.10 to 4.65, 12.68 to 19.10, 1.24 to 2.74, and 11.09 to 16.54 ppm in

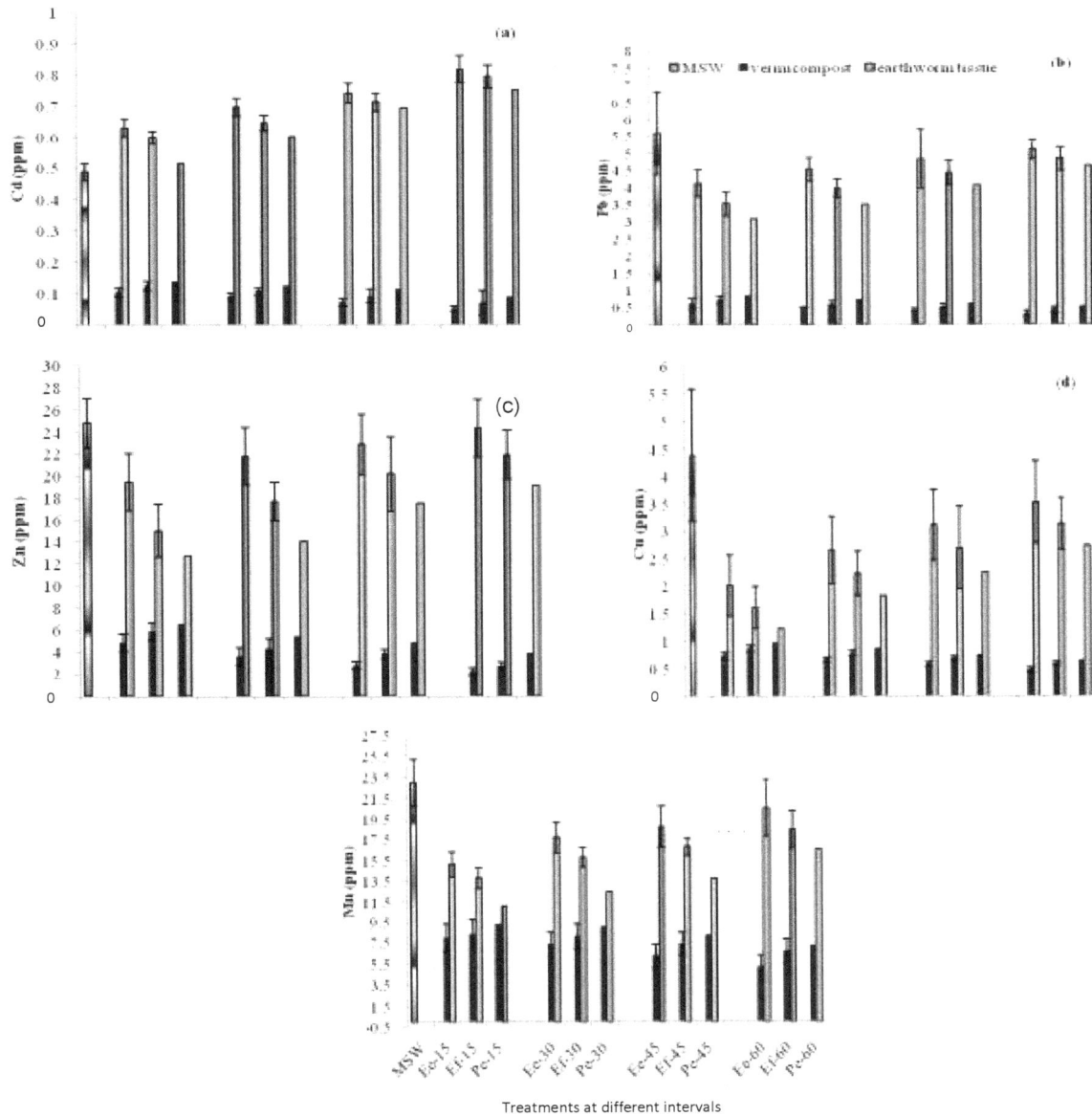

Figure 1. Accumulation of (a) Cd (ppm), (b) Pb (ppm) (c) Zn (ppm), (d) Cu (ppm) and (e) Mn (ppm) in the tissue of three earthworm species – *E. eugeniae* at 15 days (Ee-15), 30 days (Ee-30), 45 days (Ee-45) and 60 days (Ee-60), *E. fetida* at 15 days (Ef-15), 30 days (Ef-30), 45 days (Ef-45) and 60 days (Ef-60), and *P. excavatus* at 15 days (Pe-15), 30 days (Pe-30), 45 days (Pe-45) and 60 days (Pe-60) compared to that present in vermicompost produced from MSW in respect to MSW.

P. excavatus prepared from MSW (Figure 1); 0.51 to 0.72, 7.46 to 9.68, 13.00 to 17.37, 1.24 to 2.12, and 16.96 to 24.02 ppm in *E. eugeniae*, 0.44 to 0.64, 6.06 to 9.25, 10.69 to 15.90, 1.05 to 1.92, and 14.81 to 20.97 ppm in *E. fetida*, 0.40 to 0.59, 5.38 to 8.46, 8.40 to 13.94, 0.80 to 1.71, and 12.88 to 17.91 ppm in *P.*

excavatus prepared from MW (Figure 2); 0.37 to 0.60, 1.90 to 2.30, 4.16 to 6.08, 0.56 to 0.84, and 7.57 to 10.09 ppm in *E. eugeniae*, 0.30 to 0.53, 1.63 to 2.21, 3.70 to 5.56, 0.50 to 0.74, and 6.97 to 9.04 ppm in *E. fetida* and 0.21 to 0.46, 1.51 to 2.14, 2.88 to 4.81, 0.43 to 0.65, and 5.27 to 7.94 ppm in *P. excavatus* prepared

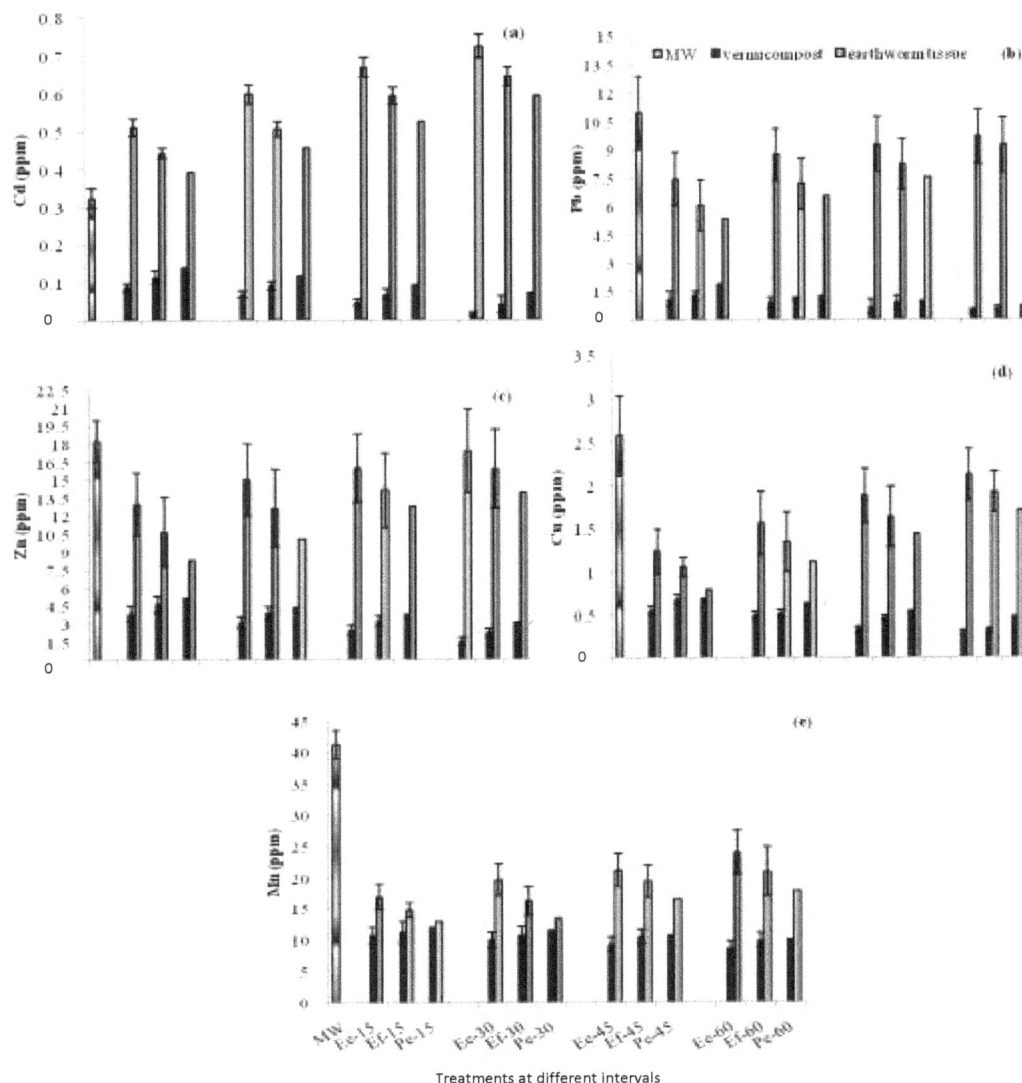

Figure 2. Accumulation of (a) Cd (ppm), (b) Pb (ppm) (c) Zn (ppm), (d) Cu (ppm) and (e) Mn (ppm) in the tissue of three earthworm species – *E. eugeniae* at 15 days (Ee-15), 30 days (Ee-30), 45 days (Ee-45) and 60 days (Ee-60), *E. fetida* at 15 days (Ef-15), 30 days (Ef-30), 45 days (Ef-45) and 60 days (Ef-60), and *P. excavatus* at 15 days (Pe-15), 30 days (Pe-30), 45 days (Pe-45) and 60 days (Pe-60) compared to that present in vermicompost produced from MW in respect to MW.

Metal accumulations in earthworm tissues

Comparison of earthworms

The comparison of heavy metal accumulation in tissues of three earthworm species used in vermicomposting of three waste substrates, i.e., vermicompost of MSW (Figure 1), MW (Figure 2) and FW (Figure 3) showed that the concentration of Cd, Pb, Zn, Cu and Mn in the tissue of *E. eugeniae* was 1.03 to 1.18, 1.04 to 1.23, 1.09 to 1.30, 1.11 to 1.24 and 1.09 to 1.21 fold higher than that of *E. fetida* and was 1.07 to 1.76, 1.07 to 1.39, 1.25 to 1.56, 1.24 to 1.62 and 1.24 to 1.46 fold higher than that of *P. excavatus*; whereas that of *E. fetida* was higher by 1.03 to1.43, 1.03 to 1.14, 1.11 to 1.28, 1.11 to 1.32 and 1.12

from FW (Figure 3), respectively.

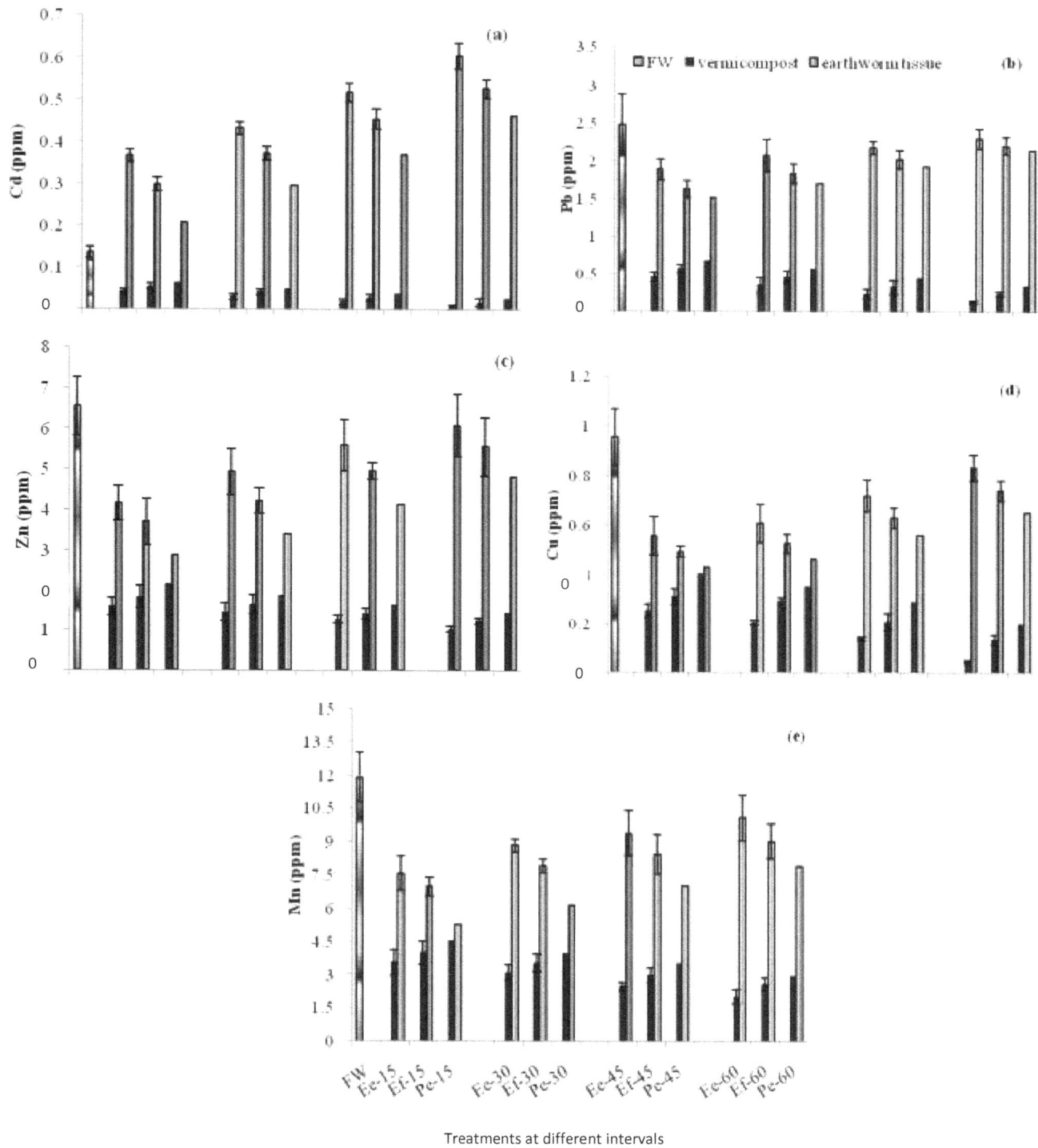

Figure 3. Accumulation of (a) Cd (ppm), (b) Pb (ppm) (c) Zn (ppm), (d) Cu (ppm) and (e) Mn (ppm) in the tissue of three earthworm species – *E. eugeniae* at 15 days (Ee-15), 30 days (Ee-30), 45 days (Ee-45) and 60 days (Ee-60), *E. fetida* at 15 days (Ef-15), 30 days (Ef-30), 45 days (Ef-45) and 60 days (Ef-60), and *P. excavatus* at 15 days (Pe-15), 30 days (Pe-30), 45 days (Pe-45) and 60 days (Pe-60) compared to that present in vermicompost produced from FW in respect to FW.

to 1.29 fold than that of *P. excavatus* at different intervals, respectively.

Removal of heavy metals from the waste substrates by earthworms

The concentration of Cd, Pb, Zn, Cu and Mn in the tissues of the three species of earthworms showed significant difference (P< 0.05) from that of their waste substrates, that is, MSW (Figure 1), MW (Figure 2) and FW (Figure 3). The removal of Pb, Zn, Cu and Mn was 9 to 26, 2 to 22, 20 to 54, and 11 to 34% by *E. eugeniae*, 13 to 37, 12 to 40, 29 to 63, and 19 to 40% by *E. fetida* and 17 to 45, 23 to 49, 38 to 72, and 28 to 52% by *P. excavatus* from MSW; it was 12 to 32, 5 to 29%, 18 to 52, and 42 to 60% by *E. eugeniae*, 16 to 45, 13 to 41 ,25 to 59, and 49 to 64% by *E. fetida* and 23 to 51, 24 to 54, 33 to 69, and 57 to 69% by *P. excavatus* from MW; and 7 to 24, 7 to 37, 13 to 41, and 15 to 36% by *E. eugeniae*, 11 to 34, 15 to 44, 22 to 48, and 24 to 41% by *E. fetida* and 13 to 39, 26 to 56, 31 to 55, and 33 to 56% by *P. excavatus* from FW at different intervals, respectively; whereas the Cd concentration was more in tissue of *E. eugeniae, E. fetida, and P. excavatus* by 29 to 68, 22 to 63, and 6 to 54% than in MSW; 58 to 123, 37 to 98, and 21 to 83% than in MW, and 170 to 344, 119 to 288 and 54 to 240% than in FW at different intervals, respectively.

The tissue concentrations of the heavy metals in earthworm species used in vermicomposting of FW were lower than those used in vermicomposting of other two substrate wastes (P <0.05). The concentration of Cd, Zn and Cu in the tissue of earthworms used in vermin-composting of MSW was higher by 1.11 to 1.35, 1.37 to 1.52 and 1.53 to 1.70 fold than that of MW and 1.36 to 2.47, 3.94 to 4.68 and 2.86 to 4.35 fold than that of FW, respectively; whereas their concentration in the tissue of earthworms used in vermicompost of MW was higher by 1.20 to 1.88, 2.85 to 3.13 and 1.83 to 2.62 fold than that of FW, respectively. The concentration of Pb and Mn in the tissue of earthworms used in vermicompost of MW was higher by 1.72 to 1.93 and 1.03 to 1.20 fold than that of MSW and by 3.56 to 4.25 and 2.06 to 2.44 fold than that of FW, respectively; whereas the increase was 2.04 to 2.22 and 1.96 to 2.08 fold in earthworms used in vermicompost of MSW compared to that in FW, respectively.

Bioaccumulation factor

The values of BAF, in the present study, showed significant variation among three earthworm species across different wastes and time intervals (P<0.05) (Table 1). The BAF for Cd was more than unity and it was 2.3 ± 0.4 in tissue of *E. eugeniae*, 2.04 ± 0.39 in that of *E. fetida*, and 1.76 ± 0.42 in that of *P. excavatus* used in vermin-composting process of urban waste (Table 1 and Figure 4). However, the ranges of BAFs were less than unity in other four metals. BAFs of Pb, Zn, Cu and Mn were 0.83 ± 0.08, 0.83 ± 0.11, 0.67 ± 0.14, and 0.68 ± 0.08 in tissue of *E. eugeniae*; 0.74 ± 0.11, 0.73 ± 0.12, 0.58 ± 0.14, and 0.61 ± 0.07 in that of *E. fetida*, and 0.68 ± 0.12, 0.62 ± 0.13, 0.50 ± 0.14, and 0.50 ± 0.08 in *P. excavatus* used in vermicomposting process of urban waste, respectively.

DISCUSSION

The present findings showed that earthworms can accumulate higher concentration of the heavy metals in their body tissues with consequent remediation in the concentration of metals in the waste substrate and consequent considerable reduction in concentration in vermicompost (Hartensein and Hartensein, 1981; Gupta et al., 2005; Suthar, 2008; Suthar and Singh, 2009). According to Edwards and Bohlen (1996) the waste substrate ingested by earthworm undergone chemical and microbial changes while passing through the alimentary canal and a great proportion of the organic fraction is converted into soluble forms that are more available to organisms. Part of the organic matter is digested, with the increase in pH and microbial activities in the gut. As a result, the possibility for metals to be bound to ions and carbonates (that is in more soluble fractions) increases in ingested material (Morgan and Morgan, 1999).

These soluble fractions can be accumulated in earthworm tissues during transit of waste through worm's gut, which results in reduction of metals in worms' excreta coming out as vermicompost and in turn, increased in the tissue of earthworm (Gupta et al., 2005; Suthar et al., 2008). Dia et al. (2004) suggested that bioaccumulation of metals in earthworms is their ability to eliminate the excess of metals. The concentrations for all the metals were found higher in the tissue of *E. eugeniae* than that of *E. fetida* and *P. excavatus*. These three species showed a considerable difference in metal concentrations in their tissues which was probably due to variation in their metabolism (Morgan and Morgan, 1992); it could be probably a species-specific feature. The variation in dietary intake of the metals could be an important factor contributing to differences in the availability of metals in their body tissues as well as vermicompost.

Hopkin (1989) stated that the earthworms have a specific capacity to regulate metals, particularly trace

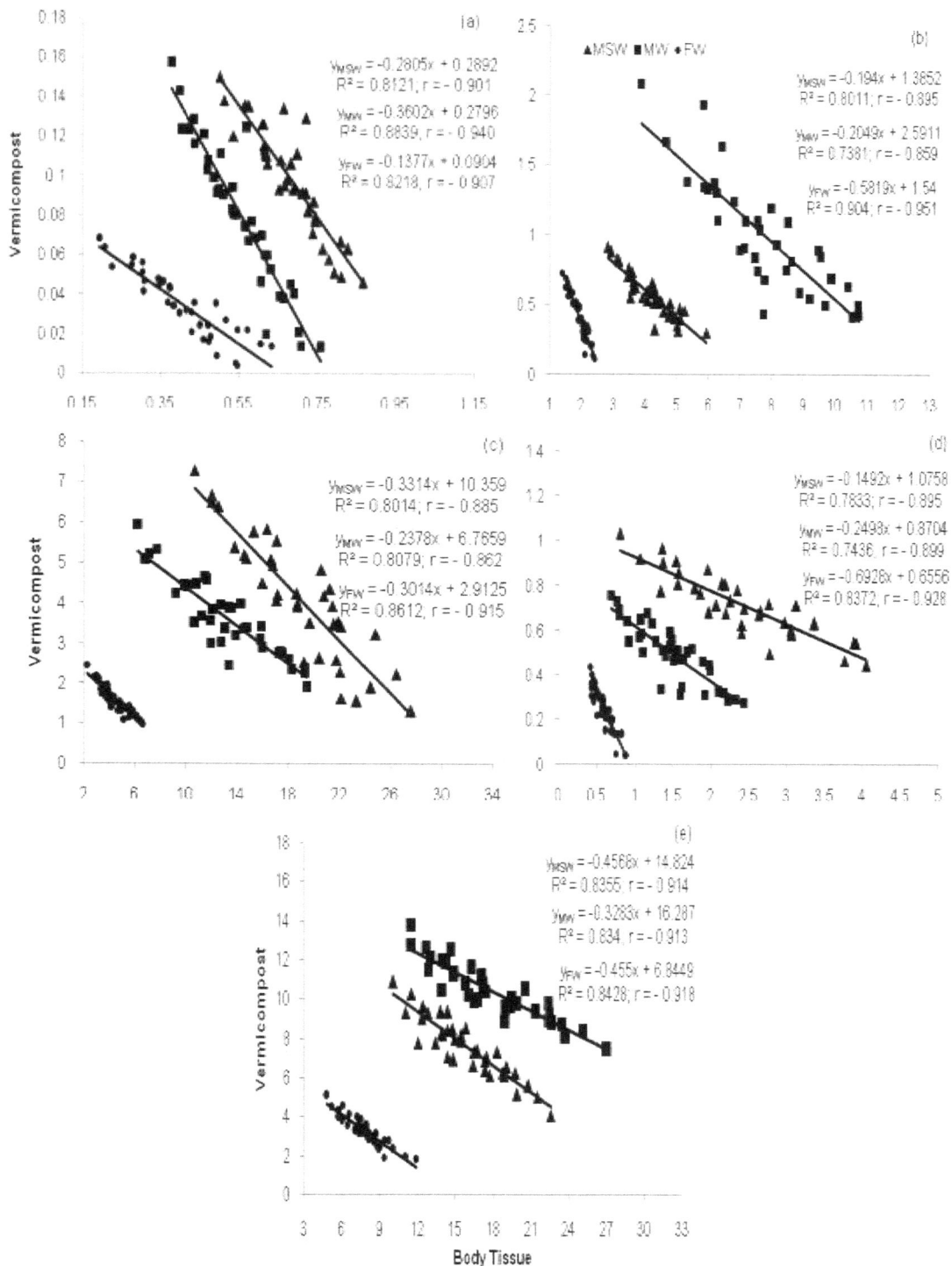

Figure 4. Regression analysis showing relationship between accumulation of (a) Cd (ppm), (b) Pb (ppm) (c) Zn (ppm), (d) Cu (ppm) and (e) Mn (ppm) in the tissue of three earthworm species – *E. eugeniae, E. fetida* and *P. excavatus* at 15, 30, 45 and 60 days with that present in vermicompost produced from MSW, MW and FW.

Table 1. Heavy metals' Bio- Accumulation Factor (BAF) of *Eudrilus eugeniae (Ee)*, *Eisenia fetida (Ef)* and *Perionyx excavatus (Pe)* during vermicomposting of MSW, MW and FW at different time intervals.

	Vermicomposting process								
	MSW			**MW**			**FW**		
	Ee	*Ef*	*Pe*	*Ee*	*Ef*	*Pe*	*Ee*	*Ef*	*Pe*
Cd									
15 Days	1.28	1.22	1.05	1.58	1.37	1.21	2.69	2.19	1.54
30 Days	1.42	1.32	1.23	1.84	1.56	1.41	3.17	2.73	2.17
45 Days	1.52	1.46	1.42	2.06	1.83	1.62	3.81	3.34	2.73
60 Days	1.67	1.62	1.54	2.23	1.98	1.83	4.44	3.87	3.39
Pb									
15 Days	0.73	0.63	0.55	0.68	0.55	0.49	0.76	0.66	0.61
30 Days	0.81	0.71	0.62	0.79	0.66	0.59	0.84	0.74	0.69
45 Days	0.86	0.79	0.73	0.84	0.75	0.69	0.88	0.82	0.78
60 Days	0.91	0.86	0.83	0.88	0.84	0.77	0.93	0.89	0.86
Zn									
15 Days	0.74	0.6	0.51	0.71	0.59	0.46	0.64	0.56	0.44
30 Days	0.84	0.71	0.57	0.82	0.69	0.55	0.75	0.64	0.52
45 Days	0.92	0.81	0.71	0.88	0.78	0.7	0.86	0.76	0.63
60 Days	0.97	0.88	0.77	0.95	0.87	0.78	0.93	0.85	0.74
Cu									
15 Days	0.46	0.36	0.28	0.48	0.41	0.31	0.59	0.52	0.45
30 Days	0.6	0.5	0.41	0.61	0.52	0.43	0.64	0.55	0.49
45 Days	0.71	0.61	0.55	0.73	0.63	0.56	0.76	0.66	0.59
60 Days	0.8	0.71	0.62	0.82	0.74	0.66	0.88	0.78	0.69
Mn									
15 Days	0.66	0.6	0.48	0.41	0.36	0.31	0.64	0.59	0.44
30 Days	0.77	0.69	0.54	0.48	0.39	0.33	0.74	0.67	0.52
45 Days	0.82	0.73	0.6	0.51	0.47	0.4	0.79	0.71	0.59
60 Days	0.89	0.81	0.72	0.58	0.51	0.43	0.85	0.76	0.67

metals such as Cu and Zn in their bodies, and the bioaccumulation and regulation mechanisms could be species-specific. He also suggested that exposure duration could be main determinant for observed differences in metal concentration in the tissue which is in corroboration with the present findings.

Suthar et al. (2008) reported that species-specific metal physiology in earthworms may alter the concentration of metals in their tissues. The amount of organic fractions in ingesting material denotes the availability of soluble forms of metals in a worm's gut. Lukkari et al. (2006) stated that binding of metals to organic matter particularly

more tightly bound fractions partly reduced the availability of metals to earthworms. The earthworm gut could modify the mobility of metals and favor their assimilation. Holmstrup et al. (2010) demonstrated in their study that cadmium, lead and copper accumulated to high concentrations in *Dendrobaena octaedra*.

The present findings clearly indicated that concentrations of Pb, Zn and Mn in tissue was higher in earthworms used for vermicomposting of MW, while that of Cu and Cd were higher in the worms used for composting of MSW; on the other hand, earthworms used in composting of FW showed the lower concentrations of all the metals. The concentrations of metals in earthworms were directly dependent on the metal concentrations of substrate waste in which they were used for composting. These findings are supported by earlier workers (Heikens et al., 2001; Lukkari et al., 2006; Kamitani and Kaneko, 2007). Gupta et al. (2005) reported that earthworm tissue metal level was directly related to their proportion in fly ash in vermibeds. Similar pattern of metal bioaccumulation was observed by Suthar et al. (2008), which further supported the hypothesis that tissue-metal level reflects the metal availability in the substrates.

The earthworms showed the greater concentrations of Cd in their tissues, than that of waste substrates; whereas the concentrations of Pb, Zn, Cu and Mn were more in waste than that in their tissue at the end of the vermicomposting process. The present findings clearly showed the earthworms accumulating and remediating metals from the waste substrates. In consistence to the present findings, Graff (1982) reported that earthworms (*E. fetida* and *E. eugeniae*) accumulated the heavy metals from the municipal garbage compost and concentrated them in their body tissues. The earthworm in general used three ways to handle the metals: (1) immobilization in fatty (chloragogen) cells of gut wall, (2) storage in waste nodules (or 'brown bodies') formed within the body cavity, and (3) excretion through the calciferous glands (Andersen and Laursen, 1982). The gut-related processes in earthworm may also increase metal availability. The earthworm chloragosomes consisting of modified epithelial cells, the eleocytes of the gut containing constituents of ion exchange compounds - phosphoric acid, carboxyl, phenolic hydroxyl and sulphonic acid groups acted as a cation exchange system capable of taking up and accumulating heavy metals (Ireland, 1978; Morgan and Morgan, 1988; Cooper, 1996). The accumulation of metals, especially Cd in earthworms is most probably due to the binding of metals by metallothioneins (Kagi and Kojima, 1987).

The low concentration of heavy metals in earthworm tissue at initial stage of composting and gradual increase with the progress of vermicomposting reaching higher concentrations at 60 days of composting was a consistent trend of higher metals accumulation in earthworm tissue. It is probably due to the metal levels in earthworm tissue being directly related to the availability of metals over different time intervals. The lower tissues metal concentration in tissues of earthworms with shorter time intervals, that is, at 15 and 30 days further supports the hypothesis. As the time passed, earthworms consume a great amount of organic waste to acquire required nutrition, and during the process metals are liberated in free forms due to the enzymatic actions in their gut

resulting in accumulation in their body tissue (Suthar, 2008). When the composting process progressed from 15 to 60 days, more metals in available forms were absorbed by the epithelial layer of gut and incorporated into the body tissue while the wastes transited through it.

The amount bioavailability of metals during the vermin-composting process can be calculated effectively using BAF or bioconcentration factors (BCFs) (Suthar and Singh, 2009). The BAFs have been used widely to quantify the bioaccumulation of pollutants in aquatic and terrestrial biota with assumption that organisms achieve a chemical equilibrium with respect to a particular medium or route of exposure (Mountouris et al., 2002; Hsu et al., 2006). The degree of BAFs mainly depended upon level of contamination and characteristics of waste, and the earthworm species used. Suthar and Singh (2009) suggested that BAFs could be an important indicator of metal bioconcentartion during the process of vermicomposting. BAF range increased as the time of vermicomposting process progressed from 15 to 60 days. The BAF was low in initial stage of 15 days and was higher at 60 days in the tissue of earthworms used in the vermicomposting of the waste substrates. The BAF was higher for *E. eugeniae* than that of *E. fetida* and *P. excavatus*. The BAF was higher for Cd and accordingly that of the other metals ranked in the order of Cd >Zn>Pb>Cu>Mn. The BAFs for all the heavy metals showed positive correlation with each other in all the three wastes, i.e., MSW (range of R^2 - 0.906 to 0.980, P<0.001), MW (R^2 - 0.956 to 0.989, P<0.001), FW(R^2 - 0.924 to 0.993, P<0.001), which clearly indicated that the heavy metals present in the wastes are accumulated in the body tissue of earthworms on which they devour. Higher BAF for Cd in all the samples clearly implied that the Cd concentrations in the tissues of earthworms exceeded many times the concentrations in the waste substrate.

Earthworms easily accumulate Cd and retain it in their body tissue. These results confirmed the findings of earlier studies (Brewer and Barret, 1995; Rebanova et al., 1995; Lapinski et al., 2002; Li et al., 2010). The BAFs of earthworms were higher in wastes that showed relatively better mineralization. Thus, the BAF values were directly related to the amount of waste assimilated by the worms (Suthar et al., 2008). The observed differences for BAFs could be related to the difference in specific metabolism and regulating mechanism of organic matter fractions of organo-metal compounds in the earthworm body (Lukkari et al., 2006; Suthar and Singh, 2009).

Suthar et al. (2008) have demonstrated higher ranges of BAFs for earthworms collected from contaminated substrates; while some earlier studies reported considerable ranges of BCFs for metals in earthworms (Dia et al., 2004; Hsu et al., 2006).

Conclusions

The present findings suggested that vermicomposting could be an appropriate technology for remediation of metals from obnoxious wastes. It was found that the vermicompost of urban wastes was not only rich in plant nutrient but also have minimum risk of environmental contamination due to lower metal concentrations availability in it. Internal metal concentrations in all the three earthworms were significantly and negatively correlated to heavy metal concentrations in their respective vermicompost.

The present study depicted that the earthworm especially, *E. eugeniae* can be utilized effectively for *ex situ* remediation of metals from urban waste. The higher BAF ranges for metals implied higher metal accumulation in earthworms' tissue, which may affect the food chain through biomagnification. It further revealed that the accumulation of metals in worm's tissue, especially uptake of metals, not only remediated the metals from the urban wastes and their vermicomposts but also improve vermicompost quality reducing the metal concentration.

ACKNOWLEDGEMENTS

The research grants for present research were sanctioned nu UGC (New Delhi) in the form of a Major Research Project (F. No. 32-634/2006(SR) dated 02-03-2007). SP availed a research fellowship in the form of a project fellow in the research project

REFERENCES

Andersen C, Laursen J (1982). Distribution of heavy metals in *Lumbricus terrestris, Aporrectodea longa* and *A. rosea* measured by atomic absorption and X-ray fluorescence spectrometry. Pedobiologia, 24: 347-356.

Brewer SR, Barret GW (1995). Heavy metal concentrations in earthworms following long-term nutrient enrichment. Bull. Environ. Contam. Toxicol., 54: 120–127.

Chattopadhyay S, Dutta A, Ray S (2009). Municipal solid waste management in Kolkata, India – A review. Waste Manage., 29: 1449–1458.

Cooper EL (1996). Earthworm immunity. In: Reinkevich B, Muller WEG (Eds.), Invertebrate immunology. Thomson Press, New York, pp. 10-95.

CPCB (2000). Management of Municipal Solid Waste, Delhi: Central Pollution Control Board.

Dia J, Becquer T, Rouiller JH, Reversat G, Bernhard-Reversat F, Nahmani J, Lavelle P (2004). Heavy metal accumulation by two earthworm species and its relationship to total and DTPA extractable metals in soils. Soil Biol. Biochem., 36: 91–98.

Edwards CA, Bohlen PJ (1992). The effects of toxicchemicals on earthworms. Rev. Environ. Contam. Toxicol., 125: 23-99.

Edwards CA, Bohlen PJ (1996). Biology and Ecology of Earthworms, third ed. Chapman and Hall, London. Pp 207-210

Forbes D (1996). Asian metropolis: Urbanization and the Southeast Asian city. Melbourne, Australia. Oxford University Press, United States, p.120.

Fries GF (1982). Potential Polychlorinated Biphenyl Residues in Animal Products from Application of Contaminated Sewage Sludge to Land. J. Environ. Qual., 11: 14-20.

Gish CHD, Christensen RE (1973). Calcium, nickel, lead and zinc in earthworms from roadside soil. Environ Sci. Technol., 7: 1060–1062.

Graff O (1982). Vergleich der Regenswurmaten *Eisenia foetida* and *Eudrilus eugeniae* Hinsichlich Ihrer Eignung zur Proteinwinnung aus Abfallstoffen. Pedobiologia, 23: 277-282.

Gupta SK, Tewari A, Srivastava R, Murthy RC, Chandra S (2005). Potential of *Eisenia foetida* for sustainable and efficient vermicomposting of fly ash. Water Air and Soil Pollut., 163: 293–302.

Gupta R, Garg VK (2009). Vermiremediation and nutrient recovery of non-recyclable paper waste employing *Eisenia fetida*. J. Hazard. Mater., 162: 430-439.

Hardoy J, Mitlin D, Satterthwaite D (2001). Environmental problems in an urbanizing world - Finding solutions for cities in Africa, Asia and Latin America. Earthscan Publications, London, p. 448.

Hartensein R, Hartenstein F (1981). Physicochemical changes affected in activated sludge by the earthworm *Eisenia foetida*. J. Environ. Qual., 10: 377-382.

Heikens A, Peijnenburg WJGM, Hendriks AJ (2001). Bioaccumulation of heavy metals in terrestrial invertebrates. Environ. Pollut., 113: 385–393.

Hobbelen PHF, Koolhaas JE, van Gestel CAM (2006). Bioaccumulation of heavy metals in the earthworms *Lumbricus rubellus* and *Aporrectodea caliginosa* in relation to total and available metal concentrations in field soils. Environ. Pollut., 144: 639–646.

Holmstrup M, Sørensen JG, Overgaard J, Bayley M, Anne-Mette B, Slotsbo S, Fisker KV, Maraldo K, Waagner D, Labouriau R, Asmund G (2010). Body metal concentrations and glycogen reserves in earthworms (*Dendrobaena octaedra*) from contaminated and uncontaminated forest soil. Environ. Pollut. In Press.

Ireland MP (1975). Metal content of Dendrobaena rubida (Oligochaeta) in a base metal mining area. Oikos 26: 74-90.

Ireland MP (1978). Heavy metal binding properties of earthworm chloragosomes. Acta Biol. Acad. Sci., H. 29: 385–394.

Kagi JHR, Kojima Y (1987). Chemistry and biochemistry of metallothionein. In: Kagi JHR, Kojima Y (Eds.), Metallothionein II. Birkhauser Verlag, Basel, Switzerland, pp. 25–61.

Kamitani T, Kaneko N (2007). Species-specific heavy metal accumulation patterns of earthworms on a floodplain in Japan. Ecotoxicol. Environ. Saf., 66: 82–91.

Kansal A (2002). Solid waste management strategies for India. Indian J. Environ. Protect., 22: 444–448.

Karthikeyan V, Sathyamoorthy GL, Murugesan R (2007). Vermi Composting of Market Waste in Salem, Tamilnadu, India. Proceedings of the International Conference on Sustainable Solid Waste Management, Chennai, India, pp. 276-281.

Katz SA, Jennis SW (1983). Regulatory compliance monitory by atomic absorption spectrometry. Verlay Chemical International, Florida.

Keener HM, Dick WA, Hoitink HAJ (2001). Composting and beneficial utilization of composted by-product materials. In: Power JF, Dick WA (Eds.), Land Application of Agricultural, Industrial, and Municipal By-products. Soil Science Society of America, Madison, West Indies, pp. 315–341.

Kruse EA, Barrett GW (1985). Effects of municipal sludge and fertilizer on heavy metal accumulation in earthworms. Environ. Pollut., (Series A) 38: 235-244.

Kumar S, Bhattacharyya JK, Vaidya AN, Chakrabarti T, Devotta S, Akolkar AB (2009). Assessment of the status of municipal solid waste management in metro cities, state capitals, class I cities, and class II towns in India: An insight. Waste Manage., 29: 883–895.

Lapinski S, Borowiec F, Pospiech N, Soltyk-Stefanska M (2002). Heavy metals accumulation in the body of earthworm Eisenia fetida (Sav.) used in animal nutrition. Ukrainian Academy of Agrarian Sciences, Anim. Biol., 4: 205–209.

Li L, Xu Z, Wu J, Tian G (2010). Bioaccumulation of heavy metals in the earthworm Eisenia fetida in relation to bioavailable metal concentrations in pig manure. Bioresource Technol., 101: 3430-3436.

Lukkari T, Teno S, Vaisanen A, Haimi J (2006). Effect of earthworms on decomposition and metal availability in contaminated soil: Microcosm studies of populations with different exposure histories. Soil Biol. Biochem., 38: 359–370.

Morgan JE, Morgan AJ (1988). Earthworms as biomonitors of Cadmium, Lead and Zinc in metalliferous soils. Environ. Pollut., 54: 123-138.

Morgan JE, Morgan AJ (1992). Heavy metal concentrations in the tissues, ingesta and faeces of ecophysiologically different earthworm species. Soil Biol. Biochem., 12: 1691–1697.

Morgan JE, Morgan AJ (1999). The accumulation of metals (Cd, Cu, Pb, Zn and Ca) by two ecologically contrasting earthworm species (Lumbricus rubellus and Aporrectodea caliginosa): implications for ecotoxicological testing. Appl. Soil Ecol., 13: 9-20.

Mountouris A, Voutsas E, Tassios D (2002). Bioconcentration of heavy metals in aquatic environments: the importance of bioavailability. Mar. Pollut. Bull. ,44 : 1136-1141.

Pare T, Dinel H, Schnitzer M, Dumonet S (1998). Transformation of carbon and nitrogen during composting of animal manure. Biol. Fert. Soils, 26: 173–178.

Pattnaik S, Reddy MV (2010). Assessment of Municipal Solid Waste management in Puducherry (Pondicherry), India. Resour Conserv. Recy., 54: 512-520

Rebanova V, Tuma V, Volakova L (1995). Earthworms of the Lumbricidae family as bioindicators of heavy metals in soil. In: Proc. Inter. Sci. Conf. "The 35TH Anniversary of Faculty Foundation, Jihočeská Univerzita, Zemědělská Fakulta", České Budějovice, Czech Republic, pp. 307–317.

Reddy MV, Pattnaik S (2009). Vermi-composting of Municipal (Organic) Solid Waste and its implications. In: Singh, S.M. (Ed.), Earthworm ecology and environment. Internation Book Distributing Co., Lucknow, India, pp. 119-113.

Rosciszewska M, Lapinski S, Borowiec F, Popek W, Drag E (2003). Relationship between the content of some heavy metals in substrate and their accumulation in the bodies of Dendrobaena octaedra (Sav.). Chem. Agric., 4: 557–562.

Shahmansouri MR, Pourmoghadas H, Parvaresh AR, Alidadi H (2005). Heavy metals bioaccumulation by Iranian and Australian earthworms (Eisenia fetida) in the sewage sludge vermicomposting. Iranian J. Environ. Health Sci., Eng. 2: 28-32.

Smith R (2009). A critical review of the bioavailability and impacts of heavy metals in municipal solid waste composts compared to sewage sludge. Environ. Int., 35: 142-156.

Suthar S (2008). Metal remediation from partially composted distillery sludge using composting earthworm Eisenia fetida. J. Environ. Monitor., 10: 1099-1106.

Suthar S, Singh S, Dhawan S (2008). Earthworm as bioindicators of metals (Zn, Fe, Mn, Cu, Pb and Cd) in soils: Is metal bioaccumulation affected by their ecological categories. Ecol. Eng., 32: 99–107.

Suthar S, Singh S (2009). Bioconcentrations of Metals (Fe, Cu, Zn, Pb) in Earthworms (Eisenia fetida) inoculated in Municipal Sewage Sludge: Do Earthworms Pose a Possible Risk of Terrestrial Food Chain Contamination? Environ. Toxicol., 24: 25-32.

Udovic M, Plavc Z, Lestan D (2007). The effect of earthworms on the fractionation, mobility and bioavailability of Pb, Zn and Cd before and after soil leaching with EDTA. Chemosphere, 70: 126–134.

Vermeulen F, Van den Brink NW, D'Havé H, Mubiana VK, Blust R, Bervoets L, De Coen W (2009). Habitat type-based bioaccumulation and risk assessment of metal and As contamination in earthworms, beetles and woodlice. Environ. Pollut., 157: 3098-3105.

Wang Q, Zhou D, Cang L, Li L, Zhu H (2009). Indication of soil heavy metal pollution with earthworms and soil microbial biomass carbon in the vicinity of an abandoned copper mine in Eastern Nanjing, China. Eur. J. Soil Biol., 45: 229-234.

Wizewardena JDH, Gunaratne SP (2004). Heavy metal contents in commonly used animal manure. Ann. Sri Lanka Dept. Agr., 6: 245-253.

11

Risk assessment of using coated mobile recharge cards in Nigeria

O. J. Okunola[1]*, Y. Alhassan[1], G. G. Yebpella[1], A. Uzairu[2], A. I. Tsafe[3], E. S. Abechi[2] and E. Apene[4]

[1]National Research Institute for Chemical Technology, Zaria, Kaduna State, Nigeria.
[2]Department of Chemistry, Ahmadu Bello University, Zaria, Kaduna State, Nigeria.
[3]Usmanu Danfodiyo University, Sokoto, Sokoto State, Nigeria.
[4]Federal College of Forestry Mechanization, P. M. B. 2273, Afaka, Kaduna, Kaduna State, Nigeria.

The risk assessment of coatings on mobile phone recharge cards on end users has been investigated using mobile phone recharge cards of three major brands (designated as A, B and C) with dominations of ₦200, ₦400 and ₦500 purchased from retail shops in Zaria, Nigeria. To appraise the health risk associated with heavy metal contamination of food by metals from the coatings, Daily Intake of Metals (DIM) and Health Risk Index (HRI) were calculated. The heavy metals was analysed using AAS. Mean concentration of metal ranged between; 12926 to 130, 554 to 294, 13175 to 7025.5, 23691 to 561, 700 to 222.5, 15230 to 5554, 2745.8 to 1429, and 75525 to 11397 µgg^{-1} for Mn, Cu, Cr, Zn, Cd, Ni, Pb and Fe respectively. Generally, high concentration of metals was found in Sample C with the exception of Cu. Also, the distribution concentration of the metals were found to be Fe > Cr > Ni > Pb > Zn > Cd > Cu > Mn, followed by Fe > Pb > Ni > Cr > Zn > Cu > Mn > Cd and Fe > Zn > Ni > Cr > Mn > Pb > Cu > Cd for Samples A, B and C respectively in that order of decreasing magnitude. Analysis of difference of means using t-test ($p < 0.05$) showed significant difference between samples for Pb, Cu and Cr among the heavy metals determined. The trend of DIMs for heavy metals in the coatings were in the order of Cd > Al > Zn > Fe > Ni > Cr > Mn > Pb > Cu, with intake from Sample B being greater than Samples C and A for Pb, Cu, Cr, Cd and Ni. Also, the HRI of metals indicated that Pb, Cd and Ni (especially Samples B and C) through adhering 'silver' coatings on mobile recharge cards on nails or under the finger contaminating food were higher than 1, which indicates that users experience relatively high health risk.

Key words: Heavy metals, recharge cards coating, health risk, AAS, Nigeria.

INTRODUCTION

During the last decades, there has been growing interest in determining heavy metal levels in foods and other common food contaminants. With the detection of toxic metals in majority of products sold in third world countries, concern to researchers and health practitioners has tremendously increased. From toxicological and environmental point of view, the determination of toxic metals in metallic products is interesting because by using the products, users could be exposed to these metals through food contamination (Ashraf, 2006).

Contamination of food by heavy metals is a serious hazard depending on the relative level of the metals. Some of these metals such as Cd and Pb, injure the kidneys and cause symptoms of chronic toxicity, including impaired organ function, poor reproductive capacity, hypertension, tumors and hepatic dysfunction (Abou-Arab et al., 1996). On the other hand, Cr, Cu, Fe, Zn and Mn are the major causes of nephritis, anuria and extensive lesions in the kidneys (Chukwujindu et al., 2007).

Metals, a major category of globally-distributed pollutants, are natural elements that have been extracted from the earth and harnessed for human industry and products for millennia. Metals are notable for their wide environmental dispersion from such activity, their

*Corresponding author. E-mail: adio4oj@yahoo.com or okunolaoj@gmail.com.

Table 1. Mean (±SD) characteristics of mobile phones rechargeable cards.

Sample	Length of coating area (cm)	Width of coating area (cm)	Weight of coating (mg)
A	4.1±0.1	0.8±0.0	6.0±0.2
B	3.8±0.0	0.8±0.0	1.0±0.0
C	3.5±0.3	0.8±0.0	0.6±0.0

tendency to accumulate in selected tissues of the human body, and their overall potential to be toxic even at relatively minor levels of exposure. It is essential to characterize the level of heavy metals in mobile phone recharge cards with 'silver' coating since it is not very uncommon to find these coatings deposited under the finger nails. Without proper washing of nails before eating, metal exposure through this route of entry could be a pathway for human intake. The objective of this study is to characterize the levels of metals (Cd, Pb, Cu, Cr, Ni, Co, Fe, Al and Zn) in "silver' coatings of major brands of mobile phone recharge cards sold in Nigeria and subsequently, to determine implications on human health.

MATERIALS AND METHODS

Glass wares, crucibles and plastic containers were washed with liquid soaps, rinsed with distilled water and soaked in 10% HNO_3 for 24 h; cleaned with distilled – deionized water to prevent contamination (Adnan, 2003). Reagents used were of analytical grades. Mobile phone recharge cards of major brands with dominations of ₦200, ₦400 and ₦ 500 were purchased from retail shops in Zaria, Kaduna State in October, 2009. The three major brands of recharge cards were designated as A, B and C. Twenty recharge cards were collected for each brand. During the analysis, the silver coatings were carefully scratched off using a stainless steel scraper into a polythene bag.

The samples were digested with a mixture of HNO_3 and $HClO_4$ for metal determination. 0.05 ± 0.0010 g of the silver coatings sample was placed in a digestion tube and predigested using 10 ml of concentrated HNO_3 at 105°C until the liquor was clear. Then 5 ml of $HClO_4$ was added and digested for 1 h until the liquor became colourless. The samples were evaporated slowly to almost dryness, cooled and dissolved in 5 ml of 1 M HNO_3. The digested samples were filtered through Whatman's No. 1 filter paper and diluted to 50 ml with distilled-deionized water. The sample solutions were analysed with graphite furnace AAS (Shimadzu AA-6800). Metal determinations were done in triplicates for each brand. Blank were also digested in the same manner with the sample. Calibration standards were made by dilution of nitrate salts of all metals determined in 0.5 M of nitric acid supplied by Scharlau Japan. A recovery test of the total analytical procedure was carried out for the metals in selected samples by the Spiking experiment. Acceptable recovery of >90% were obtained for all metals, an indication of good analytical protocol.

Health risk assessment of these metals was done according to Sajjad et al. (2009). Daily intake of metals (DIM) was determined by the following equation:

$$DIM = C_{Metal} \times D_{Intake}/B_{Average\ weight}$$

Where, C_{Metal} = Heavy metal concentration in coatings; D_{Intake} =

Intake of metals (approximate 10% of what is scratch is retained under and on the nails); $B_{Average\ weight}$ = Average body mass of an adult 65 kg

Health risk index (HRI) was calculated by using Daily Intake of Metals in food (DIM) and Reference Oral Dose (R_fD). The following formula is used for the calculation of HRI:

$$HRI = DIM/R_fD$$

If the value of HRI is less than 1 (HRI<1), then the health risk exposed to the population is considered acceptable (IRIS, 2003).

RESULTS AND DISCUSSION

The characteristics of the coatings and results of the levels of heavy metals (μgg^{-1}) are presented in Table 1 and Figure 1 respectively. Mean Mn concentration was the highest in Sample C (12926 μgg^{-1}), and the lowest in Sample A (130 μgg^{-1}). Cu concentration was the highest in Sample B (554 μgg^{-1}) and the lowest in Sample A (294 μgg^{-1}).

The highest and the lowest concentrations of Cr were 13175 and 7025.5 μgg^{-1} in Samples C and B, respectively. Zn had the highest concentration (23691 μgg^{-1}) in Sample C and the lowest (561 μgg^{-1}) in Sample A. The highest concentration of Cd (700 μgg^{-1}) was in Sample C, while Sample B had the lowest (222.5 μgg^{-1}). The highest concentration of Ni (15230 μgg^{-1}) was in Sample C and the lowest (5554 μgg^{-1}) in Sample A. Pb concentration was the highest in Sample C (2745.8 μgg^{-1}) and the lowest (1429 μgg^{-1}) in Sample A. The concentration of Fe ranged between 75525 (Sample C) to 11397 μgg^{-1} (Sample A). Generally, high concentration of metals was found in Sample C with the exception of Cu. Also, the distribution concentration of the metals were found to be Fe > Cr > Ni > Pb > Zn > Cd > Cu > Mn for Sample A, Fe > Pb > Ni > Cr > Zn > Cu > Mn > Cd for Sample B and Fe > Zn > Ni > Cr > Mn > Pb > Cu > Cd for Sample C in the order of decreasing magnitude. Analysis of difference of the concentration means using t-test (p < 0.05) showed significant differences between Samples for Pb, Cu and Cr among all the heavy metals determined.

To appraise the health risk associated with heavy metal contamination of food by metals from the 'silver' coatings of recharge cards of mobile phones, Daily Intake of Metals (DIM) and health risk Index (HRI) were calculated.

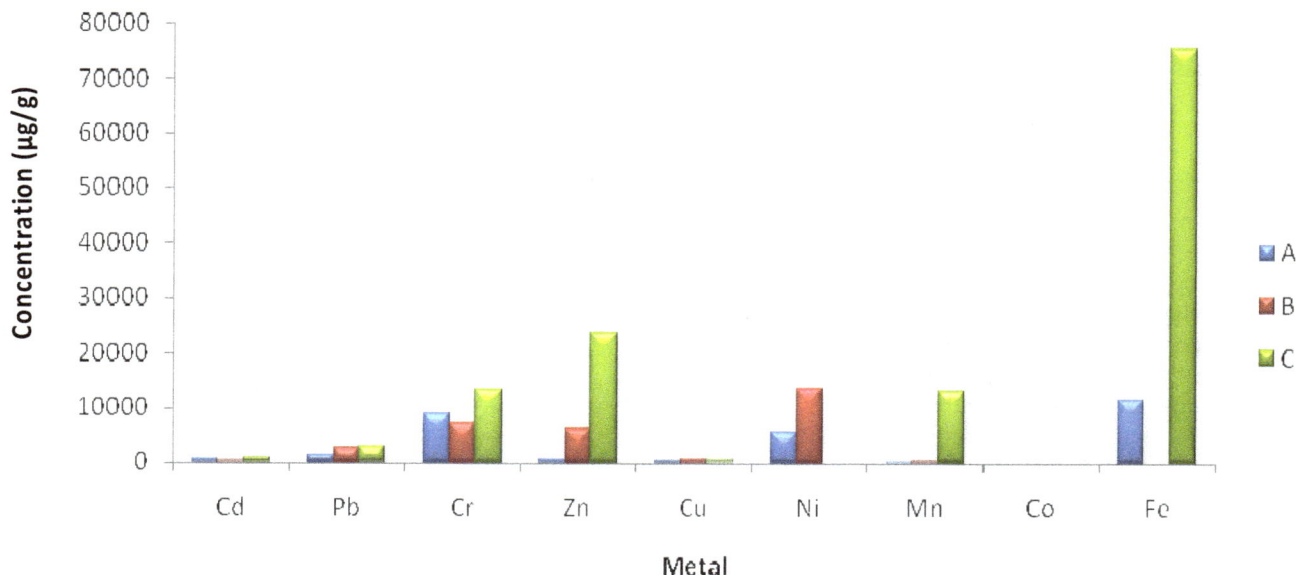

Figure 1. Mean concentration of heavy metal in mobile phones rechargeable cards coatings.

The DIM of heavy metals was estimated based on the average concentration of each heavy metal in each sample as shown in Figure 2. The estimated DIMs of heavy metals (Mn, Pb, Cr, Fe, Al, Ni, Cu, Zn, Pb and Cd) ranged between: 2.43 E-2 to 2.20E-3 µgday^{-1} for Pb, 5.11E-3 to 4.52E-4 µgday^{-1} for Cu, 6.49E-2 to 1.38E-2 µgday^{-1} for Cr, 3.18E-2 to 2.00E-4 µgday^{-1} for Mn, 1.94E-1 to 2.88E-3 µgday^{-1} for Zn, 1.86E-1 to 1.75E-2 µgday^{-1} for Fe, 2.05 to 7.66E-1 µgday^{-1} for Cd, 2.49E-1 to 4.78E-3 µgday^{-1} for Al, and 1.10E-1 to 8.55E-3 for Ni. The trend of DIMs for heavy metals in the coatings were in the order of Cd > Al > Zn > Fe > Ni > Cr > Mn > Pb > Cu, with the intake from Sample B being greater than Samples C and A for Pb, Cu, Cr, Cd and Ni.

The Health Risk Index (HRI) has been recognized as a useful indicator for evaluation of risk associated with the consumption of metals in contaminated food (Sridhara et al., 2008). The health risk assessment in this study was done for some metals (Pb, Cu, Cr, Mn, Zn, Cd and Ni) as shown in Figure 3. The HRIs of Pb, Cd and Ni (especially in Samples B and C) were higher than 1 (HRI > 1), which show that the users of mobile phones using 'silver' coating recharge cards are experiencing relatively high health risk. However, Moriguchi et al. (2004) suggested that the ingested dose of heavy metals is not equal to the absorbed pollutant dose in reality, as a fraction of the ingested heavy metals may be excreted, with the remainder accumulated in body tissues where it can affect human health.

According to Chien et al. (2002) and Sajjad et al. (2009), if the value of HRI is less than 1 (HRI < 1), the health risk to the population is considered acceptable. On the other hand, if the HRI is equal or greater than 1 (HRI ≥ 1) the population is exposed to unacceptable health risk. Based on the results, it shows that the population consuming Sample B and C recharge cards are exposed to high health risks due to Pb, Cd and Ni.

Health implications

The health implications of heavy metals refer to the harmful effects of heavy metals to the body when being consumed above the bio-recommended limits. Cadmium is toxic at extremely low levels; there is no 'safe exposure' for the human body even at minute levels. Long term exposure results in renal dysfunction, characterized by tubular proteinuria. High exposure can lead to obstructive lung disease- and cadmium pneumonitis, resulting from inhaled dusts and fumes. It is characterized by chest pain, cough with foamy and bloody sputum, and death of the lung tissues' lining because of excessive accumulation of watery fluids. Cadmium is also associated with bone defects, viz; osteomalacia, osteoporosis and spontaneous fractures, increased blood pressure and myocardic dysfunctions. Depending on the severity of exposure, the symptoms of effects include nausea, vomiting, abdominal cramps, dyspnea and muscular weakness. Severe exposure may result in pulmonary odema and death. Pulmonary effects (emphysema, bronchiolitis and alveolitis) and renal effects may occur following subchronic inhalation exposure to cadmium and its compounds (McCluggage, 1991; INECAR, 2000; European Union, 2002; Young, 2005).

Lead is the most significant heavy metal toxin and the

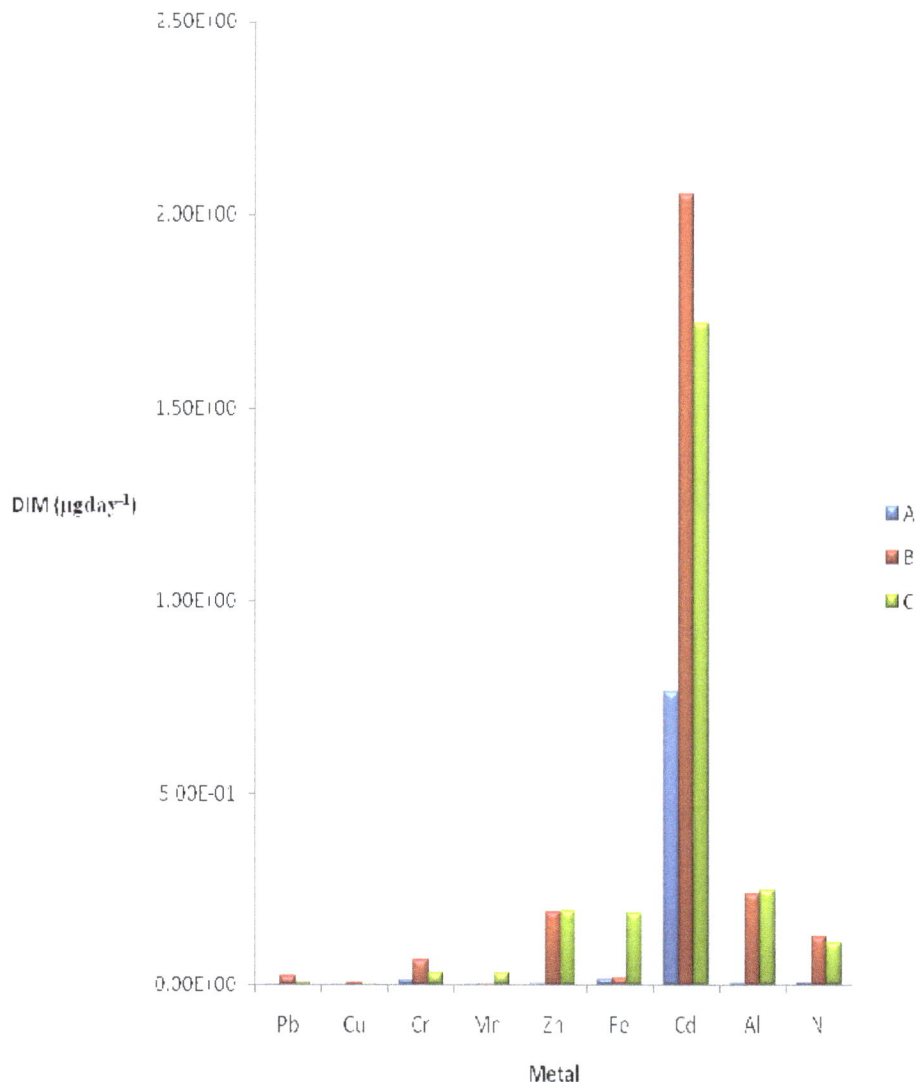

Figure 2. Daily Intake of Metals in recharge cards coatings.

inorganic forms are absorbed through ingestion (food and water) and inhalation (Ferner, 2001). A notably serious effect of lead toxicity is its teratogenic effect. Lead poisoning also causes inhibition of the synthesis of haemoglobin; dysfunctions in the kidneys, joints and reproductive systems- and cardiovascular system as well as acute and chronic damages to the Central Nervous System (CNS) and Peripheral Nervous System (PNS) (Ogwuegbu and Muhanga, 2005). Other effects include damage to the gastrointestinal tract (GIT) and urinary tract resulting in bloody urine, neurological disorder and can cause severe and permanent brain damage. While inorganic forms of lead typically affect the CNS, PNS, GIT and other biosystems, organic forms predominantly affect the CNS (McCluggage, 1991; INECAR, 2000; Ferner, 2001; Lenntech, 2004). Lead affects children by

leading to the poor development of the grey matter of the brain, thereby resulting in poor intelligence quotient (IQ) (Udedi, 2003). Its absorption in the body is enhanced by Ca and Zn deficiencies. Acute and chronic effects of lead result in psychosis.

Toxicological implications

The poisoning effects of heavy metals are due to their interference with normal body biochemistry in the normal metabolic processes. When ingested, in the acid medium of the stomach, they are converted to their stable oxidation states (Zn) and combine with the body's biomolecules such as proteins and enzymes to form strong and stable chemical bonds. Figure 4 shows

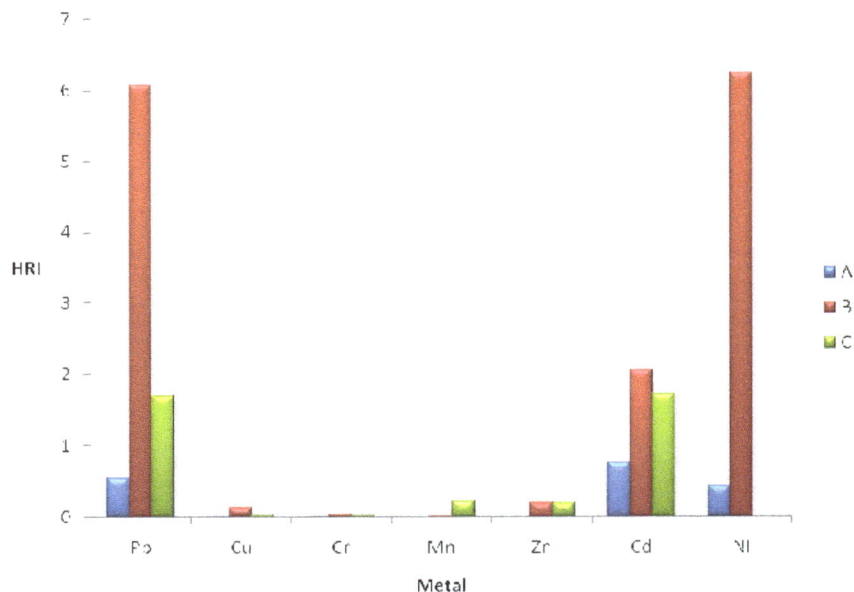

Figure 3. Health Risk Index of heavy metals.

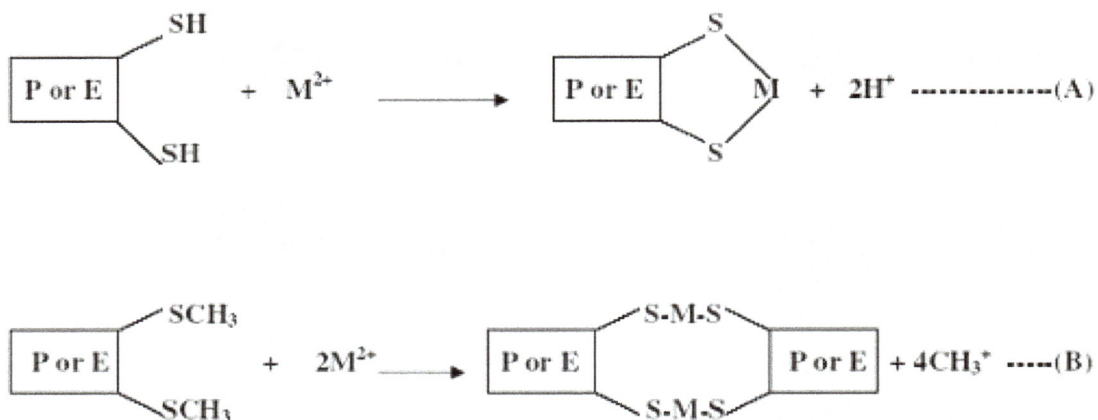

Figure 4. Biochemistry of toxicity of heavy metals (Adopted from Ogwuegbu and Ijioma, 2003). (A) = Intramolecular bonding; (B) = Intermolecular bonding; P = Protein; E = Enzyme; M = Metal The hydrogen atoms or the metal groups in the above case are replaced by the poisoning metal and the enzyme is thus inhibited from functioning, whereas the protein–metal compound acts as a substrate and reacts with a metabolic enzyme.

their reactions during bond formation with the sulphydryl groups (-SH) of cysteine and sulphur atoms of methionine (-SCH) (Ogwuegbu and Ijioma, 2003).

Conclusion

The results obtained from this study generally indicated the presence of heavy metals especially Pb, Cd, Cr and Ni in coatings on mobile phone recharge cards majorly used in Nigeria. The presence of the metals is of greater concern especially due to their health risk compared to other metals. The results of the study also suggested that population using Samples B and C recharge cards are potentially exposed to unacceptable health risks due to Pb, Cd and Ni. In order to avoid food contamination, proper washing of hands after scratching off coatings on recharge cards is a basic but important practice. Apart from that, using the printed numbers to top up the mobile phone account should be encouraged since this does not present metal exposure hazard and is economical.

ACKNOWLEDGMENT

The authors are thankful to the Chief Technologist, National Research Institute for Chemical Technology, Pastor Gandu for his assistant in running the samples. Thanks are also extended to the Management of the National Research Institute for Chemical Technology, Basawa, Zaria for their support in making this research a success.

REFERENCES

Adnan MM (2003). Determination of Cadmium and lead in different cigarette brands in Jordan. Environ. Monit. Assess., 104: 163-170.

Chien LC, Hung TC, Choang KY, Yeh CJ, Meng PJ (2002). Daily intake of TBT, Cu, Zn, Cd and As for fishermen in Taiwan. Sci. Total Environ., 285: 177-185.

Chukwujindu MAI, Godwin EN, Francis OA (2007). Assessment of contamination by heavy metals in sediments of Ase River, Niger Delta, Nigeria. Research J. Environ. Sci., 1(5): 220-228.

European Union (2002). Heavy Metals in Wastes, European Commission on Environment http://ec.europa.eu/environment/waste/studies/pdf/heavy_metalsreport.pdf.

Ferner D J (2001). Toxicity, heavy metals. eMed. J., 2(5): 1.

Institute of Environmental Conservation and Research INECAR (2000). Position Paper Against Mining in Rapu-Rapu, Published by INECAR, Ateneo de Naga University, Philippines www.adnu.edu.ph/Institutes/Inecar/pospaper1.asp.

Lenntech Water Treatment and Air Purification (2004). Water Treatment, Published by Lenntech, Rotterdamseweg, Netherlands www.excelwater.com/thp/filters/Water Purification.htm.

McCluggage D (1991). Heavy Metal Poisoning, NCS Magazine, Published by The Bird Hospital, CO, U.S.A. www.cockatiels.org/articles/Diseases/metals.html.

Ogwuegbu MO, Ijioma MA (2003). Effects of Certain Heavy Metals On The Population Due To Mineral Exploitation. In: International Conference on Scientific and Environmental Issues In The Population, Environment and Sustainable Development in Nigeria, University of Ado Ekiti, Ekiti State, Nigerian, pp. 8-10.

Ogwuegbu MOC, Muhanga W (2005). Investigation of Lead Concentration in the Blood of People in the Copper belt Province of Zambia. J. Environ., (1): 66-75.

Sajjad K, Robina F, Shagufta S, Mohammad AK, Maria S (2009). Health risk assessment of heavy metals for population via consumption of vegetables. World Appl. Sci. J., 6(12): 1602-1606.

Sridhara CN, Kamala CT, Samuel DSR (2008). Assessing risk of heavy metals from consuming food grown on sewage irrigated soils and food chain transfer.Ecotoxicol. Environ. Saf., 69(3): 513-524.

Udedi SS (2003). From Guinea Worm Scourge to Metal Toxicity in Ebonyi State, Chemistry in Nigeria as the New Millennium Unfolds, 2(2): 13-14.

Young RA (2005). Toxicity Profiles: Toxicity Summary for Cadmium, Risk Assessment Information System, RAIS, University of Tennessee rais.ornl.gov/tox/profiles/cadmium.shtml.

Preconcentration and removal of heavy metal ions from aqueous solution using modified charcoal

Jimoh, T.[1*], Egila, J. N.[2], Dauda, B. E. N.[1] and Iyaka, Y. A.[1]

[1]Department of Chemistry, Federal University of Technology, Minna, Nigeria.
[2]Department of Chemistry, University of Jos, Plateau State, Nigeria.

Levels of Pb(II), Cu(II), Zn(II), Mn(II), and Co(II) ion in aqueous solutions were determined by atomic absorption spectrometry (AAS) technique after preconcentration on diethyl dithiocarbamate supported by EDTA modified charcoal from Khaya senegalensis wood. The sorbed elements were eluted with acid and subsequently analyzed. The influence of pH, contact time and initial metal ion concentration by the modified charcoal was investigated. At 90 min contact time the uptake level of metal ions from solution was found to depend on the metal ion-substrate contact time, ion concentration and ion type. The percentage recovery of the metals ions retained by modified charcoal was highest at pH 5. The sorption recovery for the metal ions was within the range of 95.1-99.8%. It was observed that EDTA modification of charcoal significantly improved the percentage recovery of the metal ions. The results obtained from the study showed EDTA modified charcoal to be favourable for preconcentration and removal of heavy metal ions from aqueous solution.

Key words: Preconcentration, removal, heavy metals, modified charcoal, environmental pollution.

INTRODUCTION

Excessive release of heavy metals into the environment due to industrialization and urbanization has posed a great problem worldwide (Iqbal and Saeed, 2007; Jimoh, 2010). Unlike organic pollutants, the majority of which are susceptible to biological degradation, heavy metal ions do not degrade into harmless end products (Guo et al., 2008). The presence of heavy metal ions is a major concern due to their toxicity to many life forms. It therefore becomes imperative to determine these metal ion concentrations in the water bodies.

In some cases, however there are many difficulties in determining traces of heavy metals in environmental samples due to insufficient sensitivity or matrix interferences. Thus, a preconcentration and/or separation step is often recommended (Unob et al., 2005). The atomic absorption spectrometry (AAS) technique, which offers fast multi-elemental analysis, suffers from a poor sensitivity in the determination of heavy elements in environmental samples like natural water and other real samples. This drawback can be overcome by a combination of a suitable preconcentration technique with subsequent AAS determination. These preconcentration methods provide low detection limits and also help to avoid matrix interferences in the analysis of real samples (Suvardhan et al., 2003).

Consequently, many methods have been developed for the preconcentration and separation of heavy metals from various samples before being discharged into the environment. These include liquid-liquid extraction (LLE), co-precipitation, resin chelation, electrochemical deposition, ion exchange, coagulation or flocculation and solid-phase extraction (SPE) (Juang and Shiau, 2000; Amuda and Amoo, 2006; Amuda and Alade, 2006; Amuda et al., 2006). These techniques however, have disadvantages such as incomplete metal removal, high reagent and energy requirements and generation of toxic sludge or other waste products (Ahalya et al., 2003).

Thus preconcentration by solid-liquid extraction considered in this work is one of the most efficient methods used to remove heavy metals from effluents using naturally available material as an adsorbent. This technique has an edge over the other methods due to its sludge free clean operation and complete removal of toxic

---

*Corresponding author. E-mail: teejayola01@yahoo.com.

metals even from dilute solution (Ahalya et al., 2003). Preconcentration procedures of metals by dissolving metal-substrates with acid and determining the metal concentration by atomic absorption spectrometry have been reported by (Wu et al., 2000; Rhazi et al., 2002; Ngah et al., 2004; Unob et al., 2005).

The objective of this work was to use EDTA modified charcoal in simultaneous preconcentration and removal of Pb(II), Co(II), Mn(II), Cu(II) and Zn(II) ion from aqueous solutions.

MATERIALS AND METHODS

Aqueous solution preparation

All the reagents used were of analytical grade and were prepared with high quality distilled-deionised water. The aqueous solutions containing Pb, Co, Mn, Cu and Zn was prepared from $Pb(NO_3)_2$, $CuSO_4.5H_2O$, $MnCl_4$, $CoCl_2.6H_2O$, and $Zn(NO_3)_2.6H_2O$. 1000 $mgdm^{-3}$ stock solution of each of the salts was prepared. Other concentrations of 5, 10, 20, 30, 40 and 50 $mgdm^{-3}$ were prepared from each stock solution by serial dilution.

Sample collection

The charcoal used in this work was obtained from *Khaya senegalensis* tree along Government House, Minna. Sampling was done at random within this area. A composite sample was made from where the representative samples were collected for this study. The charcoal was obtained by burning the wood in a limited supply of air.

Sample pretreatment

The samples were deoiled by soaking in hot deionized water with detergent for 24 h. They were then rinsed in hot deionized water to remove all debris, after which they were air dried for a period of one week. The dried samples were ground using a clean blender and sieved through a 400 μm mesh screen. The fine powder retained on the 400 μm clean sieve was then used for the preconcentration studies.

Adsorbent preparation for preconcentration studies

Charcoal was modified by soaking with 0.5 $moldm^{-3}$ EDTA (ethylene diamine tetra acetic acid) for 24 h. This was done to significantly improve the uptake levels of these heavy metal ions.

Preconcentration procedure

1 $moldm^{-3}$ each of these metals (Lead, Zinc, Copper, Cobalt and Manganese) was extracted as metal diethyl dithio-carbamates with 2:5 acetone-chloroform by varying the pH from 1 to 8. The pH of solutions was adjusted to the desired value by using either 0.5 $moldm^{-3}$ NaOH or concentrated HCl. These metals were then adsorbed on a 2 g column of modified charcoal in 5:4:1 mixture of Tetrahydrofuran, Ethylene glycol and 6.0 $moldm^{-3}$ HCl. Succesive elution was effected with 6.0 $mol.dm^{-3}$ HCl for Co, Cu, Mn and Pb while 2.0 $moldm^{-3}$ HNO_3 was used for Zn. The elements of interest

in the eluates were analyzed by flame atomic absorption spectrophotometer, model Pye Unicam SP-9 Cambridge, UK.

Sorption experiment

The experimental procedure described by Okieimen and Okieimen, (2001) was modified as follows: 0.5 g of the pretreated powdered samples were taken and shaken with 100 cm^3 of a metal ion solution whose concentration was 5 $mgdm^{-3}$. The bottles were shaken at various time intervals of 30 to 180 min at room temperature in a reciprocating shaker at 300 rpm. At the end of each contact period the mixture was filtered using Whatman filter paper No. 42 and the filtrate was stored in sample bottles in a refrigerator prior to analysis. The filtrate was analysed for the heavy metal ions using FAAS, model Pye Unicam SP-9 Cambridge, UK.

Multimetal batch adsorption study

The experiments were carried out in the batch mode for the measurement of adsorption capabilities at 33°C using different concentrations of 10, 20, 30, 40 and 50 $mgdm^{-3}$ for the various heavy metal ions studied. 100 cm^3 of various concentrations (10.0 $mgdm^{-3}$ to 50.0 $mgdm^{-3}$) of the metal ions were placed in 250 cm^3 conical flasks, and 0.5 g of the substrate was each added. The corked conical flask was shaken in a reciprocating shaker at 300 rpm for constant metal ion-substrate contact period of 90 min at 33°C. The separation of the adsorbents and solutions was carried out by filtration with Whatman filter paper No. 42 and the filtrate stored in sample cans in a refrigerator prior to analysis. The residual metallic ion concentrations were also determined using an flame atomic absorption spectrophotometer (FAAS).

Data analysis

The adsorptive capacity of metal ions per unit adsorbent (mg metal g^{-1} dry biosorbent) was determined using the following expression:

$$ q = \frac{V \left(C_i - C_{eq} \right)}{M} \qquad \text{---------------------- 1} $$

Where q is the metal uptake (mg metal ions g^{-1} dry weight of adsorbent entrapped), V is the volume of metal solution (cm^3), C_i is the initial concentration of metal ions in the solution ($mgdm^{-3}$), C_{eq} is the final concentration of metal ions in the solution ($mgdm^{-3}$), and M is the dry weight of adsorbent. Percentage removal of heavy metal ions from initial solution concentration calculated from the following Equation.

$$ R_{em} = \frac{C_o - C_t}{C_o} \times 100\% \qquad \text{-------------- 2} $$

where C_0 is the initial metal ion concentration ($mgdm^{-3}$), and C_t is the metal ion concentration at time t ($mgdm^{-3}$).

RESULTS

The percentage recovery was calculated from the amount of metals (1 $moldm^{-3}$) in starting solution (N_s) and the amount of metals eluted from the column (N_{fl}) as in

Table 1. Recovery of metals with column preconcentration based on pH variation.

pH	Recovered amount (mgdm^{-3})									
	Pb^{2+}	% Recovery	Zn^{2+}	% Recovery	Cu^{2+}	% Recovery	Co^{2+}	% Recovery	Mn^{2+}	% Recovery
1	0.25	75.00	0.27	73.00	0.30	70.00	0.33	67.00	0.42	58.00
2	0.19	81.00	0.21	79.00	0.26	74.00	0.29	71.00	0.34	66.00
3	0.17	83.00	0.19	81.00	0.24	76.00	0.28	72.00	0.30	70.00
4	0.14	85.70	0.20	80.30	0.21	78.90	0.25	75.40	0.35	64.60
5	0.01	99.80	0.10	98.00	0.11	97.90	0.20	96.00	0.24	95.10
6	0.10	97.50	0.15	96.30	0.12	97.10	0.21	94.90	0.25	93.70
7	0.12	96.00	0.15	94.90	0.20	93.20	0.23	92.40	0.29	90.40
8	0.12	93.80	0.17	91.40	0.20	89.90	0.23	88.40	0.30	85.00

Equation 1.
The results are summarized in Table 1

$$\% \text{ recovery} = \frac{N_{fi}}{N_s} \times 100 \quad\ldots\ldots\ldots\ldots\ldots\ldots 3$$

DISCUSSION

From Table 1, at pH 5, the recovery of the metals ions is in the range of 95.1 – 99.8%. The acceptable range of percentage recovery is 80-110% as reported by many authors (Wu et al., 2000; Rhazi et al., 2002; Ngah et al., 2004; Unob et al., 2005). The obtained results therefore agree with the findings of Unob et al. (2005) who had earlier studied the preconcentration of heavy metals from aqueous solution using chitosan flake at pH 5. Similar results were also obtained by Ngah et al. (2004) who investigated the preconcentration of Zn(II) and Pb(II) using chitosan flake column in which the metals were optimally retained by chitosan at pH 5.

Effect of varying pH on the adsorption of metal ions

Metal sorption is critically linked to pH and the effect of pH of the solution is an important controlling factor in the adsorption process (Ngah et al., 2004; Faria et al., 2004). For this, the role of hydrogen ion concentration was examined at different pH. The effect of pH on the preconcentration of metals ions extracted as diethyldithiocarbamates complexes, that was adsorbed on a 2 g column packed with modified charcoal in 5:4:1 mixture of tetrahydrofuran, methyl glycol and 6 moldm^{-3} HCl was examined. It was observed that with the increase in the pH of the solution the percentage removal of metal ions increased up to the pH 5 as shown in Table 1. At pH 5 maximum recoveries were obtained for all the metal ions studied. The preconcentration experiment showed that the highest removal by modified charcoal

was 99.8% for Pb(II) ions in the aqueous solution at pH 5.

The improved removal levels of Pb(II) ion and other metal ions by EDTA modified substrates may be ascribed to the chemical reaction of the metal ions with the EDTA on the charcoal which result to the formation of the metal complexes. These metal complexes however, remain in solution and are recovered as metal ion upon addition of acids. The rate of recovery was based on their ionic radii and atomic weight (Jimoh, 2010). It should be noted that the increase in the percentage recovery was not directly proportional to the substrates suggesting that other factors play a role in the adsorption and desorption mechanism. It is quite plausible that an initial high adsorption rate (possibly by ion exchange) followed by chelation of the metal ions with the carboxyl group may have been the case. A similar behaviour was shown by column parked with chitosan flake biomass where the percentage of Pb(II) ion removal was much higher than other metal ions at pH 5 (Unob et al., 2005).

The increase in percentage removal of metal ions due to increase in pH may be explained on the basis of a decrease in competition between proton (H$^+$) and positively charged metal ion at the surface sites and also by decrease in positive charge near the surface resulting in a lower repulsion of the adsorbing metal ion. pH values of 7 to 8 suggest increase in alkalinity level and as such, there was a further decrease in the rate of adsorption by modified charcoal for Pb(II), Cu(II), Zn(II), Co(II) and Mn(II) ions in the aqueous solution. With increase in pH from 7 to 8, the degree of protonation of the adsorbent functional group decreased gradually and hence removal was decreased. A close relationship between the surface basis of the adsorbents and the anions is evident. This is similar to other findings where the interaction between oxygen-free Lewis basic sites and the free electrons of the anions, as well as the columbic interactions between the anions and the protonated sites of the adsorbent are the main adsorption mechanism (Faria et al., 2004; Oyebamiji et al., 2009; Egila et al., 2010).

The low percentage recovery at pH 1 and 2 may be ascribed to the fact that the higher mobility of H$^+$ ions

Table 2. Results of the effect of contact time on the amount of metal ions bound (mg100 cm^3) on Modified charcoal at 33°C.

Contact time(min)	Pb^{2+}	Zn^{2+}	Mn^{2+}	Cu^{2+}	Co^{2+}
30	1.5	1.7	1.8	1.8	1.9
60	2.3	2.4	2.7	2.7	2.8
90	2.7	2.9	3.3	3.4	3.0
120	2.5	2.7	3.1	2.5	2.6
150	2.2	2.3	2.5	2.2	2.3
180	1.2	1.3	1.4	1.9	2.0

present favoured the preferential adsorption of hydrogen ions compared to metal ions. This trend is not far from what was observed by Ajmal et al., (2000) and Wong et al. (2003) who independently suggested that at lower pH value, the surface of the adsorbent is surrounded by hydroxonium ions (H_3O^+), thereby preventing the metal ions from approaching the binding sites of the sorbent. This means that at higher H^+ ions concentration, the biosorbent surface becomes more positively charged such that the attraction between modified charcoal and metal cations is reduced (Saeed et al., 2004; Jimoh, 2010). However, as the pH increases, more negatively charged surface becomes available thus facilitating greater metal ions removal. It is commonly agreed that the preconcentration of metal cations increases with increasing pH as the metal ionic species become less stable in the solution (Unob et al., 2005).

Therefore, the column packed with modified charcoal is suitable for the preconcentration of these metals in aqueous solution. At pH 5, the recovery of metal ions from the column was in the range of 95.10-99.80%. This is the pH at which maximum recovery was obtained for the metal ions present in the aqueous solution. The effectiveness of the used method was presented in terms of percentage recovery of metal ions ranging from 95.10 to 99.80 at pH 5. The column parked with modified charcoal from *Khaya senegalensis* was able to recover 99.8% Pb (II) ions after successful elution with 6 moldm^{-3} HCl. This value is within the acceptable limits.

Effect of contact time on the adsorption of metal ions

Tables 2 show the percentage removal of the various metal ions by the modified charcoal. For all the metal ions present in the aqueous solution, there were a progression in the rate of adsorption but it was not linear at any time. It was also observed that the rate of adsorption increased significantly for all the metal ions present for 90 min of contact time.

This result is essential, as equilibrium time is one of the important parameters for an economical wastewater treatment system. The trend observed for the various metal ions under the same experimental conditions revealed that the pattern of adsorption is a function of the substrate as well as that of the metal ions (Egila and Okorie, 2002; Egila et al., 2010; Jimoh, 2010). The increase of contact time further from 90 to 180 min had no significant effect on the percentage sorption of all metal ions.

Further more, manganese with small ionic size was observed to have adsorbed at a faster rate than the other metal ions. The trend of adsorption in the modified charcoal is Mn(II)>Co(II)>Cu(II)>Zn(II)>Pb(II). This occurrence may be explained by considering the ionic radii which was observed for Mn(II) (0.67Å), Co(II), Cu(II) with higher rate of adsorption at equilibrium time than Zn(II), Pb(II) ions (1.20Å). According to Abia and Asuquo (2006) during sorption of metal ions, the ions of smaller ionic radii tend to move faster to potential adsorption. Similar observation had been reported by several other researchers (Mohsen, 2007; Oboh and Aluyor, 2008; Oyebamiji et al., 2009; Egila et al., 2010).

The effect of initial metal ions concentrations on the adsorption of the metal ions

The uptake level of Mn(II), Co(II), Cu(II), Zn(II) and Pb(II) ions from solutions containing various amounts of the metal ions by modified charcoal is shown in Table 3 as the metal ions concentration increases from 10 to 50 mg/100 cm^3. The obtained results show that the amount of the metal ions bound by the modified charcoal depended on the type of metal ions and its concentration. The uptake levels of metal ions followed this order Mn (II) >Co(II) >Cu(II)>Zn(II)>Pb(II). The difference in the uptake levels of the metal ions can be explained in terms of the difference in the ionic size and atomic weight of the metal ions, the mode of interaction between the metal ions and the substrate.

The initial faster sorption rate of each metal ion could be due to the availability of the uncovered surface area of the adsorbents, since adsorption kinetics depends on the surface area of the adsorbent (Qadeer and Akhtar, 2005; Egila et al., 2010; Jimoh, 2010). The trend could also be explained in terms of the progressive increase in the columbic interaction between the metal ions and the absorbent active sites. Moreover, more adsorption sites were being covered as the metal ions concentration

Table 3. Results of the effect of varying metal ions concentration on the adsorption of metal ions on modified charcoal at 33°C for 90 min.

Initial metal ions concentration (mg/100 cm³)	Amount of metal ions adsorbed				
	Pb^{2+}	Zn^{2+}	Mn^{2+}	Cu^{2+}	Co^{2+}
10	3.50	3.55	4.83	4.85	4.90
20	9.36	10.20	12.43	12.30	12.50
30	17.50	17.80	18.50	19.00	20.50
40	19.04	20.20	23.02	25.00	23.02
50	23.00	24.00	34.60	35.20	34.29

Table 4. Results of the Freundlich constants for modified charcoal.

Heavy metal ions	1/n (Freundlich Constant)	K_f (Freundlich Constant)	R^2 (Correlation coefficient)
Pb^{2+}	0.65	2.05	0.965
Cu^{2+}	0.74	2.57	0.968
Zn^{2+}	0.73	2.58	0.975
Mn^{2+}	0.76	2.73	0.976
Co^{2+}	0.75	2.68	0.975

increased (Egila et al., 2010; Jimoh, 2010), and higher initial concentrations led to an increase in the affinity of the metal ions towards the active sites (Oboh and Aluyor, 2008).

The decline in the uptake level despite increase in the initial metal ions concentration is due to the availability of smaller number of surface sites on the adsorbent for a relatively larger number of adsorbing species at higher concentrations (Somasekhara et al., 2007). The results obtained from this research work conformed to the findings by Oboh and Aluyor (2008) that studied Sour sop seeds as biosorbent to remove heavy metals ions from aqueous solution.

Adsorption isotherms

The equilibrium relationship between adsorbent and adsorbate are described by adsorption isotherms which is usually the ratio between the quantity adsorbed and that remaining in solution at a fixed temperature at equilibrium. Most often biosorption equilibria are described with adsorption isotherms of Langmuir or Freundlich types. Since the adsorption isotherms are important to describe how adsorbates will interact with adsorbents and so are critical for design purposes, therefore, the correlation of equilibrium data using an equation is essential for practical adsorption operation. When the sorption data of the metal ions investigated on modified charcoal were plotted logarithmically, they all fitted the Freundlich adsorption isotherm.

The values of correlation coefficient as contained in Table 4 indicate that the adsorption process conforms to the Freundlich isotherms (Igwe et al., 2005; Kurniawan, 2008; Oyebamiji et al., 2009). The results also show that the R^2 values for substrate was greater than 0.9 for Freundlich Isotherm, so the adsorption was multi layered and physio-sorption type. Mn(II) ion with small ionic size was observed to have the highest value for 1/n, R^2 and K_f than the other metal ions. The trend of occurrence was Mn(II)>Co(II)>Cu(II)>Zn(II)>Pb(II). This phenomenon may be explained by considering that Mn (II) ion has an ionic radius of 0.67Å while Pb(II) ion has is 1.20Å. Since Mn (II) ion has a smaller ionic radius, it tends to diffuse to the potential adsorption sites easier than other metal ions. The smaller ionic radius of Mn(II) might probably be attributed to its greater adsorptivity.

The value of 1/n which is less than 1 indicates that the metal ions are favourably adsorbed by the modified charcoal. A smaller value of 1/n indicates better adsorption mechanism and formation of relatively stronger bond between adsorbate and adsorbent (Egila and Okorie, 2002; Egila et al., 2010; Jimoh, 2010).

Conclusion

The obtained results from this work have confirmed the potential of preconcentration by solid-liquid extraction as one of the most selective method for determination of metal ions in low concentrations. It has also proved the effectiveness of heavy metal removal from aqueous solution by EDTA modified charcoal sourced from *Khaya senegalensis*. This has the potential to be used in metal ion removal from wastewater and assist in reducing environmental pollution from such metals as well to save the fund of purchasing the commercial adsorbents.

REFERENCES

Abia AA, Horsfall Jr M, Didi O (2006). The use of chemically modified and unmodified cassava waste for the removal of Cd, Cu, and Zn ions from aqueous solution. Bioresource Technol., 90(3): 345-348.

Ahalya N, Ramachandra TV, Kanamadi RD (2003). Biosorption of heavy metals. Res. J. Chem. Environ., 7: 71-78.

Ajmal M, Rao RA, Ahmad K, Ahmad R (2000). Adsorption studies on Citrus reticulate (fruit peel of orange): removal and recovery of Ni (II) from electroplating wastewater. J. Hazard. Mat., 79: 17-131.

Amuda OS, Amoo IA (2006). Coagulation/flocculation process and sludge conditioning in beverage industrial wastewater treatment. J. Hazard. Mat., 747: 778-783.

Amuda OS, Alade A (2006). Coagulation/flocculation process in the treatment of abattoir wastewater. Desalination, 796: 22- 31.

Amuda OS, Amoo IA, Ajayi OO (2006). Performance optimization of coagulation / flocculation process in the treatment of beverage industrial wastewater. J. Hazard. Mat., 129 (1-3): 69-72.

Egila JN, Dauda BEN, Jimoh T (2010). Biosorptive removal of cobalt (II) ions from aqueous solution by Amaranthus hydridus L. stalk wastes. Afri. J. Biotechnol., 9(48): 8192-8198.

Egila JN, Okorie EO (2002). Influence of pH on the Adsorption of trace metals on ecological and agricultural adsorbents. J. Chem. Soc. Nig., 27(2): 95-98.

Faria PCC, Orfao JJM, Pereira MFR (2004). Adsorption of anionic and cationic dyes on activated carbons with different surface chemistries. Water Resour., 38: 2043-2052.

Guo X, Zhang S, Shan XQ (2008). Adsorption of metal ions on lignin. J. Hazard. Mat., 151(1): 134-142.

Iqbal M, Saeed A (2007). Production of an immobilized hybrid biosorbent for the sorption of Ni (II) from aqueous solution. Process Biochem., 42: 148-157.

Jimoh T (2010). Preconcentration and Removal of Heavy Metals from Aqueous Solution by African Spinach(Amaranthus hybridus L) stalk and Pawpaw (Carica papaya) seeds (Unpublished Masters thesis submitted to Postgraduate School, Federal University of Technology, Minna, Nigeria for the Award of Master of Technology in Analytical Chemistry). pp. 67-68

Juang RS, Shiau RC (2000). Metal removal from aqueous solutions using chitosan-enhanced membrane filtrations. J. Membrane Sci., 765: 159-167.

Kurniawan TA (2008). Removal of recalcitrant contaminants from stabilized landfill leachate by a combination of advanced oxidation processes (AOP) and granular activated carbon (GAC) adsorption, PhD dissertation, pp. 121-123

Mohsen AH (2007). Adsorption of lead ions from aqueous solution by okra wastes. Inter. J. Physical Sci., 2(7): 178-184

Ngah WSW, Kamari A, Koay Y (2004). "Equilibrium and kinetics studies of adsorption of copper(II) on chitosan and chitosan/PVA beads". Int. J. Biol. Macromol., 34: 155-161.

Oboh OI, Aluyor EO (2008). The Removal of Heavy metal ions from aqueous solution using sour sop seeds as biosorbent. Afr. J. Biotechnol., 7(24): 4508-4511

Okieimen CO, Okieimen FE (2001). Enhanced metal sorption by ground nut (Arachis hypogea) husks modified with thioglycollic acid. Bull. Piere Appl. Sci., 20c: 13-20.

Oyebamiji BJ, Adesola NA, Babarinde OAP, Vincent OO (2009). Kinetic equilibrium and thermodyamics studies of Cd(II) and Pb(II) from aqueous solutions by Tallinium triangulare (water leaf). Pacific J. Sci. Technol., 10(1): 428-436.

Rhazi M, Desbrieres J, Tolaimate A, Rinaudo M, Vottero P, Alagui A, El Meray M (2002) "Influence of the nature of the metal ions on the complexation with chitosan. Application to the treatment of liquid waste". Eur. Polym. J., 38: 1523-1530.

Saeed A, Muhammed I, Waheed Akhtar M (2004). Removal and recovery of lead (II) from single and multimetal (Cd, Cu, Ni, Zn) solutions by crop milling waste (black gram husk). J. Hazard. Mat., 117: 65-73.

Somasekhara RK, Emmanuel KA, Ramaraju KA (2007). Removal of Mn(II) from aqueous solution using Bombax malabaricum fruit shell substrate. E-J. Chem. 4(3): 419-427.

Suvardhan K, Suresh KK, Reddy KM, Chiranjeevi P (2003). Determination of trace element by Atomic Absorption Spectroscopy (AAS) after Preconcentration on a support Impregnated with Coniine Dithiocarbamate. Proceedings of the Third International Conference on Environment and Health, Chennai, India, Chennai: Department of Geography, University of Madras and Faculty of Environmental Studies, York University, pp. 562– 569.

Unob F, Ketkangplu P, Chanyput P (2005). Preconcentration of Heavy metals from aqueous solution using Chitosan flake. J. Sci. Res., 30(1): 87-95.

Wong KK, Lee CK, Low KS, Haron MJ (2003). Removal of Cu and Pb by tartaric acid modified rice husk from aqueous Solutions. Chemosphere, 50: 23-28.

Wu F, Tseng R, Juang R (2000). "Comparative adsorption of metal and dye on flake- and bead-types of chitosans prepared from fishery wastes". J. Hazard. Mat., 73: 63-75.

13

Spatiotemporal assessment of metal concentration in fish and periwinkles in selected locations of Lagos Lagoon, Nigeria

Ladipo M. K.[1], Ajibola V. O.[2]* and Oniye S. J.[3]

[1]Yaba College of Technology, Lagos, Nigeria.
[2]Department of Chemistry, Ahmadu Bello University, Zaria, Nigeria.
[3]Department of Biological Sciences, Ahmadu Bello University, Zaria, Nigeria.

Fish and periwinkles, from nine locations in the Lagos Lagoon, were processed and analysed for the presence of heavy metals (chromium Cr, manganese Mn, iron Fe, nickel Ni, copper Cu, zinc Zn, cadmium Cd and lead Pb) during the dry and wet seasons. Results showed that significant differences existed in the levels of metals in periwinkle and fish caught from the different locations while the influence of the period of sample collection on elemental levels indicated no significant difference in the levels of metals accumulated between the periods. Generally, the level sequence of the elements studied was, in most cases, similar in both locations and the period of sample collection for both periwinkle and fish. The sequence was different for periwinkle and fish: for periwinkle the sequence was Fe > Mn > Zn > Cd > Ni > Cu > Cr > Pb and for fish the sequence was Fe > Zn > Cd > Cu > Mn > Ni >Pb > Cr. The observed heavy metals concentrations in these organisms were below the recommended limits for human consumption.

Key words: Fish, periwinkle, heavy metals, cluster analysis, Lagos Lagoon.

INTRODUCTION

Many pollutants, including heavy metals, which are toxic to aquatic organisms and cause their lethal or sublethal deterioration, may find their way to aquatic ecosystems through various sources located at the catchment area or at distant places. Lagoons are directly exposed to different contaminants from various sources and the Lagos Lagoon is not an exception. It is a shallow expanse of water with restricted circulation in a micro tidal environment. This aquatic resource of multiple usages receives inputs of domestic sewage, industrial waste waters, sawdust and particulate wood wastes, petroleum hydrocarbons, cooling water from a thermal power station and emissions from automobile exhaust. Moreover, the chemical quality of the marine organisms,

their contents of pollutant heavy metals, will become of major importance to resources managers and public health officials (Barbour et al., 1999). It is well documented that heavy metals have a great ecological significance due to their toxicity and accumulative behaviour. They accumulate in the tissues of aquatic animals and may become toxic when accumulation reaches substantially high levels. Accumulation levels vary considerably among metals and species (Heath, 1987). The presence of some heavy metals in aquatic environments and their accumulations in fish and periwinkles has been investigated during recent years (Hornung and Kress, 1991; Kalay et al., 1999; Canli and Atli, 2003; Ayenimo et al., 2005; Dural et al., 2006; Davies et al., 2006; Yalçın et al., 2008; Burger and Gochfeld, 2005; Chindah et al., 2009). Human beings may be contaminated by organic and inorganic pollutants associated with aquatic systems and inorganic pollutants

---
*Corresponding author. E-mail: tunjiajibola2003@yahoo.com.

associated with aquatic systems by consumption of contaminated fish and other aquatic foods from this environment (Mackay and Clarck, 1991). This fact is due to the capacity of some aquatic organisms to concentrate heavy metals up to 105 times the concentration present in the water (Singh and Narwal, 1984; Aderinola et al., 2009). Many of these metals tend to remain in the ecosystem and eventually move from one compartment to the other within the food chain. Bioaccumulation and magnification is capable of leading to toxic level of these metals in fish, even when the exposure is low. The presence of metal pollutant in fresh water is known to disturb the delicate balance of the aquatic ecosystem. Fishes are notorious for their ability to concentrate heavy metals in their muscles and since they play important role in human nutrition, they need to be carefully screened to ensure that high levels of some toxic trace metals are not being transferred to man through fish consumption (Olowu et al., 2010).

Metals, such as iron, copper, zinc and manganese, are essential metals since they play important roles in biological systems, whereas mercury, lead and cadmium are toxic, even in trace amounts. The essential metals can also produce toxic effects at high concentrations. Only a few metals with proven hazardous nature are to be completely excluded in foods for human consumptions (Yalçın et al., 2008). Due to the deleterious effects of metals on aquatic ecosystems, it is necessary to monitor their bioaccumulation in key species, because this will give an indication of the temporal and spatial extent of the process, as well as an assessment of the potential impact on the health of organisms. In order to elucidate the evidence of the pollution effects on the aquatic fauna of the Lagos Lagoon, determination of these metals in periwinkle and fish from nine locations of the lagoon was performed during the dry and rainy seasons. Further-more, considering the nutritional values of fish and the large quantities consumed in Nigeria, the results are briefly discussed in the light of some guidelines for human consumption. The comparison of the study areas in term of elemental impregnation in fish and periwinkle was performed using multivariate statistical analysis.

MATERIALS AND METHODS

Sampling and sample preparation

The choice of the benthic organism – *Tympanotonus fuscatus* (Periwinkle) is based on the fact that they have poor mobility and are typically in direct contact with sediments, they have the greatest exposure to pollutants. On the other hand, the choice of the pelagic organism - *Tilapia guineensis* (Fish) is due to their availability round the year, which makes them useful in field monitoring of pollution. These organisms occupy different community in Lagos Lagoon. And since the levels of certain elements were dependent on the size of the fish, wet weights were used to compare their levels found in each location. Data obtained from nine sampling stations in

the study area (Figure 1) during the periods of investigation were processed.

The fish species *Tilapia guineensis* was caught at each station by employing cast nets and set (gill) nets which are usually left overnight in each station in the lagoon. The fish selected for use weighed between 120 – 150 g. *Tympanotonus fuscatus* (Periwinkle) weighing between 9 – 12 g were collected at each sampling point where the water is shallow because the animal apparently had a preference for shallow waters. Five fish samples and 10 periwinkle samples were collected from each sampling location. The samples were frozen in different polythene bags and deposited at Biological Sciences Department of University of Lagos, Akoka for identification. The samples were washed thoroughly with double distilled water and air dried for about 6 h after which they were transferred to a well-ventilated oven set at 60°C for 24 h. When the samples were confirmed dry, they were pulverised whole to fine powder and stored in air tight crew-capped plastic containers.

Dry tissue samples weighing 0.5 g were digested with 6 ml of concentrated nitric acid and 1 ml of 30% hydrogen peroxide. The digestion was carried out in a microwave digester using the microwave digestion. The completely digested samples were filtered and diluted to 25 ml in volumetric flasks with distilled water. The resulting solutions were analysed for metals using Flame Atomic Absorption Spectrophotometer (Perkin Elmer AA Winlab equipped with MS Windows application software).

Statistical analysis

Statistical analysis was performed using the Graph Prism statistical package version 5.0. All statistical samples submitted to tests were first checked for normality and homogeneity of the variances by means of Shapiroe Wilk. In all cases there was departure from normality therefore non-parametric tests were used in the subsequent analysis. The significance of differences of trace element levels among the fish caught at different locations was tested by Friedman (F) tests and the influence of the period of sample collection on elemental levels was tested by means of Kruskal Wallis (KW) tests followed by Wilcoxon test for paired samples.

RESULTS AND DISCUSSION

The levels of metals in the periwinkle and fish caught during the sampling periods at the different locations of the lagoon are given in Tables 1 to 4. The concentration of Cr in the periwinkle varied between 0.01 and 0.14 µg/g during the J07 period and 0.01 and 0.13 µg/g during the F08 sampling period. The mean values for all the locations during these periods were 0.06 and 0.04 µg/g respectively. Accumulation of Cr by periwinkle was found to be higher than that accumulated by the fish which was 0.002 – 0.06 µg/g and 0.001 – 0.08 µg/g in the J07 and F08 periods respectively. The variation in concentration of Cr in fish and periwinkle caught in all the locations was significant during the sampling periods ($p > 0.005$ at 95% Confidence level).

Manganese was the second highest in the level of metals' concentrations in periwinkle from all locations. The values recorded ranged between 43 to 95 µg/g in the J07 samples and 48 to 99 µg/g in F08 samples and mean

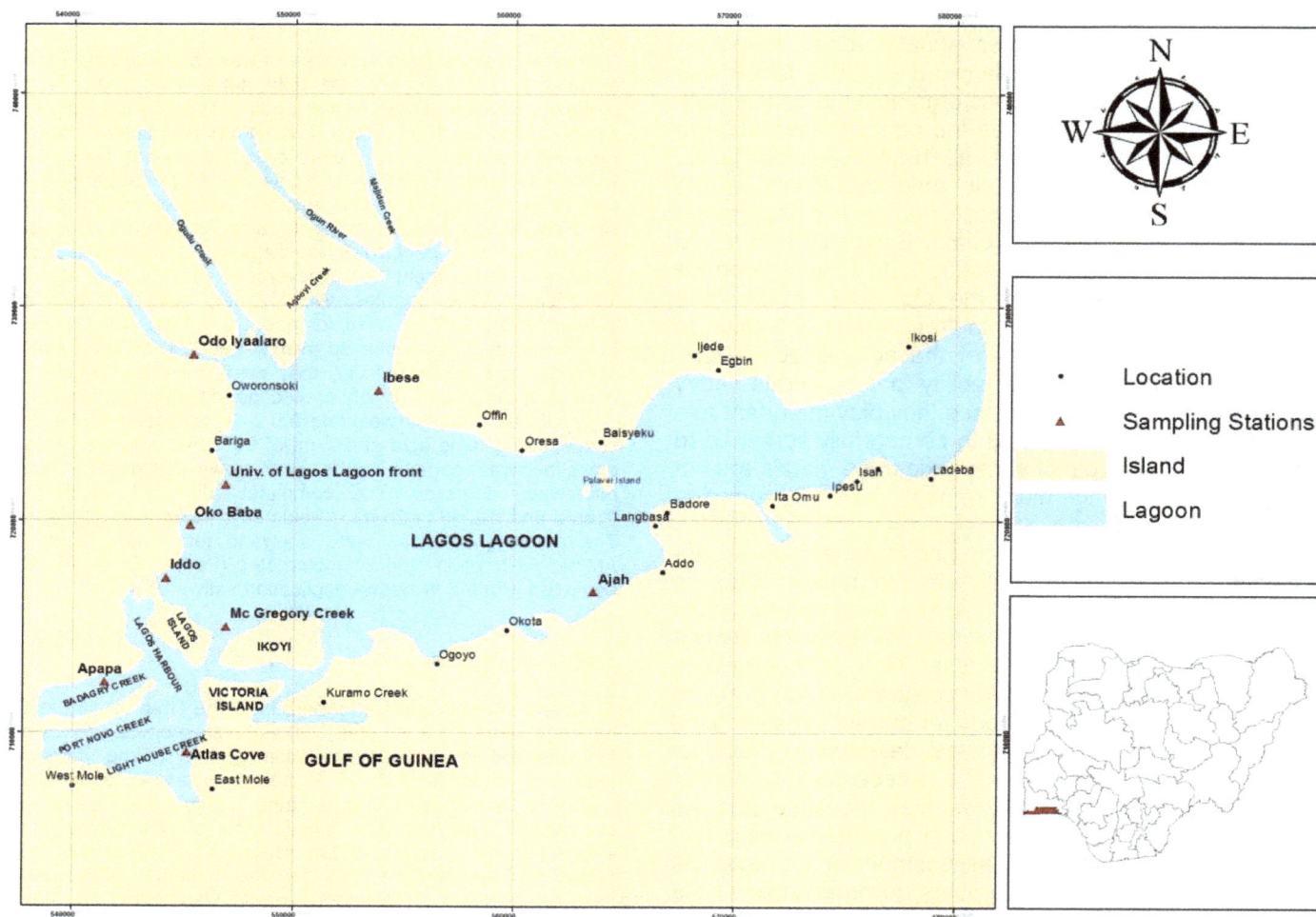

Figure 1. Map of the Lagos Lagoon showing sampling areas.

Table 1. Levels of heavy metal (µg/g) in homogenates of periwinkle *Tympanotonus fuscatus* collected from different locations of the Lagos Lagoon in the July 2007 sampling period.

Location	Cr	Mn	Fe	Ni	Cu	Zn	Cd	Pb
L1	0.01	59	4000	1.6	0.75	2.2	2.2	0
L2	0.01	53	3920	1.3	0.62	2	2.1	0
L3	0.02	47	3625	1.82	0.52	3.1	2.8	0
L4	0.03	43	1650	1.65	0.62	1.1	2.7	0
L5	0.01	62	1420	2.5	0.75	3.2	4.6	0.04
L6	0.1	64	640	2.7	0.8	2.8	3.5	0.05
L7	0.14	81	372	3.9	0.77	5.2	3.6	0.06
L8	0.12	95	3245	3.2	1.9	6.7	4.2	0.07
L9	0.14	85	2960	2.9	1.6	2.6	5.1	0.09

values of 65.44 and 70.11 µg/g respectively. The variations of Mn during these periods were not significant for all the locations ($p < 0.005$) during the periods samples were collected. The concentrations of Mn recorded in the fish caught from all the locations were relatively much lower than the levels found in the

Table 2. Levels of heavy metal (µg/g) in homogenates of periwinkle *Tympanotonus fuscatus* collected from different locations of the Lagos Lagoon in the February 2008 sampling period.

Location	Cr	Mn	Fe	Ni	Cu	Zn	Cd	Pb
L1	0.01	63	4200	1.8	0.78	2.5	2.1	0.00
L2	0.02	54	4100	1.6	0.65	2.2	2.2	0.00
L3	0.03	51	3780	2.1	0.68	4.0	3.2	0.00
L4	0.04	48	1480	1.9	0.85	2.0	3.1	0.04
L5	0.01	72	1340	2.2	0.95	5.0	5.1	0.06
L6	0.13	68	462	2.4	1.1	2.0	4.2	0.06
L7	0.01	85	301	3.8	1.2	6.0	3.8	0.07
L8	0.01	91	3568.5	3.7	1.6	9.0	4.8	0.08
L9	0.11	99	3014	3.4	1.8	2.3	5.6	0.11

Table 3. Levels of heavy metal (µg/g) in homogenates of the fish *Tilapia guineensis* collected from different locations of the Lagos Lagoon in the July 2007 sampling period.

Location	Cr	Mn	Fe	Ni	Cu	Zn	Cd	Pb
L1	0.00	1.6	72.1	0.02	2.1	6.2	2.1	0.01
L2	0.00	1.6	72	0.01	2.0	6.1	2.9	0.01
L3	0.00	2.4	60	0.03	1.8	8.5	3.9	0.02
L4	0.01	1.1	52	0.04	1.9	8.8	4.2	0.04
L5	0.01	1.0	46	0.06	2.4	14	3.8	0.04
L6	0.02	0.8	47	0.09	2.7	13	3.6	0.03
L7	0.03	0.9	46	0.08	2.5	12	3.4	0.02
L8	0.01	1.6	62	0.05	2.9	12.5	4.1	0.03
L9	0.06	3.1	57	0.06	2.7	15	4.6	0.80

Table 4. Levels of heavy metal (µg/g) in homogenates of *Tilapia guineensis* collected from different locations of the Lagos Lagoon in the February 2008 sampling period.

Location	Cr	Mn	Fe	Ni	Cu	Zn	Cd	Pb
L1	0.01	2.5	78.2	0.04	2.2	8.0	2.7	0.01
L2	0.01	2.6	76	0.04	2.2	8.0	2.8	0.02
L3	0.01	3.2	65	0.06	2.4	11	3.7	0.03
L4	0.01	2.9	57	0.06	2.6	10	3.8	0.05
L5	0.01	1.5	51	0.08	3.2	13	4.3	0.07
L6	0.00	1.2	49.5	0.07	3.5	12	4.1	0.06
L7	0.01	1.6	52.6	0.06	3.2	13	4.2	0.03
L8	0.01	2.8	70.6	0.06	2.8	15	4.4	0.05
L9	0.08	4.3	67	0.08	3.1	18	5.4	1.10

periwinkle. The values varied from 0.80 to 3.1 µg/g and 1.2 to 4.3 µg/g in J07 and 08 samples, respectively. The mean values for the locations during these periods are 1.57 and 2.51 µg/g for J07 and F08 samples respectively and the concentrations of Mn in the fish samples did not show any significant variation ($p < 0.005$ at 95% confidence level).

Fe is found in natural fresh waters and has no health-based guideline value, although high concentrations in water give rise to consumer complaints (WHO, 2004). The concentration of the metals when analysed in periwinkle and fish indicated that Fe showed the highest concentrations and mean levels relative to the other metals analysed in all locations. Comparatively, periwinkles had relatively higher mean Fe concentrations than the fish, suggestive of differences due to physiology and feeding habits and the fact that periwinkles are exposed to more Fe from sedimentary sources (Tay et al., 2005). The high concentrations of iron in the periwinkles could further be associated with the fact that this metal is naturally abundant in Nigerian soils and because the source of metal depositories are the aquatic system (Olowu et al., 2010). The concentrations found in the periwinkles sometimes exceeding the concentrations found in sediments. The concentrations of Fe found ranged from 372 to 4000 µg/g in J07 samples and 301 to 4200 µg/g in F08 samples. The mean values for all the locations were 2425.78 and 2471.72 µg/g for J07 and F08 samples respectively. The variations in the concentration of Fe in the periwinkle during these periods were not significant ($p < 0.005$). The levels of Fe recorded in the periwinkles were in most cases about five times higher than the concentrations recorded for the fish samples. The concentration of Fe was the highest of all the metals in the fish and it ranged from 46 to 72.1 µg/g and 49.5 to 78.2 µg/g in samples collected during the J07 and F08 periods respectively. This finding agreed with that reported by Olowu et al. (2010) who worked on fish caught from the Epe and Badagry sections of the Lagos Lagoon. They reported further that more of the metal concentrated in the intestine of both fishes (*Tilapia zilli* and cat fish, *Chryscichthys nigrodigitatus*) they investigated from the aquatic environment and attributed this to the function of intestine, which serves as the ultimate depository of all substance coming into the fish alimentary canal. The mean concentrations of Fe from all the locations were 57.12 and 62.99 µg/g for J07 and F08 samples, respectively. The concentration of Fe in the fish samples caught from the different locations also did not show any significant variation ($p < 0.005$).

The concentrations of Ni found in the periwinkle were higher than the values found in fish. The concentration found in periwinkle ranged from 1.3 to 3.9 µg/g and 1.6 to 3.8 µg/g for J07 and F08 samples respectively. The mean values of Ni for all the locations during the J07 and F08 sampling periods were 2.40 and 2.54 µg/g respectively and variation in the concentration for samples caught at the different locations did not vary significantly during the period of investigation ($p < 0.005$). The concentration of Ni recorded in the fish samples were much lower than those recorded for periwinkle. The concentrations in fish ranged from 0.01 to 0.09 µg/g during the J07 period and 0.04 to 0.07 µg/g during the F08 period. The mean for all the locations during the J07 period was 0.05 and 0.06 µg/g respectively; the variation in the concentration of Ni from all the locations was significant ($p < 0.005$). The concentration range of Ni reported by Olowu et al. (2010) for fish had a maximum higher than range recorded for

fish samples in this work. The level of Ni concentration reported ranged from BDL – 4.0 µg/g and from BDL – 2.30 µg/g in Epe and Badagry sections of the lagoon respectively.

Copper is one of the metals classified as essential to life due to its involvement in certain physiological processes and metabolic activities in organisms, however, elevated levels of Cu have been found to be toxic. The periwinkle in many cases concentrated less Cu compared to the fish. The concentrations recorded ranged from 0.52 to 1.90 µg/g in J07 samples and 0.78 to 1.80 µg/g in F08 samples. The concentrations of Cu in periwinkles caught from all the locations did not vary significantly and the mean values for all the locations are 0.92 and 1.07 µg/g for J07 and F08 samples. The range found in fish was 1.8 to 2.9 and 2.2 to 3.5 µg/g for J07 and F08 samples respectively. The mean values for all the locations were 2.33 and 2.80 µg/g respectively for J07 and F08; the variation of Cu in fish caught from the different locations was not significant ($p < 0.005$).

Studies have shown that, Zn could be toxic to some aquatic organisms such as fish (Alabaster and Lloyd, 1980). Although Zn has been found to have low toxicity to man, prolonged consumption of large doses could result in some health complications such as fatigue, dizziness and neutropenia (Tay et al., 2005). Fish accumulated more Zn than periwinkle in all the locations studied. The concentration of Zn in fish ranged between 6.1 and 14 µg/g in the J07 samples and 8.0 and 18.0 µg/g in the F08 samples. The mean concentrations for all the locations are 10.68 and 12.0 µg/g for J07 and F08 samples respectively; the concentration of Zn in fish samples caught in the different locations did not vary significantly ($p < 0.005$). The concentration found in the work was higher than the concentrations (0.16 - 1.95 µg/g and 0.20 - 1.32 µg/g) reported by Olowu et al. (2010) for fish caught in Epe and Badagry sections of the lagoon respectively. The level of Zn in periwinkle ranged from 1.10 to 6.70 µg/g in the J07 samples and 2.0 to 9.0 µg/g in F08 samples. The mean concentrations obtained for all the locations were 3.21 and 3.89 mg/kg for J07 and F08 samples respectively. The level of Zn like in the fish did not vary significantly from one location to another during this investigation ($p < 0.005$).

Cadmium is a highly toxic metal and its uptake from water by aquatic organisms is extremely variable and depends on the species and various environmental conditions, such as water hardness (notably the calcium ion and zinc concentration), salinity, temperature, pH, and organic matter content (Rosenberg and Costlow, 1976). Cadmium is one of the most toxic elements with reported carcinogenic effects in humans (Goering et al., 1994). High concentrations of Cd have been found to lead to chronic kidney dysfunction. Cd may bio-accumulate at all levels of aquatic and terrestrial food chains. Intestinal absorption of Cd is low and biomagnification through the food chain may not be significant (Sprague, 1986). The periwinkle and fish accumulated Cd to about the same extent. The high concentration of Cd in the periwinkle could be due to the bottom-dwelling and bottom-feeding habits which are likely to ingest and contact considerable Cd-laden sediment. However, some of the fish and periwinkle caught in some locations had Cd content higher than the WHO guideline value of 2 µg/g (Kakulu et al., 1987) while they all exceeded the European Commission recommended value of 0.1 µg/g. The concentration recorded in periwinkle ranged from 2.1 to 5.1 µg/g and 2.1 to 5.6 µg/g for J07 and F08 samples respectively. The mean concentrations for all the locations were 3.42 and 3.79 µg/g for J07 and F08 samples respectively. The level of Cd in the periwinkle caught from the different locations did not vary significantly from one location to another during this investigation ($p < 0.005$). The concentration of Cd in fish ranged between 2.1 and 4.6 µg/g in the J07 samples and 2.7 and 5.4 in the F08 samples. The mean concentrations for all the locations are 3.62 and 3.93 µg/g for J07 and F08 samples respectively; the concentration of Cd in fish samples caught in the different locations did not vary significantly ($p < 0.005$). Cd contamination in inland and coastal environments could be attributed to discharge of contaminants containing Cd. Activities which may introduce Cd into these environments include electroplating and plastic manufacturing. Cd is a constituent of some pigments and significant quantities are released during the smelting of raw sulphide ores. Atmospheric depositions from non-ferrous metal mines and refineries, coal combustion and refuse incineration are other probable sources of Cd. The levels of Cd contamination in fish and periwinkle in this study have both been observed to be relatively higher than those (0.27 µg/g for fish and 0.56 µg/g for periwinkle) reported by Aderinola et al. (2009) suggesting increased industrial and domestic activities during the period these samples were taken.

The United States Environmental Protection Agency has classified Pb as being potentially hazardous and toxic to most forms of life (USEPA, 1986). This metal has been found to be responsible for quite a number of ailments such as chronic neurological disorders especially in foetuses and children (Tay et al., 2005). The mean concentration of Pb in fish was relatively higher than the concentration found in periwinkle. This could be due to the source of lead and its mode of availability to the organisms; while most trace metals originate from land-based particulate sources, Pb is mainly of anthropogenic origin as a result of emissions from automobile exhaust fumes. Such sources of Pb could be more available to fishes (caught from coastal waters, at the bank of which, there are vast trunks of roads, on which there are heavy vehicular traffics) than to

periwinkles, to which only sedimentary sources are available. The relatively higher Pb content of the fish could also be due to the presence of small bones embedded within the muscle tissue of the fish; these bones may serve as repositories for Pb (Nord et al., 2004). The concentration in the periwinkle ranged between 0.001 to 0.09 µg/g in the J07 and F08 samples respectively. The mean concentrations obtained for periwinkle caught in all locations were 0.04 and 0.05 µg/g for J07 and F08 samples respectively. The concentrations of Pb varied significantly from one location to the other during the period of this study ($p > 0.005$). The concentration of Pb in fish ranged from 0.01 to 0.80 µg/g in J07 samples and 0.01 to 1.10 µg/g for F08 samples. The mean concentrations obtained for Fish samples caught from all locations during the sampling periods were 0.11 and 0.16 µg/g for J07 and F08 samples respectively. The variation in the concentration of Pb in fish from different locations was very significant ($p > 0.005$). In many locations, the levels of Pb in fish exceeded the guideline value (0.4 µg/g) adopted by the European Commission.

Generally, there were significant variations in trace metal levels between fish and periwinkle. This may be due to the differences in physiology and feeding habits (Tay et al., 2005). Freidman statistical test showed that significant differences existed in the levels of metals in periwinkle and fish caught from the different locations while the influence of the period of sample collection on elemental levels tested by means of Kruskal Wallis (KW) test indicated no significant difference in the level of metals accumulated between the periods. Generally, the level sequence of the elements studied was, in most cases, similar in both locations and the periods of sample collection for both periwinkle and fish (Tables 1 to 4). The sequence was different for periwinkle and fish: for periwinkle the sequence was Fe > Mn > Zn > Cd > Ni > Cu > Cr > Pb > Hg (Cd and Zn exchanged places in F08 samples), for fish the sequence was Fe > Zn > Cd > Cu > Mn > Ni > Pb > Cr > Hg. Furthermore, the sequence of non-essential metals in both periwinkle and fish is worthy of note; it was as follows: Cd > Pb > Hg. Non-essential metals do not present any function for fish and periwinkle metabolism and are by consequent not regulated by the metabolism. The concentrations of Cd, Hg and Pb in fish and periwinkle may be used as an indication of the level of metal pollution of the water and sediment respectively from which they are caught (Kojadinovic et al., 2007). However, the observed heavy metals concentrations in these animals are below the recommended limits for human consumption (Davies et al., 2006; Olowu et al., 2010). The heavy metal concentration in the fish caught at the three locations could be attributed to anthropogenic metal sources affecting their environment. The public health implication of the research seems to show no possibility of acute toxicity of the heavy metals of edible

Rescaled Distance Cluster Combine

(a)

Rescaled Distance Cluster Combine

(b)

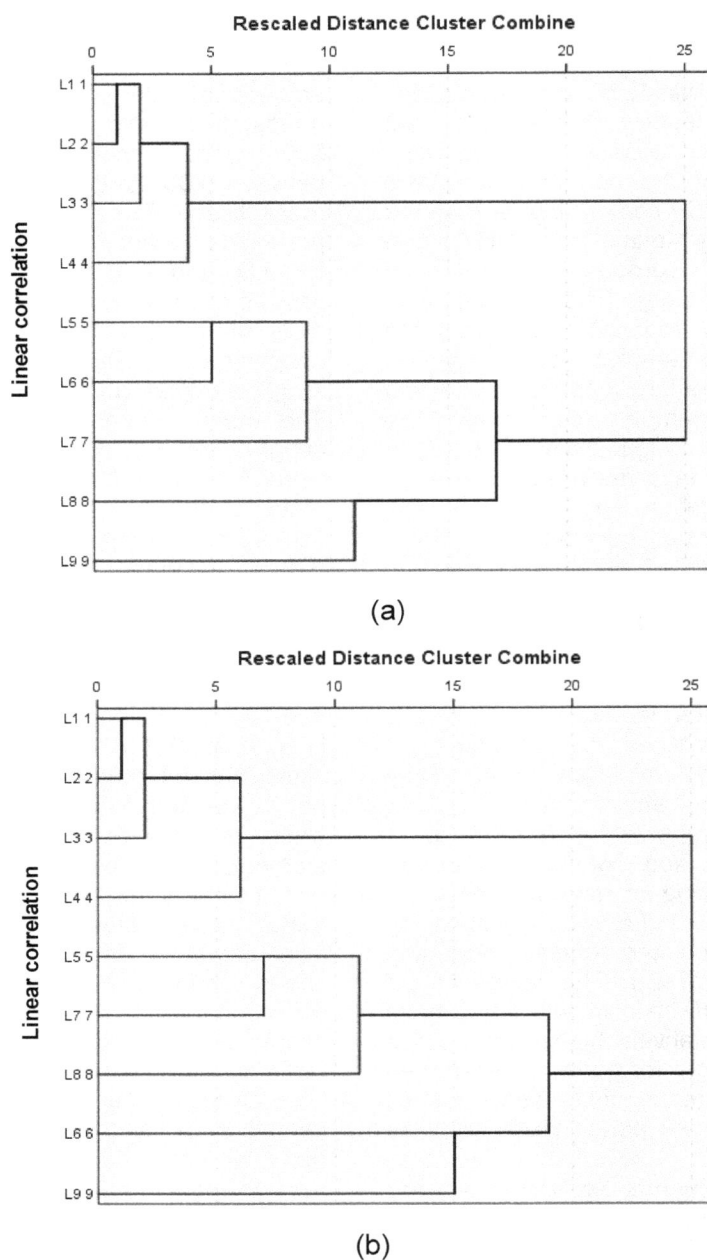

Figure 2. Dendrograms illustrating linear correlation between the heavy metals in periwinkles caught during the (a) dry season and (b) wet season.

fishes consumed (Oshisanya and Oshinsanya, 2009). Although levels of Cd, Cr and Pb, were within the normal minimum range allowable in diet of man, however, continual consumption could lead to accumulation with adverse health implications, while Cd has been linked to renal diseases and cancer (Kjellstroem, 1986) thus making periodic monitoring very imperative.

Pattern recognition and similarity in periwinkle and fish caught from the different locations during the periods of sample collection were determined by hierarchical cluster analysis based on the metals determined and presented in form of dendrograms (Figures 2). The periwinkles and fish from the different locations contained different levels of metals, however, some similarities existed between

(a)

(b)

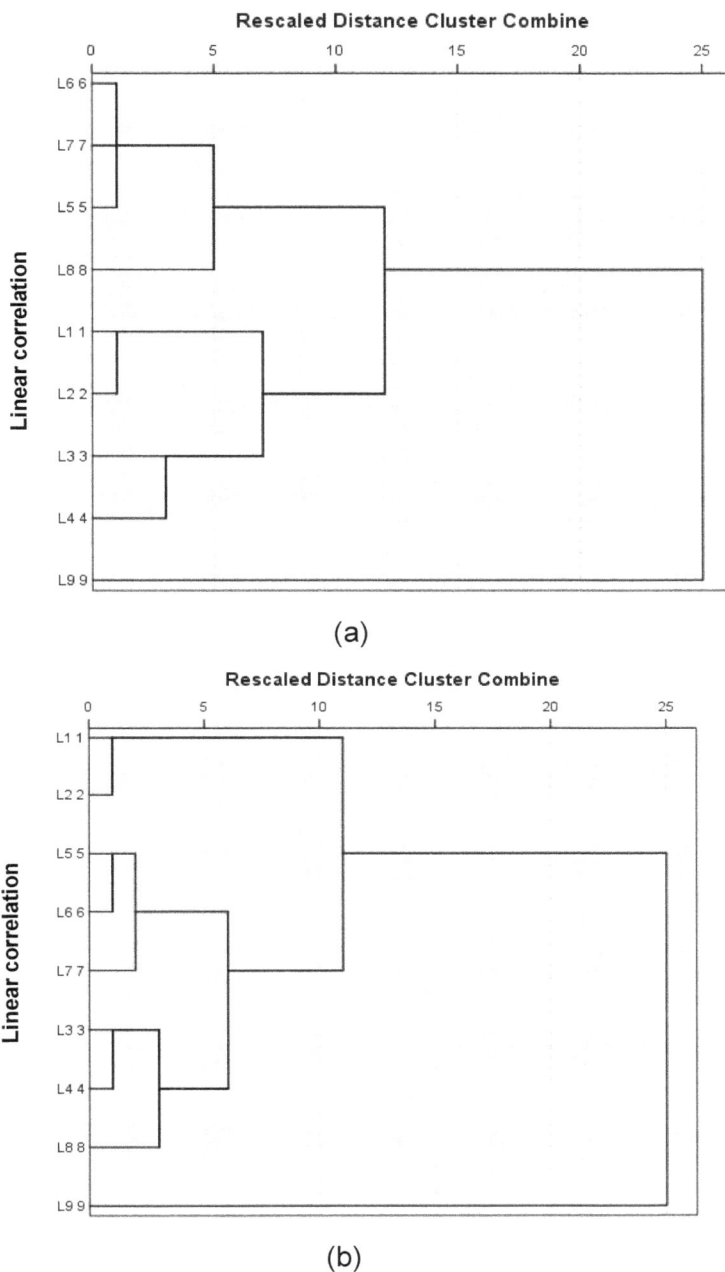

Figure 3. Dendrograms illustrating linear correlation between the heavy metals in fish caught during the (a) dry season and (b) wet season.

them. Dendrograms illustrating linear correlation between the heavy metals in periwinkles caught during the dry season and wet season (Figure 2) are made up of two clusters with greater similarities during the dry season. Cluster I consisting of periwinkles with relatively high levels of Fe and low in the other metals, while cluster II with relatively low levels of Fe and high in the other metals. Dendrograms illustrating linear correlation between the heavy metals in fish caught during the dry season and wet season consist of three main clusters (Figure 3).

The pattern was different for the different periods samples were collected; only fish from location L9 seem to be consistently different from the remaining locations. The dendrograms obtained was difficult to interpret probably due to the pelagic nature of fish.

REFERENCES

Aderinola OJ, Clarke EO, Olarinmoye OM, Kusemiju VA (2009). Heavy metals in surface, sediments, fish and periwinkle of Lagos Lagoon. Am. Eurasian J. Agric. Environ. Sci., 5(5): 609-617.

Alabaster JS, Lloyd R (1980). Water Quality Criteria for Fish, 2nd edn. Butterworths, London, pp. 35-68.

Ayenimo JG, Adeeyinwo CE, Amoo IA, Odukudu FB (2005). A preliminary investigation of heavy metals in periwinkles from Warri River, Nigeria. J. Appl. Sci., 5(5): 813-815.

Burger J, Gochfeld M (2005). Heavy metals in commercial fish in New Jersey. Environ. Res., 99(3): 403-412.

Canli M, Atli G (2003). The relationships between heavy metal (Cd, Cr, Cu, Fe, Pb, Zn) levels and the size of six Mediterranean fish species. Environ. Pollut., 121(1): 129-136.

Chindah AC, Braide SA, Amakiri J, Chikwendu SON (2009). Heavy Metal Concentrations in Sediment and Periwinkle –*Tympanotonus fuscastus* in the Different Ecological Zones of Bonny River System, Niger Delta, Nigeria. Open Environ. Pollut. Toxicol. J., 1: 93-106.

Davies OA, Allison ME, Uyi HS (2006). Bioaccumulation of heavy metals in water, sediment and periwinkle (*Tympanotonus fuscatus var radula*) from the Elechi Creek, Niger Delta Bioaccumulation of heavy metals in water, sediment and periwinkle (*Tympanotonus fuscatus var radula*) from the Elechi Creek, Niger Delta. Afr. J. Biotech., 5(10): 968-973.

Dural M, Lugalgoksu MZ, Akifözak A, Derici B (2006). Bioaccumulation of some heavy metals in different tissues of D*icentrarchus labrax*, S*parus aurata* and *Mugil cephalus* from the Amlik Lagoon of the Eastern Coast of Mediterranean (Turkey). Environ. Monit. Assess., 118: 65-74.

Goering PL, Waalkes MP, Klaassen CD (1994). Toxicology of metals. In Zinc supplement overdose can have toxic effects. (R. Hess and B. Schmidt, ed.). J. Paediatr. Haematol. Oncol., 24: 528-584.

Heath AG (1987). Water Pollution and Fish Physiology, CRC Press, Inc. Boca Rotan, Florida, p. 245.

Hornung H, Kress N (1991). Trace elements in offshore and inshore fish from the Mediterranean coast of Israel. Toxicol. Environ., 31/3: 135-145.

Kakulu SE, Osibanjo O, Ajayi SO (1987). Trace metal content of fish and shellfishes of the Niger Delta of Nigeria. Environ. Int., 13: 247-251.

Kalay M, Ay Ö, Canli M (1999). Heavy metal concentration in fish tissues from the Northeast Mediterranean Sea. Environ. Cont. Toxicol., 63: 673-681.

Kojadinovic J, Potier M, Le Corre M, Cosson RP, Bustamante P (2007). Bioaccumulation of trace elements in pelagic fish from the Western Indian Ocean (and all the references therein). Environ. Pollut., 146: 548-566.

Olowu RA, Ayejuyo OO, Adewuyi GO, Adejoro IAB, Denloye AA, Babatunde AO, Ogundajo AL (2010). Determination of Heavy Metals in Fish Tissues, Water and Sediment from Epe and Badagry Lagoons, Lagos, Nigeria. E-J. Chem., 7(1): 215-221.

Rosenberg GR, Costlow JD (1976). Synergistic effects of cadmium and salinity combined with constant and cycling temperatures on the larval development of two estuarine crab species. Mar. Biol., 38: 219-303.

Singh BR, Narwal RP (1984). Plant availability of heavy metals in a sludge-treated soil: II metal extractability compared with plant metal uptake. Environ. Qual., 13: 344-348.

Tay C, Asmah R, Biney CA (2005). Trace Metal Concentrations in Commercially Important Fishes from some Coastal and Inland Waters in Ghana. West Afr. J. Appl. Ecol., 13: 1-17.

United States Environmental Protection Agency (USEPA) (1986). Air Quality Criteria for Lead. Research Triangle Park, NC: US. Environmental Protection Agency, Office of Research and Development, Office of Health and Environmental Assessment, Environmental Criteria and Assessment Office. EPA 600/8-83-028F.

WHO (2004). Guidelines for drinking-water quality, 3rd edn. World Health Organization, Geneva. p.5-25

Yalçın T, Mustafa T, Aysun T (2008). Assessment of heavy metals in two commercial fish species of four Turkish seas. Environ. Monit. Assess., 146: 277–284.

14

Assessment of heavy metals bioaccumulation by *Eleusine indica* from refuse dumpsites in Kaduna Metropolis, Nigeria

Abdallah S.A[1]., Uzairu A[2]., Kagbu J. A[2]. and Okunola, O. J[3]

[1]Hussaini Adamu Federal Polytechnic, Kazaure, Jigawa State, Nigeria.
[2]Department of Chemistry, Ahmadu Bello University, Zaria, Nigeria.
[3]National Research Institute for Chemical Technology, Basawa, Zaria, Nigeria.

The levels of bioaccumulation of heavy metals (Pb, Cr, Zn, Cd, Mn and Cu) in *Eleusine indica* and waste soil samples from ten refuse dumpsites located in Kaduna Metropolis was assessed using atomic absorption spectrophotometer (AAS).The soil pH, electrical conductivity (EC), organic matter (OM) and available phosphorus (AP) were also determined. The results of analysis of the waste soils from the refuse dumpsites indicated that levels of the metals were in the range of; 131.93 to 205.18 mgkg^{-1} (Pb), 27.13 to 94.198 mgkg^{-1} (Cr), 259.30 to 354.708 mgkg^{-1} (Zn), 27.23 to 45.498 mgkg^{-1} (Cd), 151.68 to 227.568 mgkg^{-1} (Mn) and 42.09 to 132.11 mg kg^{-1} (Cu). The concentration recorded for *E. indica* ranged widely from 116.40 to 239.74, 6.97 to 24.84, 86.94 to 261.40, 5.10 to13.24, 90.22 to 318.51 and 41.11 to 103.84 mgkg^{-1} for Pb, Cr, Zn, Cd, Mn and Cu respectively. The results of soils pH, EC, OM and AP were found to be 7.85 to 8.60 and 0.54 to 3.22 S cm^{-1}, 2.77 to 6.32% and 42.11 to 175.55 mgkg^{-1} respectively. The bioaccumulation factors (BCF) of these metals were less than unity for all the metals except for Pb. The trend in bioaccumulation for metals in *E. indica* followed the sequence: Zn>Cu>Pb>Mn>Cd>Cr. Though the results indicated low bioaccumulation of metals by *E. indica*, however, further dumping of toxic waste could lead to toxicity to man through the food chain.

Key words: Waste soils, Refuse dump, Metals, Kaduna.

INTRODUCTION

Rapid population growth and urbanization in developing countries have led to the generation of enormous quantities of solid wastes and consequential environmental degradation (Nagendran et al., 2006). In many Nigerian cities, due to inadequate planning in the face of this rapid population growth, expectedly this put immense pressures on available public infrastructure, leading to deterioration in environmental quality and decline in the quality of life for most urban dwellers. Because of poor planning of waste collection, coupled with widespread poverty and limited awareness of the likely implications of poor waste management in the

environment by the urban populace, refuse is thrown unto roadways, and pedestrian walkways and sometimes dumped in open gutters (Okpoechi, 2007).

Furthermore, an estimated 7.6 million tons of municipal solid waste is dumped per day in developing countries (Nagendran et al., 2006). Recent studies have revealed that wastes dumpsites can transfer significant levels of toxic and persistent metals into the soil environment and eventually these metals are taken up by plants parts and transfer same into the food chain (Benson and Ebong, 2005). Consequently, higher soil heavy metal concentration can result in higher levels of uptake by plants, even though the rate of metal uptake by crop plants could be influenced by factors such as metal species, plants species, plant age and plant part (Ebong et al., 2008). Several literatures have indicated that most

---

*Corresponding author. E-mail: okunolaoj@gmail.com.

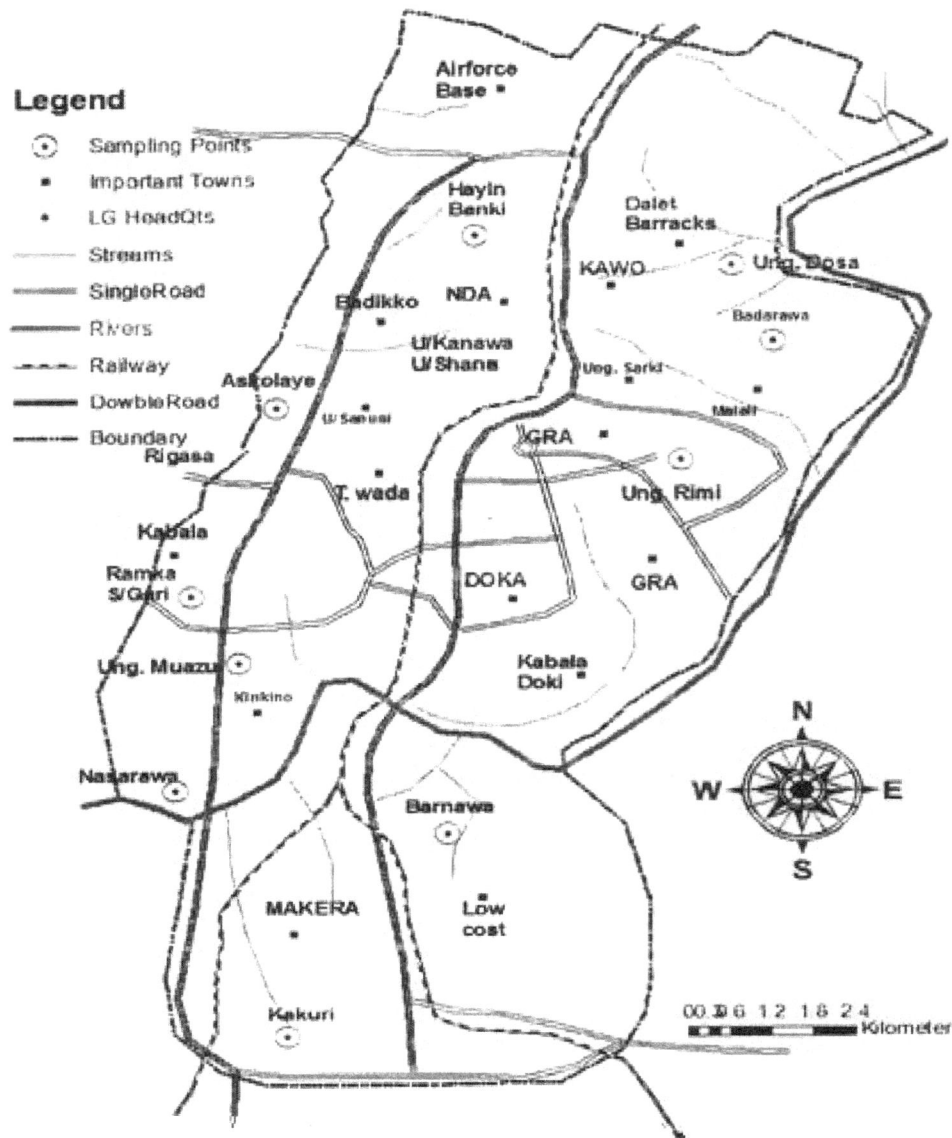

Figure 1. Map of Kaduna Metropolis showing the sampling.

of the dumpsites in the urban areas are used as fertile soils for the cultivation of some fruits and vegetables (Amusan et al., 2005; Okorokwo et al., 2005; Ebong et al., 2008). Some farmers, due to lean economic resources, resort to the use of decomposed parts of the dumpsites as manure to improve their farm yields. The cultivated plants take up these metals either as mobile ions in the soil solution through their roots or through their leaves thereby making them unfit for human consumption (Amusan et al., 2005). Based on this analogy, the present study is aimed with the objectives to determine the levels of Pb, Cr, Zn, Cd, Mn and Cu in soils and plant samples (*Eleusine indica*) from refuse dumpsites in Kaduna metropolis and evaluate the bioaccumulation of the metals in *E. indica*.

MATERIALS AND METHODS

Sampling sites

Kaduna metropolis is the capital of Kaduna State occupying the central portion of Northern Nigeria with location matrix of Latitude 10.52°N and Longitude 7.44°E. Annual temperature varies between 29 to 38.6°C. It is an administrative, industrial, a veritable commercial center and a functional urban area. It ranks second only to Kano in Northern Nigeria in terms of population (about 4,000,000 residents) (Achi et al., 2011). The study areas comprised of ten refuse dumpsites selected from four Local government areas within the metropolis which include: Kaduna South, Kaduna North, Chikun and Doka Local Government areas respectively as shown in Figure 1. The refuse dumpsites selected and their codes were: Badarawa (BD), Askolaye (AK), Ramka Sabon Gari (RSG), Unguwar Dosa (UD), Kakuri (KKR), Unguwar Rimi (UR), Nasarawa

Table 1. Mean (mgkg^{-1}±SD) % recoveries of heavy metals from spiked samples.

Metal	Soil Sample	*Eleusine indica*
Pb	90.11±0.23	86.57±0.47
Cr	89.10±0.15	95.25±0.09
Zn	91.51±0.42	97.94±0.13
Cd	106.22±0.29	101.61±0.68
Mn	96.07±0.34	104.92±0.24
Cu	101.07±1.32	103.85±0.15
Range	89.10 – 106.22	86.57 – 104.92

(NS), Hayin Banki (HB), Barnawa (BN) and Unguwar Mu'azu (UM). Control samples were equally taken 20 m away from the refuse dumpsites and the results were presented side by side with those of the impacted refuse dumpsites.

Sample Collection

The areas to be sampled in each sampling station were divided into four quadrants, each 5m^2 and five plant samples (*E. indica*) were collected by uprooting the whole plant from each quadrant in a diagonal basis following the method described by Nuonamo et al. (2002). The leafy *E. indica* in prime condition were bagged in paper bags and transported to the laboratory for pretreatment and subsequent analysis. Soil samples were collected in the same quadrants as the *E. Indica* from the dumpsites. Soil and plant samples from the background (20 m away from dumpsites) were equally collected. The background sites were located on the up-slope of each dumpsite. Three composite samples of the top soil (0 to 15 cm) were also taken from each of the quadrants. Each composite soil sample comprised 9 core soil samples (taken also in diagonal basis) bulked and homogenized before sub-sampling for laboratory analysis.

Sample preparation and pre-treatment

In the laboratory, the plant samples were thoroughly washed under running tap water to remove any attached soil particles and rinsed with double distilled water. Stainless steel scissors was used to cut the plant samples into very small portions and placed in large clean crucibles where they were oven dried at 105°C for 48 h. The dried crisp plant samples were then pulverized to powder using clean acid washed mortar and pestle and passed through 2 mm sieve. The ground plant samples were collected in labeled polythene bags and placed in desiccators to attain constant weight. The digestion of the plant sample was carried out according to the procedure described by Awofolu (2005). 0.5 g of the sieved plant sample was accurately weighed into 100 ml beaker. A mixture of 5 ml concentrated trioxonitrate (V) acid and 2 ml perchloric acid was added and this was digested on low heat using hot plate until the content was about 2 ml. The digest was allowed to cool and filtered into 50ml standard flask using Whatman Filter Paper No. 42. The beaker was rinsed with small portions of double distilled water and filtered into the flask. Triplicate digestions of each sample together with blank were carried out. The metal content was assessed using spectrophotometer (Model VGB210, Buck Scientific).

The soil samples from each site were air dried, crushed and passed through a 2 mm sieve. The fine earth fraction was retained for analysis. 5 g of the soil samples was placed in 100 ml beaker. 3 ml of 30% hydrogen peroxide was added following the procedure

described by Shriadah (1999). This was left to stand for 60 min until the vigorous reaction ceased. 75 ml of 0.5 M HCl solution was added and the content heated gently at low heat on hot plate for about 2 h. The digest was then filtered into 50 ml standard flask. Triplicate digestions of each sample together with blank were carried out. Calibration and measurement of metals were done on atomic absorption spectrophotometer, Model VGB210 Buck Scientific.

Physico-chemical parameters of soils

The pH and electrical conductivity (EC) of the soils were determined using Hanna Model No 111991000 using soil sample to distilled water 1:2 on a volume basis in accordance with the method described by Hendershot et al. (1993). Organic matter was determined based on the procedure described by Walkley and Black (1934). Available phosphorus was determined by methods described by Udo and Ogunwale (1978).

Quality assurance

Spiking experiment was conducted on the predigested soil samples and plants samples in order to ensure the reliability of the methods. This was achieved by spiking 0.5 g of the predigested *E. indica* samples with multi element standard solutions. 0.5 g each of the samples were weighed into a 100 ml beaker and spiked with 30 ml portions of multi element standards solution (0.5 mgkg^{-1}, Cd, Pb and Cr and, 5 mgkg^{-1} Zn, Cu and Mn) (Awofolu, 2005). This was then digested in triplicates. The triplicate digestion was done together with blank. Measurement of metals was done on atomic absorption spectrophotometer, VGB210 Buck Scientific model. The soil samples were similarly treated as the plant samples. Concentrations of metals in spiked and unspiked sample were used to calculate percentage recovery in order to validate the analytical procedure.

Statistical data analyses were done using SPSS version 17 (SPSS Inc. Chicago, IL, USA). All means recorded were determined considering a level of significance of less than 5% ($p < 0.05$) at 95%.

RESULTS AND DISCUSSION

Quality assurance

Table 1 indicated the percent recoveries of the heavy metals in the soil and the *E. indica* for each metal investigated. Cd showed the highest recovery in the soil (106.22 ± 0.29 mgkg^{-1}) while Cr the least (89.10 ± 0.15 mgkg^{-1}). The trend in the percent recovery in the soil was: Cd>Cu>Mn>Zn>Pb>Cr. Percent recovery was higher in *E. indica* with Mn having the highest percent recovery (104.92 ± 0.24mgkg^{-1}) and Pb having the least (86.57 ± 0.47mgkg^{-1}). The trend of metal recovery in *E. indica* was Mn>Cu>Cd>Zn>Cr>Pb. The observed good percentage recoveries indicated the validity and reliability of the digestion method and the AAS analysis adopted for this research.

Soil chemical characteristics

The results of some selected soil chemical characteristics

Table 2. Chemical parameters in waste soils of refuse dumpsites.

Sampling stations	pH	OM (%)	AP (mg/kg)	Heavy metal (mg/kg)					
				Pb	Cr	Zn	Cd	Mn	Cu
BD	8.60±0.14	3.68±0.04	91.24±2.14	143.31±0.02	29.20±0.02	354.70±0.50	30.77±0.21	179.04±0.10	70.12±0.10
AK	8.15±0.07	5.27±0.08	141.02±1.44	131.93±0.17	28.85±0.13	294.09±0.06	27.55±0.09	157.01±0.11	68.62±0.21
RSG	8.45±0.21	4.72±0.20	50.88±0.18	205.18±0.05	27.13±0.11	295.65±0.04	37.25±0.03	151.68±0.05	61.42±0.00
UD	8.10±0.14	5.04±0.25	150.00±0.47	154.25±0.05	29.61±0.04	282.50±0.23	27.23±0.01	158.02±0.02	49.37±0.02
KKR	8.1±0.15	2.87±0.11	85.86±.17	154.95±0.16	36.15±0.02	259.30v0.00	28.41±0.12	227.56±0.17	46.40±0.01
UR	7.95±0.07	4.04±0.05	77.56±0.79	148.02±0.02	28.29±0.02	283.28±0.14	38.41±0.13	168.28±0.21	42.09±0.09
NS	8.1±0.14	6.07±0.03	42.11±0.15	158.76±0.45	35.35±0.28	299.09±0.10	28.36±0.08	166.66±0.06	53.94±0.12
HB	7.85±0.07	6.32±0.08	82.06±0.08	176.00±0.12	48.19±0.02	287.39±0.13	45.49±0.28	193.37±0.14	132.11±.08
BN	8.05±0.07	4.89±0.28	175.55±0.78	161.41±0.14	48.19±0.07	289.55±0.18	31.26±0.05	170.10±0.00	88.50±0.10
UM	8.45±0.79	2.77±0.01	104.05±1.35	140.02±0.26	94.19±0.03	304.23±0.06	33.41±0.07	165.66±0.00	72.27±0.14
Mean ± sdv	8.18±0.24	4.57±1.22	100.03±43.17	157.38±20.82	40.52±20.43	294.98±24.32	32.81±5.93	173.74±22.32	68.48±26.41
Range	7.85-8.60	2.77-6.32	42.11-175.55	131.93-205.18	27.13-94.19	282.50-354.70	27.23-45.49	151.68-227.56	42.09-132.11
CONTROL	7.19±0.62	2.60±1.16	41.74±22.88	7.95±5.47	6.95±5.13	64.07±18.35	5.19±2.17	86.12±29.74	53.18±18.94
USEPA (1986) mgkg⁻¹	-	-	-	30 - 300	100	300	3.0	100-300	250

were summarized in Table 2. The pH values in the study areas ranged from 7.85 to 8.60 suggesting that all the soils were basically alkaline in nature. The control areas were basically alkaline in nature. The control areas depicted mean pH value of 7.19. Soil pH, nature of soil and climatic changes affect the rate of uptake of metals by plants (Alloway and Ayres, 1997). Metal mobility has been shown to decrease with increasing soil pH due to precipitation of hydroxides, carbonates or formation of insoluble organic complexes (Smith et al., 1996). Hence, high pH values observed in the present study could lead to decreased mobility of metals in the soil.

The electrical conductivity of the soils ranged from 0.54 to 3.22 Scm⁻¹ The control areas have EC mean value of 1.10 Scm⁻¹. The organic matter content of waste soils from the dumpsites ranged from 2.77 to 6.32% while the mean organic matter

for the control areas was 2.60%. The waste soils from the refuse dumpsite have higher organic matter content than the control areas. This observation agrees with the work of Bamgbose et al. (2000). The organic matter content of the waste soils from the refuse dumpsites ranged from 2.77 to 6.32% while the mean organic matter for the control areas was 2.60%. Researches have shown that among other factors such as presence of dolomite and phosphates, organic matter in soils reduce the concentration of metals by precipitation, adsorption and complexation (Mench et al., 1994, Chen and Lee, 1997) and thus making them unavailable to the plants. In the present study, the refuse dump soils were observed to have higher levels of organic matter (2.77 to 6.32%) and available phosphorus (phosphates) (42.11 to 175.55 mgkg⁻¹) which

reduced the concentrations of the metals absorbed by plant refuse dump soils than the control areas.

The levels of concentration of the six metals investigated in waste soils from the refuse dumpsites as shown in Table 2 revealed the level of Pb in the study area ranged from 131.93 – 205.18mgkg⁻¹ while Cr and Zn ranged from 27.13 to 94.19 and 259.30 to 354.70 mgkg⁻¹ respectively. Cd has the least concentrations compared to the other metals with a range of 27.23 to 45.49 mgkg⁻¹. Other ranges of metal concentrations were 151.68 to 227.56 mgkg⁻¹ for Mn and 42.09 to 132.11 mgkg⁻¹ for Cu. The mean concentrations of the metals from the study and control areas (side by side) respectively were: Pb: 157.38, 40.52, Cr: 7.95, 6.95, Zn: 294.98, 64.07, Cd: 32.81, 5.19, Mn: 173.74, 86.12 and Cu: 68.48,

Table 3. Correlation matrix between metals in waste soils of refuse dumpsites.

Metal	Pb	Cr	Zn	Cd	Mn	Cu
Pb	1					
Cr	-0.210	1				
Zn	-0.190	0.030	1			
Cd	0.507*	0.162	-0.040	1		
Mn	-0.060	0.039	-0.320	0.068	1	
Cu	0.238	0.334	0.140	0.612*	0.126	1

*indicate significant at p < 0.05

Table 4. Concentrations of heavy metals (mgkg^{-1} dw) in *E. indica* in refuse dumpsites.

Sampling stations	Metals					
	Pb	**Cr**	**Zn**	**Cd**	**Mn**	**Cu**
BD	181.64±0.19	18.03±0.03	138.39±0.06	13.24±0.04	90.22±0.01	66.93±0.09
AK	173.18±0.06	16.32±0.04	216.35±0.01	8.02±0.02	108.01±0.02	61.05±0.05
RSG	148.03±0.03	17.42±0.14	261.40±0.01	13.08±0.03	212.06±0.17	51.63±0.13
UD	239.74±0.20	24.84±0.17	136.43±.02	5.10±0.01	160.16±0.05	41.11±0.12
KKR	128.34±0.11	15.67±0.11	103.33±0.04	12.94±0.06	318.51±0.05	60.00±0.11
UR	116.4±0.09	16.38±0.05	143.84±0.17	10.32±0.09	129.26±0.19	64.99±0.10
NS	163.48±0.02	9.67±0.06	86.94 ±0.16	8.02±0.01	213.28±0.05	103.84±0.17
HB	184.92±0.08	12.95±0.05	150.67±0.34	6.89±0.11	102.5±0.13	65.34±0.05
BN	192.42±0.03	6.97±0.03	229.85±0.22	8.10±0.01	194.64±0.02	60.39±0.05
UM	131.34±0.11	15.94±0.06	177.09±0.02	8.72±0.05	122.23±0.05	46.84±0.17
Mean±SD	165.95±36.82	15.42±4.86	164.43±56.13	9.44±2.84	165.09±70.60	62.21±16.92
Range	116.4-239.74	6.97-18.03	86.94-261.40	5.10-13.24	90.22-318.51	41.11-103.48
Ctrl	34.91±10.98	4.83±3.25	73.79±31.15	3.72±0.86	64.64±41.11	34.33±15.33
USEPA mgkg^{-1} (1986)	0.2–20*	0.03–14*	1-400*	0.1–2.4*	-	5 -20*

*Radojevic and Bashkin (2006).

53.18 mgkg^{-1}. This indicated that the dumpsites had significantly higher heavy metals burden than the control areas. This agrees well with the findings of past researchers (Bamgbose and Odukoya, 2000; Eddy et al., 2006; Ebong et al., 2008 and Uba et al., 2008). The profile of metal abundance in the study areas was: Zn>Mn>Pb>Cu>Cr>Cd. Among the six metals investigated, only Cd exceeded the EC (1986) permissible limit (2.0 - 3.0mg/kg).

Nickel-cadmium batteries, cadmium-pigments, ceramics, glasses, paints and enamels, cadmium coated ferrous and non-ferrous products, cadmium stabilized polyvinyl chloride (pvc) products, cadmium alloys, cadmium electronics or electronic compounds are among anthropogenic sources of Cd in the environment (Baldini et al., 2000). Values of 950 mg/kg Mn, 100 mg/kg Cr, and 75 mg/kg Zn were reported for uncontaminated soils (Sparks, 2003). However, the values of concentrations of metals from waste soils reported by Uba et al. (2008) were higher than those reported in this study. Statistical analysis performed using Pearson correlation as shown

in Table 3 indicated positive correlations among the metals except Pb vs Cr, Zn and Mn and Zn vs Cr and Mn which were negatively correlated. Significant correlations (p < 0.05) were observed between Pb vs Cd and Cd vs Cu. Positive correlation among the metals in soils samples could indicate common sources of the metals which could be related to known geochemical associations among them while negative or insignificant positive correlation between the metals indicate that the appearance of local high concentration for one metal by possible contamination does not necessarily indicate high values for other metals. It may also indicate different sources or biogeochemical behaviours (Okunola et al., 2008).

Heavy metal levels in plants

The mean levels of metals in *E. indica* are shown in Table 4. Pb ranged from 116.40 to 239.74 mgkg^{-1} while Cr and Zn ranged from 6.97 to 24.84 mgkg^{-1} and 86.94 to

Table 5. Correlation between metals in plant and waste soil from refuse dumpsites.

Parameter	Pb-soil	Cr-soil	Zn-soil	Cd-soil	Mn-soil	Cu-soil
Pb-plant	0.017	-0.229	0.161	-0.246	-0.268	0.288
Cr-plant	-0.110	-0.234	0.061	-0.161	-0.171	-0.400
Zn-plant	0.360	0.067	0.064	0.199	-0.527*	0.249
Cd-plant	0.209	-0.221	0.264	0.035	0.327	-0.309
Mn-plant	0.340	-0.162	-0.600*	-0.336	0.474	-0.421
Cu-plant	-0.038	-0.232	0.162	-0.066	0.110	-0.010

*Indicate significant at $p < 0.05$.

Table 6. Correlation between metals in plant from refuse dumpsites.

Parameter	Pb-plant	Cr-plant	Zn-plant	Cd-plant	Mn-plant	Cu-plant
Pb-plant	1					
Cr-plant	0.233	1				
Zn-plant	0.038	-0.079	1			
Cd-plant	-0.603*	0.018	0.082	1		
Mn-plant	-0.249	-0.191	-0.181	0.320	1	
Cu-plant	-0.163	-0.604	-0.482	0.032	0.121	1

*Indicate significant at $p < 0.05$.

261.40 mgkg^{-1} respectively. Cd had the least concentrations among the metals investigated in the plant. It had values that fluctuated between 5.10 to 13.24 mgkg^{-1}. Mn had range of values from 90.22 to 318.51 mgkg^{-1}g and Cu 41.11 to 103.84 mgkg^{-1}.

The relative abundances of the heavy metals as detected in the *E. indica* samples from the refuse dump sites followed the sequence: Pb>Mn>Zn>Cu>Cr>Cd. The levels of the metals in the plant were generally lower than those of the waste soils from the refuse dumpsites. It has been reported that high soil pH can stabilize soil toxic elements resulting in decreased leaching effects of the soils toxic elements (Li et al., 2005). This may explain the low absorbability of the elements from the soil solution and translocation into plant tissues.

The higher level of heavy metals in the dumpsite plants may also be attributed to the large amount of wastes disposed at the dumpsite. Studies have shown that crops harvested from soils of the refuse dump sites presented higher levels of the metals when compared to the those crops from the control sites. This may be interpreted that if the level of these metals in soils is significantly increased, the plants have the potential of showing increased uptake of the metals. This is also supported by studies that plants grown in soils possessing enhanced metal concentrations have increased heavy metal ion content (Alloway and Ayres, 1997; Amusan et al., 2005). High levels of heavy metals as revealed in this study; Pb (15.28 to 76.92 mgkg^{-1}), Cd (1.96 to 9.80 mgkg^{-1}), Zn (45.37 to 237.96 mgkg^{-1}) Mn

(84.00 to 132.00 mgkg^{-1}) and Cu (2.14 to 48.00 mgkg^{-1}) had already been reported in soils in roadside soils in Kaduna metropolis (Okunola *et al.*, 2008). Correlation calculations between metals in plants and waste soil as shown in Table 5 revealed mild positive correlation between Pb-plant vs Pb, Zn, Cu in soil, Cr-plant vs Zn-soil, Zn-plant vs Pb, Cr, Zn, Cd and Cu in soil, Cd-plant vs Pb, Zn, Cd and Mn in soil, Mn-plant vs Pb and Mn in soil, and Cu-plant in Zn and Mn in soil. It can therefore be inferred that with an increase in the amount of metals in soil due to percolation, the uptake of metals by plants also increases (Aksoy et al., 2000). Also, the positive relationships among both plants and soil content of metals might be a cause of metal toxicities to both plants and animals through their entry into food chain (Awofolu, 2005).

Also, correlation coefficients between the metals in plants are presented in Table 6. A perusal of the table revealed mild positive correlation between the different metals in plant. The positive correlation could indicate common sources of metals in plants.

Bioaccumulation factor (BCF)

The Bioaccumulation factor (BCF) is a competent technique developed to assess the level of the metal in the plant as a fraction of the soils total. Previous studies have indicated that the uptake of metals by plants differs from one metal to another, from one plant species to

Table 7. Bioaccumulation factor of heavy metals in *E. indica* from refuse dumpsites.

Sampling Sites	Metal					
	Pb	**Cr**	**Zn**	**Cd**	**Mn**	**Cu**
BD	0.86	0.19	0.75	0.30	0.78	1.24
AK	0.29	0.20	0.93	0.18	0.42	0.47
RSG	0.22	0.12	1.06	0.19	0.68	0.45
UD	1.27	0.41	0.48	0.08	0.18	0.41
KKR	0.54	0.10	0.48	0.37	1.06	0.63
UR	0.60	0.12	0.82	0.34	0.43	0.73
NS	0.37	0.19	0.48	0.26	0.13	1.11
HB	0.51	0.20	0.83	0.13	0.29	0.83
BN	0.96	0.10	1.12	0.12	0.25	0.26
UM	0.53	0.08	0.51	0.18	0.36	0.41
Control	0.60	0.50	1.00	0.62	0.70	0.34

another and from one dumpsite to the other (Amusan et al., 2005; Agyarko et al., 2010). The results from this study as shown in Table 7 indicated that the uptake of each metal differs from one dump site to another. This variation might be explained in terms of the available salts present in the soil which differ from one dumpsite to the other. The ranges of BCF for *E. indica* – soil were as follows: Pb: 0.22 to 1.27, Cr: 0.08 to 0.41, Zn: 0.48 to 1.12, Cd: 0.08 to 0.37, Mn: 0.13 to 1.06 and Cu: 0.26 to 1.24 respectively.

The general sequence of bioaccumulation of the metal in *E. indica* was: Zn>Cu>Pb>Mn>Cd>Cr. The mean BCF values for the soil - *E. indica* in the control areas were: 0.60, 0.50, 1.00, 0.62, 0.70 and 0.34 for Pb, Cr, Zn, Cd, Mn and Cu respectively. The mean BCF value for Zn in the control area was unity, which indicated that the concentration of the metal in the *E. indica* was equal to that of the soil. Generally, the BCF of the heavy metals in the control area were higher than those of the refuse dumpsites. This depicted higher bioavailability of the metals in the control areas than the refuse dumpsite. Amusan et al. (2005) reported that other factors apart from total soil metal concentration could influence the bioaccumulation factor (soil plant transfer ratios). Chambers and Sidle (1991) found that metal levels in plants highly vary when related to soil metal levels and according to Fleming and Parle (1977), the uptake of heavy metals varies widely depending on the plant species being studied. They also found out that metal uptake was controlled by such variables like pH, organic matter content and soil type. Generally most of the heavy metals were less available to plants under alkaline conditions than under acid conditions (Hess and Schmid, 2002). The results showed that the BCF value for Cu at BD and NS were greater than unity. This indicated that plant uptake of this metal at the sites were not restricted by pH or other parameters. Zinc also depicted BCF greater than unity at RSG and BN, Pb at UD and Mn at

KKR. The high level of these metals in the plant at these sites might be due to direct deposition and foliar absorption more than the translocation from roots to the upper part of the plant.

Conclusion

The results of this study have revealed high levels of heavy metals in the waste soils and *E. indica* from the metropolitan refuse dumpsites. The metals were however within the concentration limits set by USEPA (1986) except Cd which exceeded the set limit (3.0 mg/kg) in all the dumpsites. The result also depicted the concentrations of the metals in the waste soil samples to be higher than the plant samples. The soil-plant transfer ratios (BCF) were less than unity except for Pb (1.27) in UD, Zn (1.06 and 1.12) in RSG and BN, Mn (1.06) in KKR and Cu (1.24 and 1.11) in BD and NS respectively. The BCF for the investigated metals followed the sequence: Zn>Cu>Pb>Mn>Cd>Cr. Soil factors such as pH, electrical conductivity, organic matter and available phosphorus might have contributed to the low absorbability of the metals. The practice of using dumpsites for cultivation of crops should be avoided since plants can bioaccumulate heavy metals into food chain.

REFERENCES

Achi MM, Uzairu A, Gimba CE, Okunola OJ (2011). Chemical fractionation of heavy metals in soils around the vicinity of automobile workshops in Kaduna Metropolis, Nigeria. J. Environ. Chem. Ecotoxicol., 3(7): 184-194.

Agyarko K, Darteh E, Berlinger B (2010). Metal levels in some refuse dump soils and plants in Ghana. Plant Soil Environ., 56(5): 244-251.

Aksoy A, Sahcen U, Duman F (2000). *Robina pseudo-acacia L.* as a possible biomonitor of heavy metal pollution in Kayseri. Turk. J. Bot., 24: 279-284.

Alloway BJ, Ayres DC (1997). Chemical Principles of Environmental Pollution. 2nd edu. Black Academic and Professional Publication. Pp. 190-217.

Amusan AA, Ige DV, Olawale R (2005). Characteristics of soil and crops uptake of metal in municipal waste dumpsites in Nigeria. J. Hum. Ecol., 17: 167-171

Awofolu OR (2005). A survey of trace metals in vegetation, soil and lower animals along some selected major roads in metropolitan city of Lagos. Environ. Monit. Assess., 105: 431-447.

Baldini M, Stacdhini P, Cubadele F, Miniero R, Parodi P, Facelli P (2000). Cadmium in Organs and Tissues of Horses, Slaughtered in Italy. Food Addit. Contam., 17(8): 676-687.

Bamgbose O, Odukoye O, Arowolo TOA (2000). Earthworms as bio-indicators of metal pollution in dumpsites of Abeokuta, Nigeria. Revista de Biologia Tropical v. 48 n. J. San. Jose.

Benson NU, Ebong GA (2005). Heavy metals in vegetables commonly grown in a tropical garden ultisol. J. Sustain. Trop. Agric. Res., 16: 77-80.

Chambers J, Sidle P (1991). Fate of heavy metals in an abandoned lead-zinc tailings   pond: 1, Vegetation. J. Environ. Qual., 20: 745-758.

Chen ZS, Lee DY (1997). Evaluation of remediation techniques on two cadmium polluted soils in Taiwan. In: Iskandar A., Adriano D.C. (eds.): Remediation of Soils Contaminated with Metals. Science Reviews, Northwood. Pp. 209-223.

Ebong GA, Akpan MM, Mkpenie VN (2008). Heavy metal contents of municipal and rural dumpsite soils and rate of accumulation by *Carica papaya* and *Talinum triangulare* in Uyo, Nigeria. E-J. Chem., 5(2): 281-290.

Eddy NO, Odoemelen SA, Mbaba A (2006). Elemental composition of soil in some dumpsites. Electronic J. Environ. Agric. Food Chem., 5(3): 1349-1365.

European Commission (EC) (1986). European Commission, office for Official Publications of the European Communities, Luxemburg, Council Directive 66/27/EEC on the protection of environment and in particular of soil, when sewage sluge is used in Agriculture.

Fleming G, Parle P (1977). Heavy metals in soils, herbage and vegetables from an industrialized area west of Dublin city. Irish. J. Agric., 16: 35-48.

Hendershot WH, Lalande H, Duquette M (1993). Soil reaction and exchangeable acidity. In: carter. M. R. (Ed), Soil sampling and Methods of Analysis for Canadian Society of Soil Sciences. Lewis Boca Raton, FL, pp. 141-145.

Hess R, Schmid B (2002). 'Supplement overdose can have toxic effects'. J. Pediat. Heamatol. Oncol., 24: 582-584.

Li X D, Wong CSC, Thornton I (2005). Urban environmental geochemistry of trace metals. Environ. Pollut., 129: 113-124.

Mench M, Didier VL, Loffler M, Gomez A, Masson P (1994). A mimicked *in-situ* remediation study of metal-contaminated soils with emphasis on cadmium and lead. J. Environ. Qual., 23: 58-63.

Nagendram R, Joseph K, Esakku S, Visvanatha C, Norbu T (2006). Municipal Solid Waste Dumpsites to Sustainable Landfills. Development Insight October – December. Pp. 65-68.

Nuonomo I, Yemefack M, Tchienkuoa M, Njougang R (2002). Impact of Natural Fallow Duration on Top Soils Characteristics in Southern Cameroon. Nig. J. Res., 3: 52-57.

Okorokwo NE, Igwe JC, Onwuchekwe EC (2005). Risk and health implication of polluted soils for crops production. Afr. J. Biotechnol., 4(13): 1521-1524.

Okpoechi CU (2007). Municipal solid wastes management and its implications for sustainable development in Nigeria. Int. J. Environ., 5(1 &2): 124-133.

Okunola OJ, Uzairu A, Ndukwe GI, Adewusi SG (2008). Assessment of Cd and Zn in roadside surface soils and vegetations along some roads of Kaduna Metropolis, Nigeria. Res. J. Environ. Sci., 2(4): 266-274.

Shriadah MMA (1999). Heavy metals in mangrove sediments of United Arab Emirates shoreline (Arabian Gulf). Water Air Soil Pollut., 116: 523-534.

Smith CJ, Hopmans P, Cook FJ (1996). Accumulation of Cr, Pb, Cu, Ni, Zn and Cd in soil following irrigation with untreated urban effluents in Australia. Environ. Pollut., 94 (3): 317-323.

Sparks DL (2003). Environmental Soil Chemistry 3rd Edition. Academic Press.

Uba S, Uzairu A, Harrison GFS, Balarabe ML, Okunola OJ (2008). Assessment of heavy metals bioavailability in dumpsites of Zaria Metropolis, Nigeria. Afr. J. Biotechnol., 7(2): 122-130.

Udo EJ, Ogunwale JA (1978). Laboratory manual for the analysis of soils, plants and water samples. Department of Agronomy University of Ibadan. Pp. 45.

United States Environmental Protection Agency (USEPA) (1986). Quality Criteria for Water. United States Environmental Protection Agency. Office of Water Regulations and Standards. Washington D.C. 20460

Walkley A, Black IA (1934). An examination of the Detjare method for Determining soil organic matter and a proposed modification of the chromic Acid titration. Soil Sci., 37: 29-36.

15

Study of multi-resistance to heavy metals, antibiotics and some hydrocarbons of bacterial strains isolated from an estuary basin

K. Lyamlouli, K. Kharbouch, A. Moutaouakkil and M. Blaghen*

Laboratory of Microbiology, Pharmacology, Biotechnology and Environment, Faculty of Sciences, Université Hassan II Ain-CHock, route Eliadida, B. P. 5366, Casablanca, Morocco.

Microorganisms contained in estuary water samples (Deltebre, Spain) have been the subject of several tests of resistance against various chemicals, such as heavy metals, hydrocarbons, and antibiotics. Isolates were plated (cultivated) on Trypticase Soy Agar plates and purified for further screening. The strains were extremely resistant to heavy metals, with peculiarly, high average minimal inhibitory concentration (18700 µmol/l for arsenic and 10600 µmol/l for lead), and they also showed that they were able to grow in the presence of significant concentrations of sodium chloride (more than 50 g/l), and an interesting resistance to hydrocarbons, and antibiotics. Results showed that the most resistant strains to all the tested pollutants belong to *Pseudomonas putida* and *Stenotrophomonas maltophilia*. The kinetics of growth in the presence of certain heavy metals (Arsenic (9600 µmol/l), Cobalt (1200 µmol/l), and Lead (4600 µmol/l)), showed that the isolates had a great ability to multiply in presence of such growth inhibitors, even in high concentrations. The study of growth of the isolated strains in the presence of aromatic hydrocarbons (Benzene (4 mmol/l), Toluene (4 mmol/l), Naphthalene (6 mmol/l)) as the sole carbon source was also carried out. Isolates showed a significant sensitivity in the presence of high concentrations of hydrocarbons however, the proliferation was surprisingly fast in the presence of naphthalene. The isolated strains have shown that they can be of considerable significance, regarding the remediation of some heavy metals and aromatic compounds in heavily polluted sites.

Key words: Heavy metals resistant bacteria, antibiotics resistance, hydrocarbons resistance, salt tolerance, bioremediation, seawater.

INTRODUCTION

Industrialization and human activities in general, shoulder a great responsibility for turning our environment into dumping sites for waste materials. As a result many water resources have been rendered unwholesome and hazardous to man and various ecosystems (Bakare et al., 2003). Many chemical substances are water soluble and therefore easy to gain access to various water systems forming a threat for the fauna and flora in these systems. Pollutants can be transported by water at all stages of the water cycle. The fauna and flora are also affected by the accumulation of pollutants in the tissues of plants and other terrestrial and aquatic animals, and more generally along the food chain, needless to say that it is seriously prejudicial to the natural balance of various ecosystems. In consequence, human welfare will be affected directly.

Estuaries are large areas where mass exchanges are done between drainage basins and the sea, and thus they have been greatly the focus of scientific research (Muxika et al., 2005; Sarkar et al., 2007; White and Wolanski, 2008; Wolanski et al., 2008; Wolf et al., 2009). They are among the most exposed areas to different types of pollution, especially pollution due to oil spills (Anupama and Padma, 2009), whether it is a direct or indirect contribution due to industrialization and urbanization. This is why living beings that inhabit these environments are generally exposed to multi-elemental pollution (hydrocarbons, antibiotics, dyes and heavy

*Corresponding author. E-mail: karim-2l@hotmail.fr.

metals); For example, a large number of textile dyes are sold each year, with approximately 2% released directly in various water sources and 10% lost during the coloring process (Pearce et al., 2003). Color is one of the most obvious indicators of water pollution, and discharge of highly colored synthetic dye effluents can cause damage to the receiving water bodies (Nigam et al., 1996).

In view of the fact that several authors have attested of the strong antibiotics, hydrocarbons and heavy metals resistance concerning bacteria isolated from a natural environment (Cohen, 1992; Levy, 1998; Levy and Marshall, 2004), the aim of our study was to contribute to the research and the isolation of bacteria with halophilic properties, in a relatively sensitive natural environment, and study the tolerance in the presence of several types of pollutants.

METHODS

Sampling sites

The description of sampling sites is represented in Appendixes A, B and C. Deltebre is located between Barcelona and Valencia, near the town of Tortosa in the south of the Tarragona province (southern part of Catalonia). This place is very special because it is part of a delta, where the Ebre (Spain's main river) flows into the Mediterranean Sea (hence the name Deltebre).

The sampling was carried out, more specifically, in the old abandoned Deltebre saline; characterized primarily by their high content of sodium chloride (NaCl) and their high biological activity (swamp).

Samples processing and bacteria isolation

Seawater and sediment samples were collected aseptically from an estuary basin in Spain. The samples were put on sterile tubes and conserved at 4°C.

The collected samples were diluted with sterile Distilled water, sown in TSA (Trypticase Soy Agar, Bio-born) and incubated at 37°C for 48 h; then a pure culture was obtained by successive isolation of colonies in the same media. Bacterial identification was done by biochemical analysis according to the standardized micromethod API 20E and 20 NE (Biomeriaux) (Filali et al., 2000).

Minimal inhibitory concentration (MICs)

Heavy metals

The liquid medium (Nutrient Broth), non-amended (controls) or amended by adding the metal element at different concentrations from stock solutions, was inoculated with uniform volume (100 µl) of cell suspensions of preculture of one night of each strain diluted to 1%, the minimal inhibitory concentration (MIC) is defined as the lowest concentration that causes no visible growth (Jennifer, 2001). For more accuracy, the MIC was determined using a Shimadzu UV-1800 spectrophotometer, set at a wavelength of 600 nm.

The resistance to heavy metals was tested using a concentration range, from 18.75 to 19.200 µmol/l.

The chosen dilution factor was half; nevertheless, to obtain results as accurate as possible, other concentrations were tested. According to the first obtained MIC value, we tested concentrations slightly lower, without using the dilution factor. For example, if the

first obtained MIC was 19200 µmol/l, the following concentration to be tested will be 17200 µmol/l, then 1520 0µmol/l, so on. Metals used includes: $AgNO_3$, $HgCl_2$, $CdCl_2$, $Pb(NO_3)_2$, $CoCl_2$, $CuSO_4$, KH_2AsO_4, $Ni(NO_3)_2$, $FeCl_3$, $Bi(NO_3)_3$, $ZnCl_2$ and $BaCl_2$.

Following the obtained results, our interest has focused on both the bacteria showing the highest resistance degree, in the presence of the tested heavy metals, in order to study the growth in the presence of antibiotics, hydrocarbons, and dyes.

Hydrocarbons and sodium chloride

The MIC determination was also carried out for the sodium chloride. Tubes containing 5 ml of nutrient broth with different concentrations of NaCl were inoculated with uniform volume (100 µl) of preculture of one night of each isolated strain. The tested concentrations were 5, 6, 8, 10, 15, 20, 25, 30, 40 and 50 g/l. Incubation was done at 37°C for 24 and 72 h. The MIC concerning Naphthalene, one of the most studied aromatic hydrocarbons due to its relatively high solubility in water and especially the ease of isolating, bacteria involved in its biodegradation (Mrozik et al., 2003), was also determined by using the same method already mentioned.

A minimum medium of the following composition: ($MgSO_4$ (0.1 g/l); KH_2PO_4 (1.36 g/l); $(NH4)_2SO_4$ (0.6 g/l); $CaCl_2$ (0.0 2 g/l); $MnSO_4$ (1.1 mg/l); $CuSO_4$ (0.2 mg/l); $ZnSO_4$ (0.2 mg/l); $FeSO_4$ (0.14 mg/l); NaCl (0.5 g/l), was used instead of Nutrient Broth. The same study was carried out concerning, Toluene and Benzene. The following concentrations (in mmol/l), for each hydrocarbon, were tested: Naphthalene: 0.25, 0.5, 1, 2, 3, 4, 7, 8, 14.5, 15, 16 mmol/l. Toluene and benzene: 0.5, 1, 1.75, 3, 3.5, 5, 6.5, 7, 7.5, 8, 14 mmol/l.

Antibiotics and dyes

The resistance evaluation (assessment) of the isolated strains, to various antibiotics was carried out on liquid medium (Mueller Hinton). Tubes containing 5 ml of liquid medium and various antibiotics at different concentrations (between 0.0625 and 16 mmol/l using a dilution factor of 1/2) were inoculated with 100µl of preculture of each isolated strain.

Tested antibiotics were: Oxacillin, Ceftriaxone, Ceporine, Erythromycin, Ampicillin, Fosfomycin, Rifampycin, Carbenicillin and Amphotorecin. To assess the ability of the isolated strains regarding the degradation and the discoloration of synthetic dyes, bacterial growth was followed in the presence of; one anthraquinonic (Cibacron blue) and two azoic ones (Azorubin and Blue trypan). Isolated strains were sowed into plates containing 15 g/l of agar supplemented with 100 mg/l of each tested dyes. Incubation was done at 37°C for 4 days.

Growth rate of the isolates

The growth rate of each isolated strain in the presence of heavy metals was also determined. A set of Erlenmeyer flasks containing 100 ml of nutrient broth and different concentrations of tested heavy metals, was inoculated with 1 ml of preculture of each strain.

A control for each isolate was carried out under the same conditions, without heavy metals addition.

Erlenmeyer flasks were incubated in a shaker incubator at 37°C and at 70 U/min. Bacterial growth was followed (attended) by measuring absorbance at 600 nm during 48 h. Only three metals were tested: lead, cobalt and arsenic.

The growth of the bacterial strains in the presence of aromatic hydrocarbons as sole carbon source, has also undergone extensive tests, we proceed by inoculation of Erlenmeyer flask containing 100 ml of minimal medium with the following composition: ($MgSO_4$.

(0.1 g/l); KH_2PO_4 (1.36 g/l); $(NH4)_2SO_4$ (0.6 g/l); $CaCl_2$ (0.02 g/l); $MnSO_4$ (1.1 mg/l); $CuSO_4$ (0.2 mg/l); $ZnSO_4$ (0.2 mg/l); $FeSO_4$ (0.14 mg/l); NaCl (0.5 g/l) in the presence of the following hydrocarbons concentrations:

S1: benzene (4 mmol/l), toluene (4 mmol/l), naphthalene (6 mmol/l).
S2: benzene (4 mmol/l), toluene (4 mmol/l) naphthalene (6 mmol/l).

A control blank containing the minimum medium supplied with 2% of glucose was prepared; the inoculated flasks were incubated at 37°C and agitated on a rotary shaker (150 rev/min) for 10-15 days. Absorbance was measured at 600nm.

RESULTS AND DISCUSSION

Bacterial diversity

Following the bacteriological analysis and microscopic observation, a total of 28 strains could be isolated. Bacteria that were the most tolerant in the presence of heavy metals (S1 and S2) were selected to undergo a battery of tests (resistance to antibiotics and hydro-carbons, and salt tolerance). API 20 analysis has been able to reveal the presence of two different strains: *Pseudomonas putida* (S1) and *Stenotrophomonas maltophilia* (S2).

Minimal inhibitory concentration of heavy metals, hydrocarbons and antibiotics

The obtained minimal inhibitory concentrations concerning heavy metals are represented in Table 1. The results, supported by the different MIC values, indicate that the strains are highly resistant regarding the tested heavy metals, compared to those listed in the literature (Blaghen et al., 1993; Seralathan et al., 2006), with respectively an average MIC of 18700 ± SD µmol/l and 10600 ± SD µmol/l for Arsenic and Lead, and 200 ± SD µmol/l for mercury. Naphthalene, benzene and toluene MIC are also reported in Table 1.

According to the obtained results, strains resistance toward hydrocarbons seems to be low. However, we have registered an average MIC of 15 ± SD mmol/l regarding naphthalene for S1. The obtained MICs regarding NaCl showed that all the strains are able to grow in the presence of large quantities of this element. We have obtained a value of 15 ± SD g/l for S2, and for S1 this value has exceeded 50 g/l. The MIC concerning antibiotics on liquid medium (Mueller Hinton) are represented in Table 1. All the strains were resistant to the selected antibiotics; the first signs of tolerance were clearly visible after only 18 h of incubation, the results were not surprising, since *Pseudomonas putida* and *Stenotrophomonas maltophilia* are known to exhibit a high resistance to various types of antibiotics. Microbial resistance to antibiotics is generally associated with a reduced penetration of these substances in the cell.

In most cases, genes assembled in plasmids protect bacteria against antibiotics, however, there are bacteria resistant to antibiotics and do not contain any plasmids, in this case the resistance depends more on mobile genetic elements called transposons. There are four mechanisms of resistance specified by plasmids: inactivation, impermeability, bypasses and altered target site; all occur in aquatic environments (Mudry, 2002).

Several studies reported that there would be some association between resistance to heavy metals and antibiotics, which was demonstrated by the analysis results. In fact, under conditions of metal stress, resistance to these two types of compound would help the microorganisms to adapt faster by the spread of resistance factors than by mutation and natural selection (Edward et al., 2009).

Dyes effect regarding strains growth, on solid medium after 3 days of incubation, showed that there wasn't any discoloration or bacterial multiplication; we have concluded that all the strains have no discoloration activity, neither degradation ability regarding the tested dyes.

Growth studies of the isolated bacteria

The growth curves of the strains in the presence of different concentrations of lead, arsenic and cobalt are shown in Figures 1 and 2. Figures 1 and 2 also showed that bacterial growth for S2 seems not much influenced and we assisted to a short lag phase, indicating that there is no effect on the growth properties. In presence of 9600 µmol/l of arsenic, we obtained a delayed lag phase for S1.

As reported by some authors, the results may suggest that the isolated bacterial strains are likely to be involved in the redox cycling of arsenic (Inskeep et al., 2007; Quéméneur et al., 2008). The oxidation of arsenite can either produce usable energy, or simply be a step in an eventual process of detoxification, nevertheless further study is necessary, in order to affirm this hypothesis.

The study of the isolated strains capacity to utilize aromatic hydrocarbons as an energy source was also done; the results are represented in Figures 3 and 4. Bacteria with aromatic hydrocarbons degradation pro-perties are generally isolated from soil samples, most of them belong to the genus *Pseudomonas*. The conducted study concerning the ability of the bacterial strains to grow in the presence of the selected hydrocarbons as sole carbon source showed that the strains could grow promptly in the presence of naphthalene, which was evident, referring to the obtained short lag phase concerning *Pseudomonas putida* (S1).

This relatively fast growth may be due to the use of naphthalene as a source of energy, implying a possible biodegradation of this elements, several studies have indicated that the bacteria involved in this degradation

Table 1. Minimal inhibitory concentration of NaCl, and the tested heavy metals, antibiotics, and hydrocarbons.

| Compounds | Bacterial strains | |
	S1	S2
As^{2+}	18700 ± 1000	18200 ± 1154,701
Pb^{2+}	10600 ± 1154,701	9100 ± 1000
Cu^{2+}	2300 ± 200	4800 ± 0
Co^{2+}	2400 ± 0	2400 ± 0
Ag^{2+}	625 ± 28,867	575 ± 50
Hg^{2+}	112,5 ± 25	200 ± 212,5
Cd	17,812 ± 1,875	150 ± 0
Ni^{2+}	2400 ± 0	562,5 ± 25
Cr^{2+}	287,5 ± 25	600 ± 0
Fe^{2+}	2400 ± 0	2400 ± 0
Ba^{2+}	4800 ± 0	4800 ± 0
Ceporine	7,875 ± 0,25	4 ± 0
Ceftriaxone	**	3,75 ± 0,5
Ampicillin	7,375 ± 0,478	O,625 ± 0,25
Oxacillin	10,25 ± 0,5	3,875 ± 0,25
Amphotorecin	**	4 ± 0
Erythromycin	11,625 ± 0,216	4 ± 0
Fosfomycin	3,75 ± 0,5	4 ± 0
Rifampycin	0,14 ± 0,07	0,156 ± 0,0625
Carbenicillin	3,25 ± 0,5	2 ± 0,408
Naphthalene(mmol/l)	15 ± 0,866	7,66 ± 0,577
Benzene (mmol/l)	3,5±0	1,5 ± 0,433
Toluene (mmol/l)	7,33±0,763	1,75 ± 0
NaCl (g/l)	More than 50	15 ± 0

The results are Means ± SD of quadruplicate tests concerning heavy metals and antibiotics, and triplicate tests regarding NaCl and hydrocarbons.
** The bacteria are sensitive to that antibiotic.

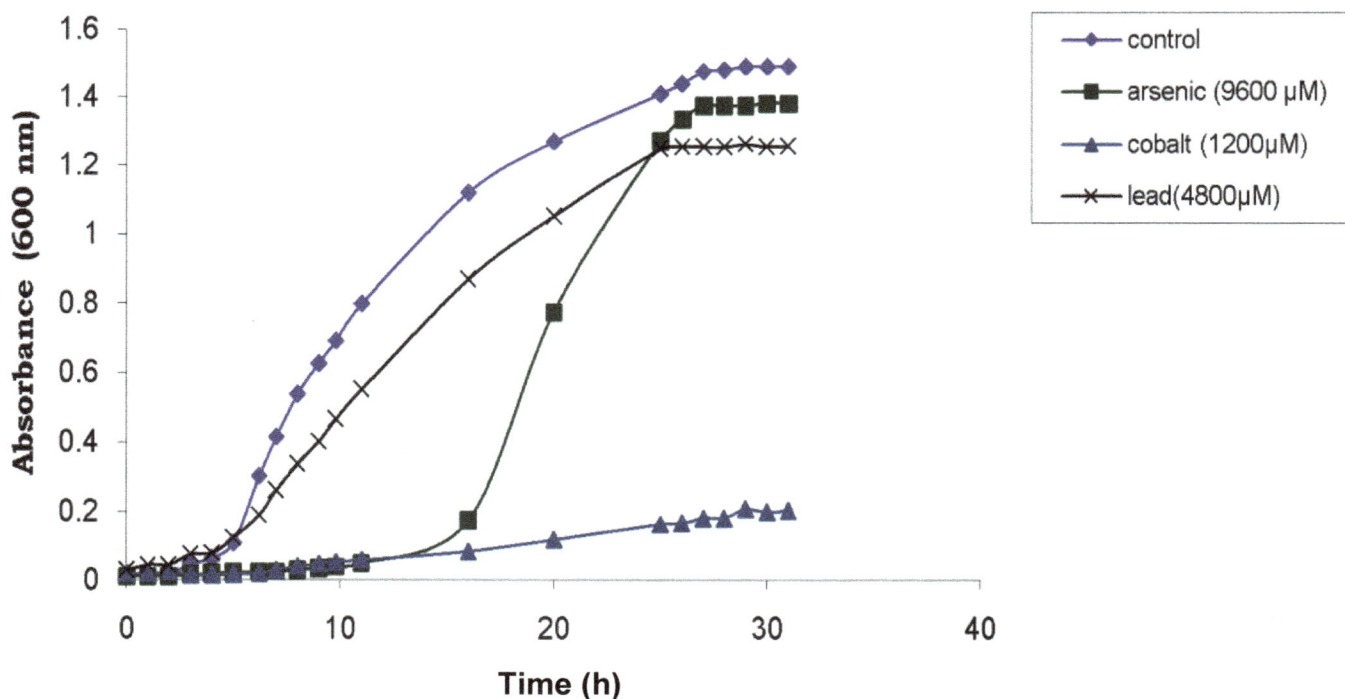

Figure 1. Arsenic, lead and cobalt effect on (S_1) growth, Incubation was carried out in aerobic conditions in a shaker incubator at 37°C, and at 70 U/min. Results are Means ± SD of triplicate tests.

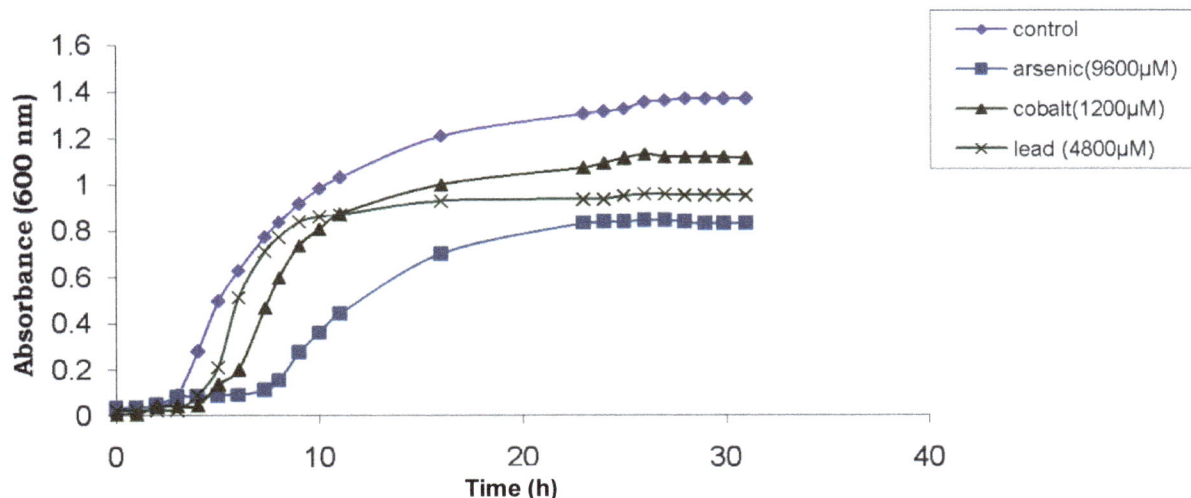

Figure 2. Arsenic, lead and cobalt effect on (S$_2$) growth Incubation was carried out in aerobic conditions in a shaker incubator at 37°C, and at 70 U/min. Results are Means ± SD of triplicate tests.

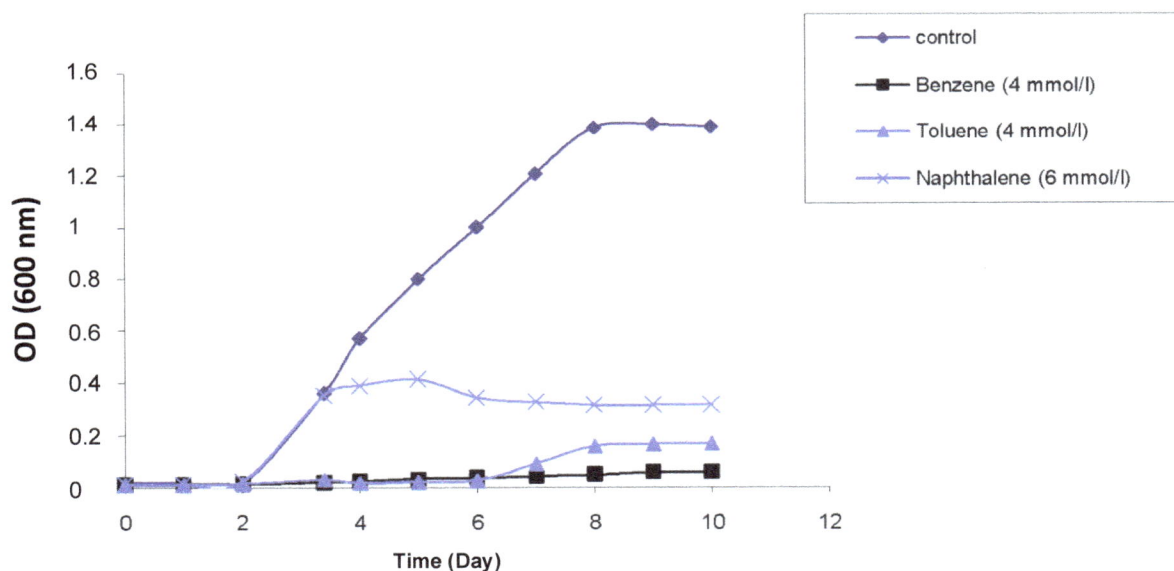

Figure 3. Growth of (S1) in the presence of hydrocarbons (naphthalene, toluene and benzene) as sole carbon source. Bacterial strain was incubated in 100 ml of minimum media containing the tested hydrocarbon, in a shaker incubator at 37°C, and at 150 Rev/min. Results are Means ± SD of triplicate tests. Results are Means ± SD of triplicate tests.

process, oxidize initially naphthalene by incorporating of both molecular oxygen into the aromatic molecule to form cis-1,2-dihydroxy-1,2-dihydronaphthalene.

In the case of *Pseudomonas putida* the Naphthalene dioxygenases acts as as multicomponent enzyme systems which are responsible for naphthalene cis-dihydrodiol formation (Mrozik et al., 2003).

In the presence of toluene, a low growth rate was reported concerning S1 after 6 days of incubation. In

order to treat wastewater, several studies have been carried out, using a variety of bacteria, among them *Pseudomonas putida* is certainly the most popular, however it appears that the degradation potential is limited by the concentration, which in our case may explain the relatively slow growth rate, in fact beyond certain values, bacterial growth tends to drop, due to the toxicity caused by the high concentration of substrate (Bordel et al., 2007).

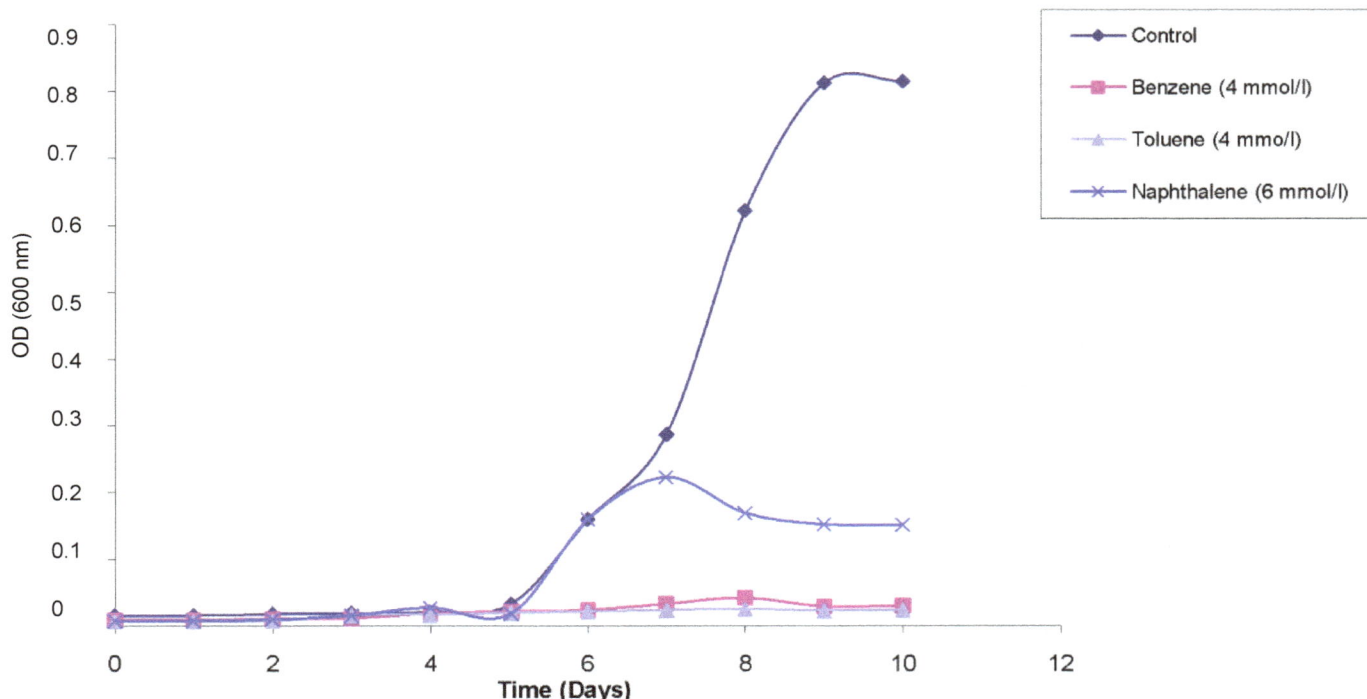

Figure 4. Growth of (S2) in the presence of hydrocarbons (naphthalene, toluene and benzene) as sole carbon source. Bacterial strain was incubated in 100 ml of minimum media containing the tested hydrocarbon, in a shaker incubator at 37°C, and at 150 Rev/min. Results are Means ± SD of triplicate tests.

Conclusion

From estuary basin water, the assorted strains have manifested by the experiments results that they were able to grow in the presence of high concentrations of heavy metals and NaCl, but also have showed an interesting degrees of tolerance against antibiotics, and some selected hydrocarbons.

The identification of these strains revealed the presence of *Pseudomonas putida* (S1) and *Stenotrophomonas maltophilia* (S2). Water is a precious commodity, more and scarcer, vital for the survival of humanity, and more generally for all species that inhabit our planet, its conservation is a duty that must be the top priority. The situation is critical, however, we should by no means deny, that efforts are being made constantly in order to propose methods and techniques more and more effective to fight against water pollution by various pollutants.

Our researches are included in this framework, with the aim of nurturing the current literature. The ability of the isolated strains to grow in the presence of heavy metals, could prove to be extremely useful in the treatment of waste water, where microorganisms are directly involved in the decomposition of organic matter, since in this biological process the heavy metals have often an inhibitory effect (Filali et al., 2000), thus the applied treatments can be optimized in order to incur a finer yield.

REFERENCES

Anupama M, Padma S (2009). Isolation of hydrocarbon degrading bacteria from soils contaminated with crude oil spills. Ind. J. Exp. Biol., 47: 760-765.

Bakare AA, Lateef A, Amuda OS, Afolabi RO (2003). The aquatic toxycity and characterization of chemical and microbiological constituents of water samples from Oba River, Odooba, Nigeria. Asia j. Microbiol. Biotechnol. Environ., Sci., 5: 11-17.

Blaghen M, Lett MC, Vidon DJM (1993). Mercuric reductase activity in a mercury resistant strain of yersinia enterocolitica. FEMS. Microbiol. Lett., 19: 93-96.

Bordel S, Muñoz R, Díaz LF, Villaverde S (2007). New insights on toluene biodegradation by *Pseudomonas putida* F1: Influence of pollutant concentration and excreted metabolites. Appl. Microbiol. Biot., 74:857-866

Cohen ML (1992). Epidemiology of drug resistance: implications for a post-antimicrobial era. Science, 257: 1050–1055.

Edward RC, Selvam GS, Kiyoshi O (2009). Isolation, identification and characterization of heavy metal resistant bacteria from sewage. International Joint Symposium on Geodisaster Prevention and Geoenvironment in Asia: JS-Fukuoka, pp. 205-211

Filali BK, Taoufik J, Zeroual Y, Dzairi FZ, Talbi M, Blaghen M (2000). Waste water bacterial isolates resistant to heavy metals and antibiotics. J. Currr. Microbiol., 4: 151- 156.

Inskeep WP, Macur RE, Hamamura N, Warelow TP, Ward SA, Santini JM (2007). Detection, diversity and expression of aerobic bacterial arsenite oxidase genes. Environ. Microbiol., 9:934-943.

Jennifer M (2001). Determination of minimum inhibitory concentrations. J. Antimicrobial. Chem., 48: Suppl. S1, 5-16.

Levy SB (1998). Multidrug resistance – a sign of the times. New Engl. J. Med., 338:1376–1378

Levy SB, Marshall B (2004). Antibacterial resistance worldwide: Causes, challenges and responses. Nat. Med., 10: S122–S129.

Mrozik A, Piotrowska-Seget Z, Labuzek S (2003). Bacterial Degradation and Bioremediation of Polycyclic Aromatic Hydrocarbons. Pol. J. Environ. Stud., 12(1): 15-25.

Mudry ZJ (2002). Antibiotic Resistance among Bacteria Inhabiting Surface and Subsurface Water Layers in Estuarine Lake Gardno, Pol. J. Environ. Stud., 11(4): 401-406

Muxika I, Borja A, Bonne W (2005). The suitability of the marine biotic index (AMBI) to new impact sources along European coasts. Ecol. Ind., 5: 19–31.

Nigam P, Banat IM, Singh D, Marchant R (1996). Microbial process for the decolourization of textile effluent containing azo, diazo and reactive dyes. Process Biochem., 31: 435-442.

Pearce CI, Lioyd JR, Guthrie JT (2003). The removal of colour from textile wastewater using whole bacterial cells: a review. Dyes Pigm., 58: 179-196.

Quéméneur M, Heinrich-Salmeron A, Muller D, Lièvremont D, Jauzein, M, Bertin, PN, Garrido, F and Joulian, C (2008). Diversity surveys and evolutionary relationships of aoxB genes in aerobic arsenite-oxidising bacteria. Appl. Environ. Microbiol., 74: 4567-4573.

Sarkar SK, Saha M, Takada H, Bhattacharya A, Mishra, P, Bhattacharya B (2007). Water quality management in the lower stretch of the river Ganges, east coast of India: an approach through environmental education. J. Clean. Prod., 15: 1559–1567.

Seralathan KK, Mahadevan S, Krishnamoorthy R (2006). Characterization of a mercury-reducing Bacillus cereus strain isolated from the Pulicat Lake sediments, south east coast of India. Arch. Microbiol., 185(3): 202-211.

White L, Wolanski E (2008). Flow separation and vertical motions in a tidal flow interacting with a shallow-water island. Estuar. Coast. Shelf Sci., 77(3): 457–466.

Wolanski E, Newton A, Rabalais N, Legrand C (2008). Coastal zone management. Encycl. Ecol., pp. 630–637.

Wolf B, Kiel, E, Hagge A, Krieg HJ, Feld, CK (2009). Using the salinity preferences of benthic macroinvertebrates to classify running waters in brackish marshes in Germany. Ecol. Indic., 9: 837–847.

Appendix A. Characteristics of the sampling site.

Continent	Europe
Country	Spain
Region	Catalonia
Locality	Deltebre
Latitude	40°43'18" N
Longitude	0°43'23" E
Altitude	6 m

Appendix B. Geographical localization of the sampling site.

Appendix C. Sampling site "Deltebre" shown in aerial photo.

16

Assessment of heavy metals in urban highway runoff from Ikorodu expressway Lagos, Nigeria

O. Ukabiala Chinwe[1]*, C. Nwinyi Obinna[2], Abayomi Akeem[3] and B. I. Alo[3]

[1]Department of Chemistry, School of Natural and Applied Sciences, College of Science and Technology, Covenant University, Km 10 Idiroko Road, P. M. B. 1023, Ota, Ogun State, Nigeria.
[2]Department of Biological sciences, School of Natural and Applied sciences, College of Science and Technology, Covenant University, Km 10 Idiroko Road, P. M. B. 1023, Ota, Ogun State, Nigeria.
[3]Department of Chemistry, University of Lagos, Akoka, Yaba, Lagos State, Nigeria.

The distribution of heavy metals in the urban high way run off from Ikorodu expressway of Lagos was studied between March to May, 2004.The heavy metals studied include Pb, Cu, Cr, Zn and Cd. The levels of these selected heavy metals were determined using Atomic Absorption Spectrophotometer (M-scientific 200 Model). Trends in the heavy metal from the runoff showed significant variations between the months were values recorded in the month of April showed high values. Statistical analyses showed different mean levels of these heavy metals assessed at the five collecting points. The distribution shows Zn > Pb > Cu > Cr > Cd. Zn recorded the highest concentration levels between (53.4 ± 35.5 - 107.5 ± 80.4 µg/l), while Cd levels (ND - 6.00 µg/L) were the lowest. However, the results obtained falls within the permissible limits of FMENV effluents limits, FHWA and WHO standards of water for domestic use.

Key words: Heavy metals, Onipanu, Obanikoro, palmgroove, road runoff.

INTRODUCTION

Water is an essential resource for living systems, agricultural production, industrial processes and domestic use. However mans' activity towards urbanization (Road and highway construction) and industrialization has chiefly led to the pollution of water by runoff flows. Road runoffs are characterized by mixture of toxicants that are released without prior treatment into the receiving water bodies. Typical pollutants include: suspended solids, heavy metals, hydrocarbons and bacteria of animal origin (Hvitved–Jacobson and Yousef, 1991).

The fate and magnitude of these pollutants found in highway runoff are site-specific and are affected by the volume of traffic, design of the road way, climate and surrounding land use. Roadways with higher average–daily traffic (for instance 30,000 vehicles per day) may produce runoffs with two to five times higher pollutant level than is found in rural highways. Highway runoff pollutants exert significant impacts that could lead to degradation of aesthetic, recreational, biological, physical and chemical qualities of the receiving waters.

According to Maltby et al., (1995) particulate materials tend to be a major constituent in road runoffs, with tendency of accumulation in the sediments of the receiving water bodies.

According to (Ademoroti, 1996a, b) metals with densities greater than 5 gcm^{-3} are referred to as heavy metals. In urban runoffs, metals such as copper, zinc, lead and cadmium have been reported with high level concentration. These heavy metals may have gained access to the runoffs through natural and anthropogenic sources (Duzgoren-Aydin et al., 2006; Florea and Busselberg, 2006). The continuous increase in heavy metal contamination of estuaries and coastal waters is a cause for concern as these metals have the ability to bioaccumulate in tissues of various biotas and may also affect the distribution and density of benthic organisms (Griggs et al., 1977).

Many heavy metals adsorb to particulates and sediments. These settle out and reduce the metals availability for biological uptake (Hung and Hsu, 2004).

The occurrences of elevated levels of heavy metals

*Corresponding author. E-mail: getchinwe@gmail.com.

Figure 1. Map showing the sampling site (Ikorodu road in the Lagos metropolis, Nigeria).
Source: www.Google maps.com.

especially in the sediments can be a good indication of man induced pollution rather than natural enrichment of the sediment by geological weathering (Davies et al., 1991; Udosen, 1992; Forner and Wittman, 1993). Several studies have reported the impacts of heavy metal (lead and Cadmium). The deleterious impacts include: reduced growth and development, cancer, organ damage in males (sterility), nervous system damage (Yilmaz, 2005; Asuquo et al., 2004; Riba et al., 2003; Ademoroti, 1996a). Vehicular movements contribute significantly to the impact of heavy metals either directly or indirectly. Vehicles part (brake lining) is known to contain copper which provide mechanical strength and assist in heat dissipation. However, wears of these parts occur as a result of abrasion and corrosion processes of tyres, brakes and clutch linings. Thus contribute to the elevated level of heavy metals in run off flows. In this study, the authors seek to assess and document the traffic related pollutant (heavy metals) loadings in runoff waters and the potential environmental impacts using the Ikorodu expressway of Lagos as a case study.

MATERIALS AND METHODS

Study area

Ikorodu road is one of the major roads in the Lagos metropolis Nigeria (Figure 1). The road is usually busy with a traffic volume of about 30,000 vehicles per day. It is a four lane dual carriage way

which stretches from Jibowu to Ikorodu town linking other roads in the city (The mainland and island areas of Lagos). The road is of flat topography, impervious in nature and with asphaltic and bituminiuous coatings which accelerate free flow of runoffs through underground ducts and receiving channels down to the lagoon.

Sample collection

Samples were collected between March and May, 2004. Replicate runoff water samples were collected at five different points along Ikorodu road of Lagos. These sampling points were located at Onipanu, Palmgroove, Obanikoro and Anthony points. The fifth is a collecting stream through which runoff flows into the Lagos Lagoon.

Sample preparation

Samples were collected in 2 liter plastic containers and 200 cm^3 reagent bottles properly cleansed with distilled deionized water prior to usage. Collection was carried out by careful immersion of the sample containers deep inside the water and sealing with tight fitting corks and stoppers after collection, in order to avoid air bubbles. Samples were transferred to a refrigerator (4°C) prior to analysis.

Preparation of standards

Instrumental calibration was carried out prior to metal determination by using standard solutions of metal ion prepared from their salts. Commercial analar grade 1000 ppm stock solutions of Pb^{2+}, Zn^{2+}, Cd^{2+}, Cu^{2+}, Cr^{2+} were diluted in 25 cm^3 standard flask and made up to the mark with deionized water to obtain working standard and

Table 1. Distribution of heavy metals in the first runoff samples in week two of March, 2004.

Heavy metal	Onipanu (µg/l)	Palmgroove (µg/l)	Obanikoro (µg/l)	Anthony (µg/l)	Receiving link (µg/l)
Pb	20.0	22.0	11.0	22.0	17.0
Cu	15.0	13.0	11.0	12.0	22.0
Zn	65.0	72.2	67.9	69.7	56.6
Cr	3.0	3.0	3.0	11.0	4.0
Cd	ND	ND	ND	ND	ND

Table 2. Distribution of heavy metals in the second runoff samples in week two of April, 2004.

Heavy metal	Onipanu µg/l	Palmgroove µg/l	Obanikoro µg/l	Anthony µg/l	Receiving link µg/l
Pb	68.0	12.0	74.0	62.0	26.0
Cu	29.0	21.0	48.0	27.0	41.5
Zn	125.4	75.0	200.0	77.9	117.5
Cr	5.0	3.0	16.0	7.0	7.0
Cd	ND	ND	0.6	ND	ND

Table 3. Distribution of heavy metals in the third runoff samples in week two of May, 2004.

Heavy metal	Onipanu µg/l	Palmgroove µg/l	Obanikoro, µg/l	Anthony µg/l	Receiving link µg/l
Pb	55.0	ND	ND	3.0	ND
Cu	26.0	12.0	13.0	37.0	23.5
Zn	48.0	59.0	54.6	12.7	46.1
Cr	1.0	1.0	3.0	3.0	3.0
Cd	ND	ND	ND	ND	ND

solution of 2.0, 3.0 and 4.0 ppm of each metal ion.

Heavy metal determination

The run off water samples were digested using concentrated nitric acid HNO_3 and concentration of Lead (Pb), Zinc (Zn), Cadmium (Cd), copper (Cu) and Chromium (Cr) measured on a M-scientific 200 model atomic absorption spectrophotometer (AAS) (Williams et al., 2007; Essien et al., 2006; Adekoya et al., 2006) at ROTAS SOIL LAB. Ibadan, Nigeria. The essence of the digestion before analysis was to reduce organic matter interference and convert metal to a form that can be analyzed by AAS.

RESULT AND DISCUSSION

The monthly distribution profiles of selected heavy metals in the run off water samples collected at five different points along the Ikorodu road of Lagos are presented in Tables 1 - 3, while their concentration in the three monthly collections are reported as mean and standard deviation and highlighted in Table 4. Trends in heavy metal in urban runoff waters revealed monthly variation in the samples investigated. At 95% confidence interval, the degree of distribution of Zn, Cu, Pb and Cr were signify-cantly different in the runoff samples at the different sampling points, while the distribution of cadmium was not. Zinc and copper were the most predominant in all the runoff samples.

The mean ± standard deviation recorded are 79.5 ± 40.7; 68.8 ± 8.7; 107.5 ± 80.4; 53.4 ± 35.5; 73.4 ± 38.6 µg/L for zinc and 23.3 ± 7.4; 15.3 ± 4.9; 24.0 ± 20.8 ; 25.3 ± 13.8; 29.0 ± 10.9 µg/L for copper . The high level of zinc recorded agrees to previous reports on road run offs that tyre wears resulting from abrasion and friction on impervious surfaces contributes significantly to the increase in the incidence of zinc in highway runoff.

40% detection was observed for lead in runoff samples collected in May 2004 as only 2 samples gave detection values (55.0 and 3.0 µg/l). The low level of lead (Pb) detection could be attributable to reduction in the use of leaded fuel and possibly the diverse nature of road runoffs within this period of assessment. Furthermore, the effect of lead (Pb) to the receiving links (water bodies) cannot be underestimated. This is due to risk posed to life particularly the aquatic organisms which serve as source of food for man (Young and Blevins, 1981). In the runoff samples collected in the month of April, the concentration of the heavy metals were very significant in the order Zn > Pb > Cu >Cr (200 > 74 > 48 > 16 µg/l). This result is of concern particularly when considering

Table 4. Metal concentration (µg/l) reported as mean and standard deviation.

Heavy metal	Onipanu µg/l	Palmgroove µg/l	Obanikoro µg/l	Anthony µg/l	Receiving link µg/l
Pb	47.7 ± 24.8	17.00 ± 2.2	42.5 ± 44.6	29.00 ± 30.1	21.50 ± 6.4
Cu	23.3± 7.4	15.3 ± 4.9	24.0 ± 20.8	25.3 ± 13.8	29.0 ± 10.9
Zn	79.5 ± 40.7	68.8 ± 8.7	107.5 ± 80.4	53.4 ± 35.5	73.4 ± 38.6
Cr	3.0 ± 2.0	2.3 ± 1.2	7.3 ± 9.3	7.0 ± 4.0	4.7 ± 2.1
Cd	ND	ND	ND	ND	ND

the toxic effects to the aquatic ecosystem, although zinc (Zn) is not a human carcinogen but excessive intake through contaminated food chain could lead to vomiting, dehydration, vomiting, abdominal pain, lethargy and dizziness (ATSDR, 1994). The sample collected from Obanikoro sampling point in April showed the presence of cadmium at 0.6 µg/l. In other monthly samples collected, cadmium was not detected in all the runoffs. This low concentration could be due to low level of cadmium in street dust as reported by Perdikaki and Madison (1999). It may also be due to first flush effect when concentrations of runoff are not yet diluted by the rain. The receiving channel showed no significant difference in the concentration of heavy metals when compared to other points where samples are collected. This may be due to runoff flow between other sampling points to the receiving channel (E). Thus the concentration levels of the metals fall almost within the same range as metals from other samples. (3 - 200 µg/l for samples from other points and 3 - 117.5 µg/l for receiving channel). Statistical analyses revealed different mean levels of heavy metals from runoffs in the different sampling points (Onipanu, Palmgroove, Obanikoro, Anthony and Receiving link), the distribution followed the same sequence Zn > Pb > Cu > Cr > Cd.

Generally the average concentration of the metals analyzed in the runoff compares well with results obtained for similar highway runoff carried out in United Kingdom highway. Also the values obtained for these heavy metal falls within the permissible limits of FMENV effluents limits, FHWA and WHO standards of water for domestic use (Clark, 1986).

REFERENCES

Adekoya JA, Williams AB, Ayejuyo OO (2006). Distribution of heavy metals in sediments of Igbede, Ojo and Ojora rivers of Lagos, Nigeria. Environmentalist 26: 277-280.

Ademoroti CMA (1996a). Environmental Chemistry and Toxicology, Foludex Press Ltd., Ibadan pp. 171- 204.

Ademoroti CMA (1996b). Standard Methods for water and effluents Analysis, Foludex press Ltd, pp. 111-120.

ATSDR (1994). Toxicological profile for zinc. US Department of Health and Human Services, Public Health services, 205-88-0608.

Asuquo FE, Ewa-Oboho I, Asuquo EF, Udoh PJ (2004). Fish species used as biomarkers for heavy metals and hydrocarbon contaminations for the cross river, Nigeria. Environmentalist 24: 29-37.

Clark RB (1986). Arine Pollution. 3rd Edn, Reinhold Publisher. United States of America, pp. 110 -121.

Davies CA, Tomlinson K, Stephenson T (1991). Heavy metal in river trees estuary sediments. Environ. Technol. 12: 961-972.

Duzgoren-Aydin NC, Wong Z, Song A, Aydin X, Li You M (2006). Fate of heavy metal contamination in road dusts and gully sediments in Guangzhou, SE China: A chemical and mineralogical assessment. Human Ecol. Risk Assess. 12: 374-389.

Essien DU, Nsikak UB (2006). Spatio- Temporal Distribution of Heavy Metals in sediments and surface water in Stubbs creek, Nigeria. Trends Appl. Sci. Res. 1(3): 292-300.

Florea A, Busselberg D (2006). Occurrence, use and potential toxic effects of metal and metal compounds. Biometals 19: 419-427.

Forner UA, Wittman GW (1993).Metal Pollution in Aquatic Environment, Springer Verlag: Berlin, 486 pp.

Griggs GB, Grimanis AP, Grimani MV (1977). Bottom sediments in a polluted marine environment, Upper Saronikos Gulf, Greece. Environ. Geol. 2(2): 97-106.

Hvitved-Jacobson T, Yousef YA (1991). Highways run off quality, environmental impacts and control. In Highway Pollution, eds.R.S Hamilton and R.M Harrison, Elsevier, London. pp. 165-208.

Hung JJ, Hsu K (2004). Present state and historical change of trace metals in coastal sediments off southwestern Taiwan. Marine Poll. Bull. 49: 986-998.

Maltby L, Boxall ABA, Forrow DM, Calow P, Betton CI (1995). The effects of motorway run – off on freshwater ecosystems. 1 Field study. Environ. Toxicol. Chem. 14: 1079-1092.

Perdikaki K, Mason CF (1999). Impact of road run off on receiving streams in Eastern England, Water Res. 33(7): 1627-1633.

Riba I, Garcia-Lugue E, Blasco J, DelVallsi TA (2003). Bioavailability of heavy metals bound to estuarine sediments as a function of pH and salinity values. Chem. Speciation Bioavailab. 15: 101- 114.

Udosen ED (1992). Aqua-terrestrial Environmental pollution Studies of Inorganic Substances from two industrial firms in Akwa Ibom State, Ph.D Thesis, Department of Chemistry University of Calabar, Calabar Nigeria 577 pp.

Yilmaz AB (2005). Comparison of heavy metals of grey mullet (M. cephalus L.) and Sea bream (S. aurata L.) caught in Iskenderun Bay (Turkey). Turk. J. Vet. Anim. Sci. pp. 257-262.

Williams AB, Ayejuyo OO, Adekoya JA (2007). Trends in trace metal burdens in sediment, Fish species and filtered water of Igbede Rivers, Lagos, Nigeria. J. Appl. Sci. 7(13): 1821-1823.

17

Heavy metals in sediments from River Ngada, Maiduguri Metropolis, Borno State, Nigeria

J. C. Akan[1]*, F. I. Abdulrahman[1], O. A. Sodipo[2], A. E. Ochanya[1] and Y. K. Askira[1]

[1]Department of Chemistry, University of Maiduguri, P. M. B. 1069, Maiduguri, Borno State, Nigeria.
[2]Department of Clinical Pharmacology and Therapeutics, College of Medical Sciences, University of Maiduguri, Maiduguri, Borno State, Nigeria.

The objective of this research was to determine the degree of heavy metal contamination in River Ngada and the extent to which the sediment quality of the river had deteriorated. In this study, metals such as Cu, Zn, Co, Mn, Mg, Fe, Cr, Cd As, Ni and Pb in the sediments were determined using Perkin-Elmer Analyst 300 Atomic Absorption Spectrophotometer. The extent of sediment quality deterioration was observed in all the sampling points to be higher with respect to all the metals studied. The levels of the above metals increased with an increase in distance from point S_1 to S_8. The metals also increased with increasing sediment depth, indicating age-long accumulation of heavy metals from anthropogenic sources. The study revealed that the levels of all the metals studied were higher than the WHO's standard sediment guideline limits. If this trend is allowed to continue unabated, it is most likely that the food web in this study environment might be at highest risk of induced heavy metal contamination.

Key words: Heavy metals, sediments, deterioration, food web, pollution.

INTRODUCTION

Heavy metals are elements having specific gravity greater than 4.0 that is, at least 5 times that of water. Heavy metals exist in water, in colloidal, particulate and dissolved forms (Adepoju-Bello et al., 2009). Their occurrence in water bodies are either of natural origin (eroded minerals within sediments, leaching of ore deposits and volcanism-extruded products) or of anthropogenic origin (solid waste disposal, industrial or domestic effluents, harbour channel dredging) (Marcovecchio et al., 2007). Some of the metals such as calcium, magnesium, potassium and sodium are essential to sustaining life and must be present for normal body functions. While others such as cobalt, copper, iron, manganese, molybdenon and zinc are needed at low levels as catalyst for enzyme activities (Adepoju-Bello et al., 2009). Excess exposure to these essential metals can however, be toxic. Water has unique chemical properties due to its polarity and hydrogen bonds which makes it is able to dissolve, absorb, adsorb or suspend many different compounds (WHO, 2007).

Thus, in nature, water is not pure as it acquires contaminants from its surrounding and those arising from humans and animals as well as other biological activities (Mendie, 2005).

Heavy metals can cause serious health effects with varied symptoms depending on the nature and quantity of the metal ingested (Adepoju-Bello and Alabi, 2005). The most common heavy metals that humans are exposed to are aluminium, arsenic, cadmium, lead and mercury. Aluminium has been associated with Alzheimer's and Parkinson's disease, senility and presenile dementia. Arsenic exposure can cause among other illnesses or symptoms, cancer, abdominal pain and skin lesions. Cadmium exposure produces kidney damage and hypertension. Lead is a cumulative poison and a possible human carcinogen (Bakare-Odunola, 2005) while for mercury, toxicity results in mental disturbane and impairment of speech, hearing, vision and movement (Hammer and Hammer, 2004). In addition, lead and mercury may cause the development of autoimmunity in which a person's immune system attacks its own cells. This can lead to joint diseases and ailment of the kidneys and circulatory system and neurons. At higher concentrations, lead and mercury can cause irreversible brain damage.

Sediments represent significant sources of heavy metal

pollution in the aquatic environment as a result of changes in pH, redox potential, diagenisis or physical perturbations within their primary sedimentary sinks. The occurrences of enhanced concentrations of heavy metals especially in sediments may be an indication of human-induced perturbations rather than natural enrichment through geological weathering (Davies et al., 1991; Binning and Baird, 2001; Eja et al., 2003). Heavy metals are non-biodegradable and they persist in the environment and may become concentrated up the food chain (Eja et al., 2003), leading to enhanced levels in liver and muscle tissues of fishes (Eja et al., 2003), aquatic bryophytes (Mouvet et al., 1993) and aquatic biota (Ramos et al., 1999). It therefore means that the biota and water quality of an aquatic ecosystem could determine the quantitative and qualitative levels of heavy metals in tissues of fauna and flora of the ecosystem. Heavy metal distribution and bioavailability in both sediments and the overlying water column have to be considered to obtain a better understanding of inter-actions between the organisms and their environment.

Sediments are ecologically important components of the aquatic habitat, which play a significant role in maintaining the trophic status of any water body (Singh et al., 1997). Sediments near urban areas commonly contain high levels of contaminants (Cook and Wells, 1996; Lamberson et al., 1992) which constitute a major environmental problem faced by many anthropogenically impacted aquatic environments (Magalhaes et al., 2007). Sediments in rivers do not only play important roles at influencing the pollution, they also record the history of their pollution. Sediments act as both carrier and sources of contaminants in aquatic environment (Shuhaimi, 2008). The contamination of sediments with heavy metals, may lead to serious environmental problem (Loizidou et al., 1992). Heavy metals may adsorb onto sediments or be accumulated by the benthic organisms; their bioavailability and toxicity depend upon the various forms and amount bound to the sediment matrices (Chukwujindu et al., 2007). Additionally, pollutants released to surface water from industrial and municipal discharges, atmospheric deposition and run off from agricultural, urban and mining areas can accumulate to harmful levels in sediments (Chukwujindu et al., 2007). In view of growing concern over the use of River Ngada by resident along the river banks and the fauna and flora of the ecosystem, this study is aimed at assessing the quality of sediment from River Ngada.

MATERIALS AND METHOD

Study area

River Ngada is located in Maiduguri Metropolis, Borno State, Nigeria. The river is used for various human activities including fishing, vegetables irrigation, bricks making and by residences along the river banks for bathing, washing and as drinking water by animals. The river originates from Rivers Yedzram and Gombole

which meet at a confluent at Sambisa both in Nigeria and flows as River Ngada into Alau Dam and stretches down across Maiduguri Metropolis then empties into Lake Chad. The river receives copious amounts of wastes from residential houses and abattoirs sited along its course. Urban waste management and garbage disposal practices in the city are very poor. Treated water from the Municipal waste and Abattoir located near the river contains large amounts of heavy metals, which when in super abundance may cause disruption to the ecological balance of the river (Akan et al., 2011).

Sample area and collection points

Sediment samples from River Ngada were collected and labelled S_1 to S_8 along the river course corresponding to the points where notable discharge of wastewater into River Ngada occurs (Figure 1). Samples were collected from Alau Dam bridge (S_1); the point of discharge of wastewater from the water treatment plant (S_2) 100 m away from water treatment plant (S_3), the Lagos bridge (S_4), the Gwange bridge (S_5), the Custom bridge (S_6); the point of discharge of Abattoir wastewater into River Ngada (S_7) is 2 km after discharge of Abattoir wastewater into River Ngada (S_8). At each sampling point, sediments were collected using Van-Veen grab sampler at a depth of 0-5, 5-10, 10-15, and 15-20 cm respectively. Sediment samples were then preserved in plastic bags, transported to the laboratory and kept frozen in the refrigerator pending analyses.

Sample preparation

All the sediment samples were oven-dried at 80 to 100 °C. The dry samples were gently crushed and sieved to collect the < 63 μm grain size. Two grammes of each sediment sample were weighed into a pre-weighed dry crucible. The crucibles were covered and placed in a muffle furnace and the temperature gradually increased to 500 °C for 2 h (Radojevic and Bashkin, 1999). The samples were removed and cooled to room temperature in desiccators. The ash samples were placed into acid-washed glass beakers. Sediment samples were digested by the addition of 10 ml of 0.25 M HNO_3, heated to dryness and thereafter 10 ml of 0.25 M HNO_3 and 3 ml of $HClO_3$ were added. The solution was then heated in a fume chamber. Sample solutions were obtained by leaching the residues with 4.0 ml of HCl and thereafter filtered and diluted with distilled water to 100 ml mark (Radojevic and Bashkin, 1999). Blank solutions were handled as detailed for the samples. All samples and blanks were stored in plastic containers. Determination of Cu, Zn, Co, Mn, Mg, Fe, Cr, Cd As, Ni and Pb were made directly on each final solution using Perkin-Elmer Analyst 300 Atomic Absorption Spectrophotometer (AAS).

Statistical analyses

Data collected are presented as mean and standard deviation and were subjected to Pearson's correlation analysis, while one-way analysis of variance (ANOVA) ($P < 0.05$) was used to assess whether heavy metals varied significantly between sampling depths and points. All statistical analysis was performed with SPSS 9.0 for Windows (Ozdamar, 1991).

RESULTS

The concentrations of heavy metals in sediment at different depths at point S_1 are presented in Table 1. The concentrations of Cr range from 28.87±0.02 to 45.14±0.02 μg/g; Pb from 54.33±0.01 to 71.23±0.11μg/g;

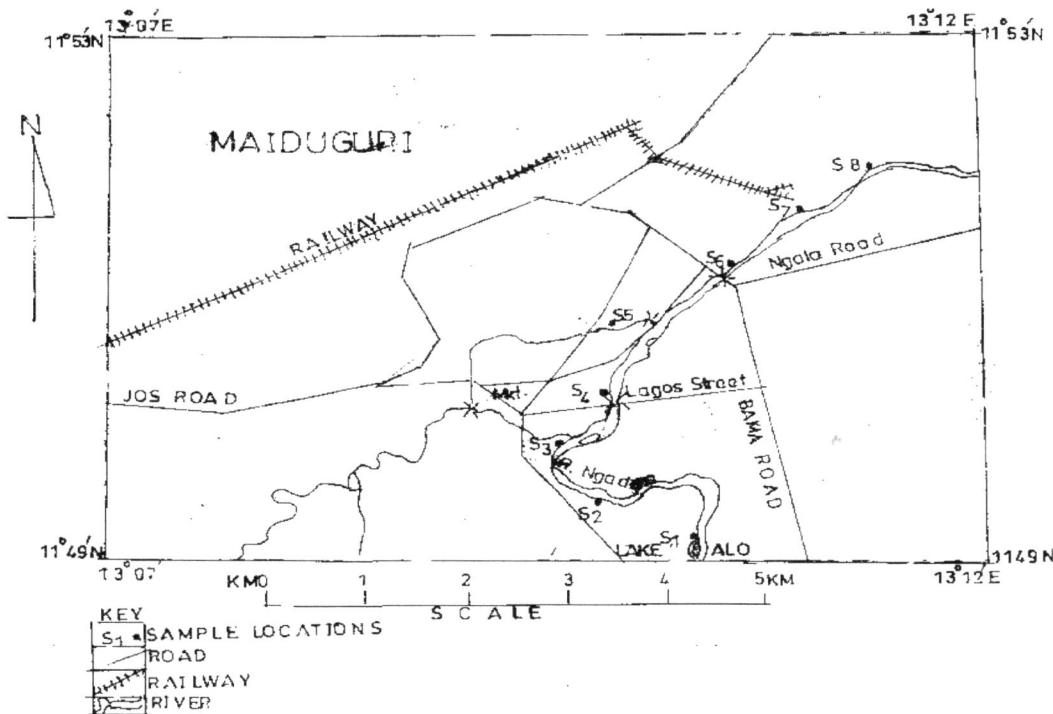

Figure 1. Map of the study area showing sample location.

Cu from 26.32±0.02 to 51.32±0.01 μg/g; Fe from 31.87±0.21 to 58.34±1.01 μg/g; Ni from 24.31±0.22 to 39.45±0.41 μg/g; Co from 12.45±0.05 to 28.34±0.21 μg/g; Mn from 43.11±0.11 to 154.34±0.10 μg/g; Cd from 7.34±0.02 to 19.34 ±0.09 μg/g; As from 12.23±0.01 to 40.67±0.76 μg/g and Zn from 132.03±0.20 to 163.45±0.06 μg/g. For point S_2, S_3, S_4 and S_5, the concentrations of heavy metals in sediment samples at different depths are as presented in Figures 2, 3, 4 and 5. The levels of Cr for points S_2, S_3, S_4 and S_5 at different depths varied between 36.00±0.67 and 82.00±3.56 μg/g; 64.00±4.56 and 97.00±0.87 μg/g Pb; 30.00±3.54 and 79.00±5.66 μg/g Cu; 43.00±1.34 and 88.00±7.56 μg/g Fe; 32.00±7.89 and 79.00±2.41 μg/g Ni; 18.00±4.67 and 64.00±6.12 μg/g Co; 54.00±4.54 and 21.00±2.01 μg/g Mn; 12.00±0.91 and 44±4.12 μg/g Cd; 15.00±0.21 and 80.00±2.81 μg/g As and 143.00±1.53 and 230.00±3.76 μg/g Zn. The concentrations of heavy metals in sediment samples at different depths for points S_6 and S_7 are as presented in Tables 2 and 3. The values of Cr ranged from 69.34±0.32 to 94.34±0.50 μg/g; 81.76±0.22 to 110.33±0.60 μg/g Pb; 48.34±0.05 to 82.55±0.06 μg/g Cu; 74.45±1.43 to 107.44±0.07 μg/g Fe; 52.23±1.01 to 91.44±0.92 μg/g Ni; 36.21±1.00 to 72.22±1.05 μg/g Co; 84.34±0.28 to 236.33±0.88 μg/g Mn; 25.23±0.55 to 56.54±0.22μg/g Cd; 33.24±0.22 to 97.67 ±0.11μg/g As and 185.45±0.87 to 248.45 ±0.87 μg/g Zn. Similarly, the levels of Cr for point S_8 ranged (Figure 6) between 74.00±0.56 and 98.00±1.03 μg/g; 89.00±7.73 and

116.00±10.56 μg/g Pb; 56.00±2.05 and 89.00±6.06 μg/g Cu; 86.00±0.87 and 118.00±2.07 μg/g Fe; 65.00±0.54 and 104.00±1.77 μg/g Ni; 53.00±3.90 and 78.00±0.88 μg/g Co; 103.00±2.63 and 243.00±2.88 μg/g Mn; 34.00±0.21 and 64.00±1.29 μg/g Cd; 45.00±0.22 and 107.00±0.66 μg/g As and 218.00±2.12 and 257.00±7.81 μg/g Zn.

DISCUSSION

Of the various sampling sites, point S_1 was found to be least polluted in relation to the heavy metals analysed. This was attributed to the fact that there were less anthropogenic activities at this point. For point S_2, the increase in concentrations of all the metals was due to the discharge of wastewater from the water treatment plant located close to it. During water processing and treatments, wastewater from the treatment plant is discharged to River Ngada, thereby increasing the levels of metals at this point. point S_3 was located 200 m away from point S_2 and had higher concentration of heavy metals when compared to point S_1 and S_2. The high levels of heavy metals at point S_3 are due to the discharged of wastewater from residential areas. Point S_4 had higher metal concentrations than points S_1 to S_3. This is due to the fact that this point is located near Lagos bridge and roads where several anthropogenic activities and discharge of sewage sludge from residential areas

Table 1. Mean concentration of heavy metals in sediment samples by depth for point S_1 of River Ngada.

Sampling Depth (cm)	Concentrations (µg/g)									
	Cr	Pb	Cu	Fe	Ni	Co	Mn	Cd	As	Zn
0-0.5	28.87a±0.02	54.33a±0.01	26.32a±0.02	31.87a±0.21	24.31a±0.22	12.45a±0.05	43.11a±0.11	7.34a±0.02	12.23a±0.01	132.03a±0.20
5-10	32.89b±0.33	58.98b±0.12	34.13b±b0.01	36.97b±0.02	29.23b±0.10	23.76b±0.04	78.23b±0.21	12.56b±0.11	30.65b±0.11	138.23b±0.65
10-15	38.43c±0.23	63.22c±0.03	41.54c±0.06	43.34c±1.03	33.34c±0.01	18.23c±0.12	126.76c±0.54	16.76c±0.43	32.23c±0.32	154.56c±0.12
15-20	45.14d±0.02	71.23d±0.11	51.32d±0.01	58.34d±1.01	39.45d±0.41	28.34d±0.21	154.34d±0.10	19.34d±0.09	40.67d±0.76	163.45d±0.06

Within column, mean with different letters are statistically significant p<0.05.

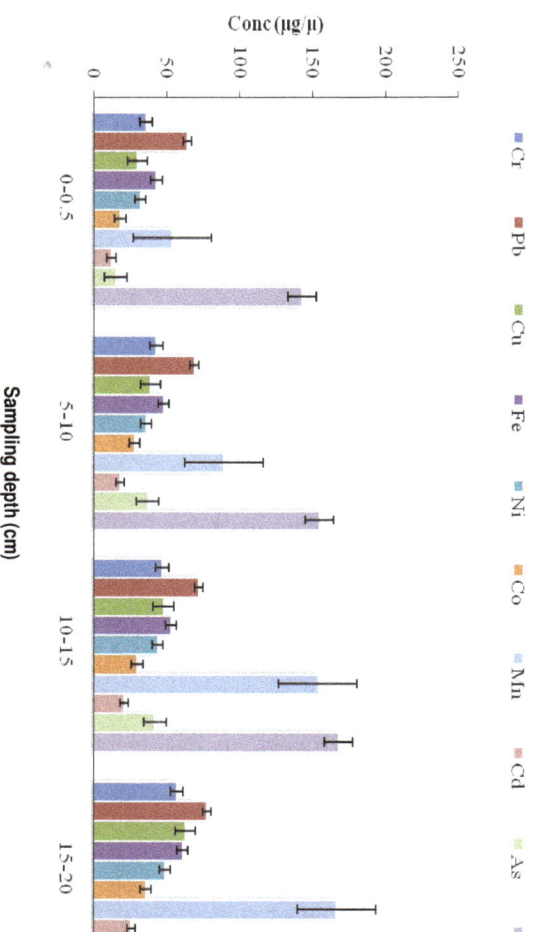

Figure 2. Mean concentration of heavy metals in sediment samples by depth for points S_2 of River Ngada.

sludge from residential areas take place. For point S_5, the levels of all the metals in the sediment samples are likely due to the discharge of wastewater from residential areas and urban garbage. Point S_6 was located at the Custom Bridge and within the market areas where there is

heavy discharge of wastewater from both residential and market areas into it. Also large quantity of garbage from the Custom market is also discharged at this point which eventually contribute to high concentrations of all the metals compared to point S_1 to S_5. Also for point S_7, the

concentrations of all the metals were high, which is due to the fact that this point is located at the immediate discharge of abattoir waste water into River Ngada. For all the metals studied, Zn was the highest and Mn the second highest. Pb the third highest, Fe was the fourth while Cd showed

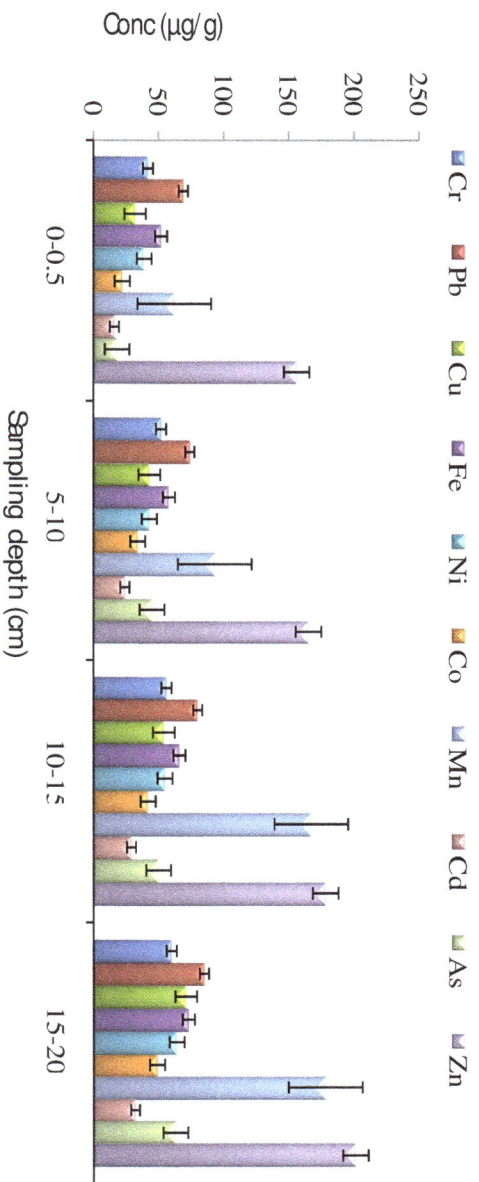

Figure 3. Mean concentration of heavy metals in sediment samples by depth for point S₃ of River Ngada.

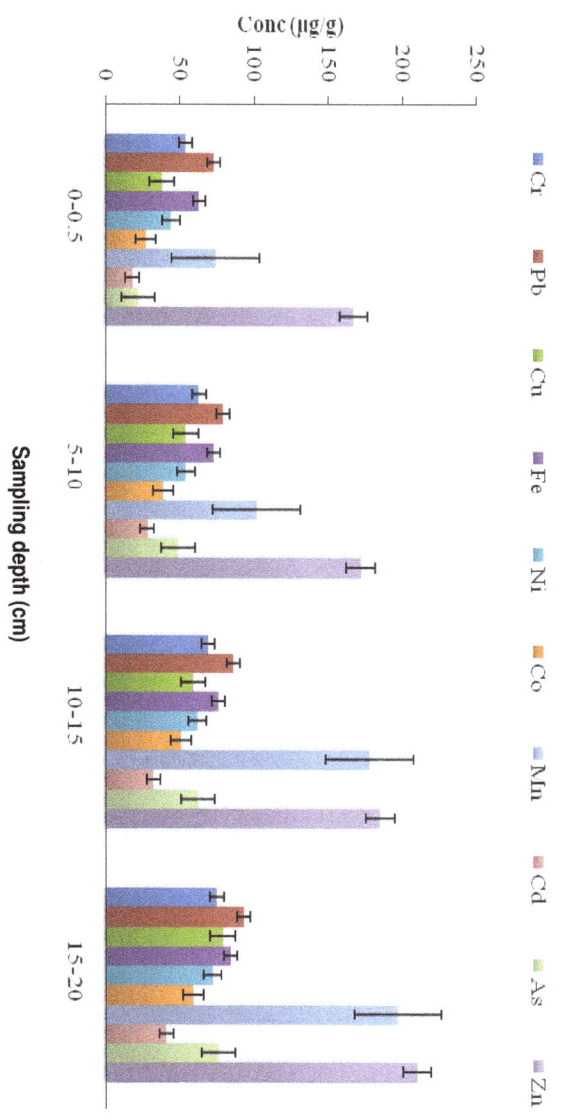

Figure 4. Mean concentration of heavy metals in sediment samples by depth for point S₄ of River Ngada.

Table 2. Mean concentration of heavy metals in sediment samples by depth for point S_6 of River Ngada.

Sampling Depth (cm)	Concentrations (μg/g)									
	Cr	Pb	Cu	Fe	Ni	Co	Mn	Cd	As	Zn
0-0.5	69.34a±0.32	81.76a±0.22	48.34a±0.05	74.45a±1.43	52.23a±1.01	36.21a±1.00	84.34a±0.28	25.23a±0.55	33.24a±0.22	185.45a±0.87
5-10	78.34b±0.11	86.65b±0.63	61.34b±0.22	88.45b±2.04	46.34b±0.12	123.02b±1.02	38.33b±0.12	56.11b±0.32		193.74b±0.43
10-15	81.87c±0.05	97.45c±0.24	68.78c±0.22	92.98c±0.47	61.12c±0.23	193.23c±2.32	42.45c±1.00	73.21c±0.11		224.74c±2.11
15-20	87.45d±0.33	103.45d±0.45	79.45d±0.12	99.34d±0.09	65.08d±0.21	221.34d±0.07	49.85d±0.12	85.65d±0.03		237.45d±0.06

Within column, mean with different letters are statistically significant p<0.05.

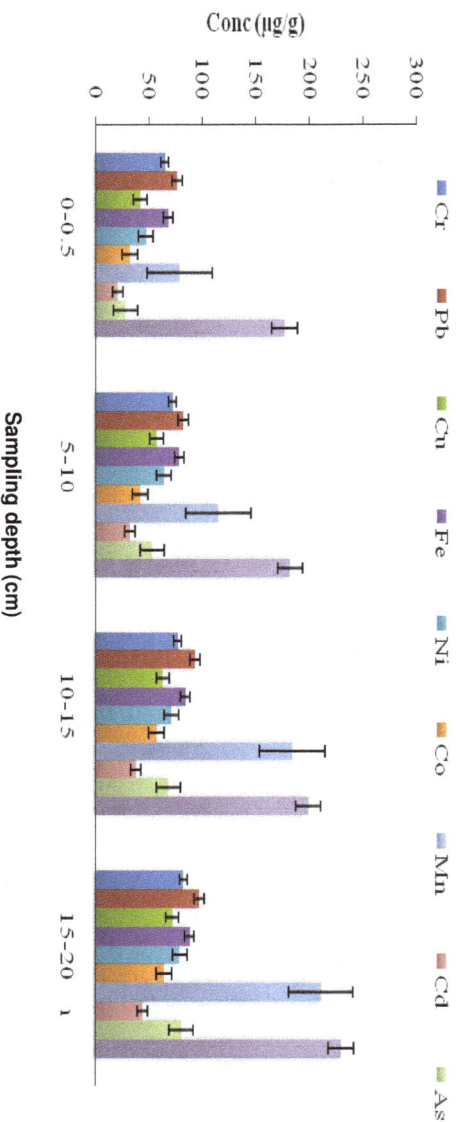

Figure 5. Mean concentration of heavy metals in sediment samples by depth for point S_5 of River Ngada.

Table 3. Mean concentration of heavy metals in sediment samples by depth for point S_7 of River Ngada.

Sampling Depth (cm)	Concentrations (μg/g)									
	Cr	Pb	Cu	Fe	Ni	Co	Mn	Cd	As	Zn
0-0.5	84.34a±0.53	53.45a±0.23	79.45a±0.02	58.45a±0.11	42.55a±0.06	92.87a±0.55	30.33a±1.05	37.01a±0.01		205.23a±0.73
5-10	92.34b±0.54	65.89b±0.72	93.56b±0.57	74.23b±0.65	53.87b±0.05	136.65b±0.43	45.45b±0.06	64.34b±0.19		212.34b±0.23
10-15	104.23c±0.54	74.34c±0.54	97.32c±0.03	86.15c±0.03	64.45c±0.11	204.45c±0.33	47.76c±0.13	87.34c±0.58		231.66c±0.16
15-20	110.33d±0.60	82.55d±0.06	107.44d±0.07	91.44d±0.92	72.22d±1.05	236.33d±0.88	56.54d±0.22	97.67d±0.11		248.45d±0.87

Within column, mean with different letters are statistically significant p<0.05.

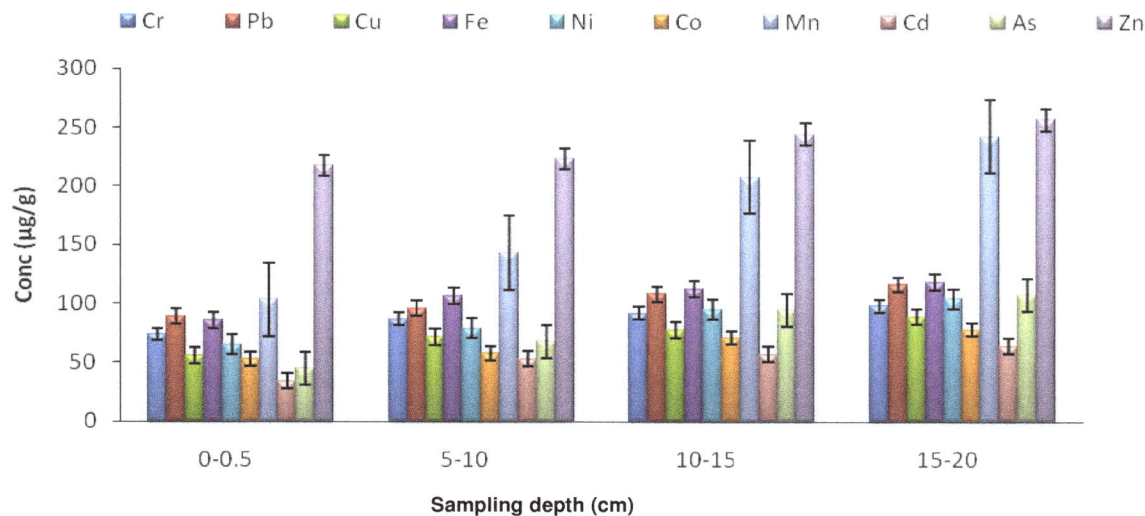

Figure 6. Mean concentrations of heavy metals sediment sample by depth for point SR of River Ngada.

the least concentrations. The high levels of Zn, Mn, Pb and Fe might be due to the disposal of solid waste from residential areas which might contain higher levels of these metals. The increase in concentrations of heavy metals with depth might be due to leaching effect, since sediments usually serve as repository of elements in aqueous environment. This conforms to report by Stephen et al. (2001) that sediment could act as sink for a wide range of contaminants including heavy metals from various sources. Based on the results obtained from this study, the concentrations of heavy metals in sediment sample for point S_1 was slightly polluted, while the levels for other points were heavily polluted. Statistical analysis using analysis of variance shows that, there was significant variation in the levels of heavy metals from points S_1 to S_8 (p< 0.05).

The highest chromium level of 98.00 µg/g was observed in point S_8 (15-20 cm depth), while the least level of 28.87 µg/g was also observed at point S_1 (0.5-5.0 cm depth). Generally, it was observed that the levels of chromium increased significantly with increase in depth. Also there was a significant increase in concentrations of chromium from point S_1 to S_8. Chromium exists in four valency states viz: Cr (II), Cr (III), Cr (IV) and Cr (VI) and reaches water supplies primarily from the discharge of industrial wastes. While Cr^{3+} is considered a minor problem, Cr^{6+} is very toxic and carcinogenic (ATSDR, 2000). Eczematous dermatitis due to trivalent Cr compounds has also been reported. Chromium and its compounds are known to cause cancer of the lung, nasal cavity and paranasal sinus and suspected to cause cancer of the stomach and larynx (ATSDR, 2000). Chromium (III) had been described as an essential nutrient that helps the body use sugar, protein, and fat (Hati et al., 2005). However, under certain environmental conditions (Awan et al., 2003) and certain metabolic transformations,

chromium (III) may readily be oxidized to chromium (VI) compounds that are toxic to human health (ATSDR, 2000). According to WHO/USEPA guideline value for sediment, the concentration of 25 µg/g, Cr is acceptable (Radojevic and Bashkin, 1999). For concentrations exceeding 25 µg/g, a condition known as allergic dermatitis could result (EPA, 1999). From the results of these analyses, the concentrations of chromium in the sediment samples exceeded the regulating limits, indicating severe contamination of sediments of River Ngada by chromium.

The highest concentration of 116 µg/g for lead was observed at point S_8, and the lowest level of 54.33 µg/g at point S_1. Lead is also used in the production of lead acid batteries, solder, alloys, cable sheathing, pigments, rust inhibitors, ammunition, glazes and plastic stabilizers. Tetraethyl and tetramethyl lead are important because of their extensive use as antiknock compounds in petrol (Abbasi et al., 1998; Sharma and Pervez, 2003; WHO, 2004). Lead toxicity leads to anaemia both by impairment of haemobiosynthesis and acceleration of red blood cell destruction. Both are dose related. Lead also depresses sperm count (Anglin-Brown et al., 1995). In addition, Pb can also produce a damaging effect on the kidney, liver and nervous system, blood vessels and other tissues (Anglin-Brown et al., 1995; Sharma and Pervez, 2003). Lead is toxic to humans and may originate in water from contact with the ground, industrial wastes and from water piping itself. Lead is a cumulative poison and has been known to cause 'plumbism' or lead poisoning at a concentration of 10 g/day, It produces damaging effects on the organs and tissues to which it comes into contact. The consequences of excess lead in the human body range from low intelligent quotient in children and high blood pressure in adults by (Ottaway, 1978). The levels of lead in the analyzed sediment samples showed that the limiting values by USEPA of 40 µg/g (EPA, 1999) was exceeded, indicating contamination of River Ngada sediment and this may pose a hazard to the aquatic biota.

The maximum concentration of 89.00 µg/g for copper was observed for point S_8, while the minimum level of 26.32 µg/g was detected in point S_1. Copper is widely used in electrical wiring, roofing, various alloys, pigments, cooking utensils, piping and in the chemical industry. Copper is present in amunitions, alloys (brass, bronze) and coatings. Copper compounds are used as or in fungicides, algicides, insecticides and wood preservatives and in electroplating, azo dye manufacture, engraving, lithography, petroleum refining and pyrotechnics. Copper compounds can be added to fertilizers and animal feeds as a nutrient to support plant and animal growth. Copper compounds are also used as food additives (Abbasi et al., 1998; Eaton, 2005; WHO, 2004). In addition, copper salts are used in water supply systems to control biological growths in reservoirs and distribution pipes and it forms a number of complexes in natural waters with inorganic and organic ligands (WHO, 2004). Copper is an essential substance to human life, however, in high concentrations, it can cause anaemia, liver and kidney damage, stomach and intestinal irritation (Turnland, 1988). Copper is generally remobilised with acid-base ion exchange or oxidation mechanism (Gomez et al., 2000). The levels of copper in the sediment samples were above the (WHO, 2004) standard values of 25 µg/g for the survivor of aquatic organism.

The maximum concentration of 118.00 µg/g for iron was observed for point S_8, while the minimum level of 31.87 µg/g was detected in point S_1. The USEPA guideline value (30 µg/g of Fe in sediment is acceptable (Radojevic and Bashkin, 1999). Above 30 µg/g, a condition known as haemo-chromatosis could result. From the result of this study, the concentration of iron in the sediment samples exceeded the guideline limit indicating severe pollution of River Ngada.

The level (104.00 µg/g) of nickel in the sediment samples was highest at point S_8, while point S_1 showed the least concentration (24.31 µg/g). Nickel is used mainly in the production of stainless steels, non-ferrous alloys and super alloys. Other uses of nickel and nickel salts are in electroplating, as catalysts, in nickel-cadmium batteries, in coins, in welding products and in certain pigments and electronic products. It is estimated that 8% of nickel is used for household appliances (WHO, 2004). Nickel is also incorporated in some food supplements, which can contain several micrograms of nickel per tablet (Abbasi et al., 1998; Eaton, 2005; WHO, 2004). The WHO guideline value of 20 µg/g showed that the concentrations of Ni in River Ngada is very high and indicate possible pollution. However, nickel limiting levels were exceeded and River Ngada could be said to be contaminated by nickel.

The maximum concentration for manganese (243.00 µg/g) was observed for point S_8, while the minimum level (43.11 µg/g) was detected in point S_1. The common

aqueous species found in water is predominantly Mn^{2+} and Mn^{4+}. Manganese is essential for plants and animals. Manganese dioxide and other manganese compounds are used in products such as batteries, glass and fireworks. Potassium permanganate is used as an oxidant for cleaning, bleaching and disinfection purposes. Other manganese compounds are used in fertilizers, varnish and fungicides and as livestock feeding supplements. Manganese can be adsorbed onto soil, the extent of adsorption depends on the organic content and cation exchange capacity of the soil. It can bioaccumulate in lower organisms (e.g., phytoplankton, algae, molluscs and some fish) but not in higher organisms; biomagnification in food chains is not expected to be very significant (Abbasi et al., 1998; Eaton, 2005; WHO, 2004). The levels of manganese in the sediment samples exceeded the USEPA limit of 30 µg/g,

The highest concentration (64.00µg/g) for cadmium was observed in point S_8, while the minimum level (7.34µg/g) was detected in point S_1. Cadmium metal is used mainly as an anticorrosive and electroplated on steel. Cadmium sulphide and selenide are commonly used as pigments in plastics. It is also used in electric batteries and in various electronic components and inorganic fertilizers produced from phosphate ores which constitute a major source of diffuse cadmium pollution (Anglin-Brown et al., 1995; Eaton, 2005). Moreover, when ingested by humans, cadmium accumulates in the intestine, liver and kidney (WHO, 2004). The kidney cortex is regarded as the most sensitive organ. Cadmium adsorbs strongly to sediments and organic matter (Sanders et al., 1999). Cadmium has a range of negative physiological effects on organisms such as decreased growth rates and negative effects on embryonic development (Newman and McIntosh, 1991). The levels of cadmium in the sediment samples were above the WHO, 2004 standard value of 6 µg/g.

Arsenic is a highly toxic metalloid element (Rodriguez et al., 2003; Pizzaro et al., 2003). The maximum concentration (107.00 µg/g) for arsenic was observed in point S_8, while the minimum level (12.23µg/g) was detected in point S_1. The levels of arsenic in the analyzed sediment samples exceeded the WHO, 2004 standard limit of 27 µg/g, indicating contamination of River Ngada.

Excessive intake of Zn may lead to vomiting, dehydration, abdominal pain, nausea, lethargy and dizziness (ATSDR, 1994). The maximum concentration (257.00 µg/g) for zinc was observed in point S_8, while the minimum level (132.03 µg/g) was detected in point S_1. Zinc is used in a number of alloys including brass and bronze, batteries, fungicides and pigments. Zinc is an essential growth element for plants and animals but at elevated levels it is toxic to some species of aquatic life (WHO, 2004). In addition, Zn is involved in a variety of enzyme systems which contribute to energy metabolism, transcription and translation. Zinc is also potentially hazardous and excessive concentrations in soil lead to phytotoxicity as it is a weed killer (Anglin-Brown et al., 1995;

Abbasi et al., 1998; WHO, 2004). Zinc is used in galvanizing steel and iron products. Zinc carbonates are used as pesticides (Anglin-Brown et al., 1995). The levels of zinc in the sediment samples exceeded the WHO guideline value of 123 µg/g.

Conclusion

From the above observations, it is clear that the concentrations of all the metals showed pronounced levels of pollution. The study therefore indicates the increasing levels of all the metals from points S_1 to S_8. The study also revealed a significant increase in the levels of all the metals with sediment depth. The results of this study proved that the activities within the metropolis might have been responsible for the elevated levels of all the metals in the sediment samples. The levels of all the metals in the sediment samples were higher than the sediment guideline limits. If this trend is allowed to continue unabated, it is most likely that the food web in this study environment might be at highest risk of induced heavy metal contamination.

REFERENCES

Abbasi SA, Abbasi N, Soni R (1998). Heavy Metals in the Environment, 1st Edn. Mittal Publications, ISBN: 81-7099-657-0, p.314.

Adepoju-Bello AA, Ojomolade OO, Ayoola GA, Coker AAB (2009). Quantitative analysis of some toxic metals in domestic water obtained from Lagos metropolis. The Nig. J. Pharm., 42 (1): 57-60.

Adepoju-Bello AA, Alabi OM (2005). Heavy metals: A review. The Nig. J. Pharm., 37: 41-45.

Akan JC, Abdulrahman FI, Mamza PT, Aishatu N (2011). Effect of Environmental Pollution on the Quality of River Ngada, Maiduguri Metropolis, Borno State, Nigeria. Global Science Books. Terrestrial and Aquatic Environ. Toxicol. 6 (1): 40-46.

Anglin-Brown B, Armour-Brown A, Lalor GC (1995). Heavy metal pollution in Jamaica 1: Survey of cadmium, lead and zinc concentrations in the Kintyre and hope flat districts. DOI: 1 0.1007/BF00146708. Environ. Geochem. Health, 17: 51-56.

ATSDR, Agency for Toxic Substances and Disease Registry (1994). Toxicological profile for zinc. US Department of Health and Human Service, Public Health Service, (205):-88-0608.

ATSDR, Agency for Toxic Substances and Disease Registry (2000). Toxicological Profile for Chromium. Atlanta, GA: U.S. Department of Health and Human Service, Public Health Service. 1600 Clifton Road N.E, E-29 Atlanta, Georgia 30333 (6-9): 95-134.

Awan AM, Baigl MA, Iqbal J, Aslam MR, Ijaz N (2003). Recovery of chromate form tannery waste water. Electron. J. Environ. Agric. Food Chem. 2 (5): 543-548.

Bakare-Odunola MT (2005). Determination of some metallic impurities present in soft drinks marketed in Nigeria. The Nig. J. Pharm., 4(1): 51-54.

Binning K, Baird D (2001). Survey of heavy metals in the sediments of the Swatkops River estuary, Port Elizerbeth South Africa. Water SA., 24: 461-466.

Chukwujindu MA, Godwin EN, Francis OA (2007). Assessment of contamination by heavy metals in sediment of Ase-River, Niger Delta, Nigeria. Res. J. Environ. Sci., 1: 220-228.

Cook NH, Wells PG (1996). Toxicity of Halifax harbour sediments: an evaluation of Microtox Solid Phase Test. Water Qual. Res. J. Canada., 31 (4): 673-708.

Davies CA, Tomlinson K, Stephenson T (1991). Heavy metals in River Tees estuary sediment. Environ. Technol., 12: 961-972.

Eaton AD (2005). Standard Methods for the Examination of Water and Waste Water. 21st Edn. American Public Health Association, Washington, ISBN. 0875530478. pp. 343-453.

Eja CE, Ogri OR, Arikpo GE (2003). Bioconcentration of heavy metala in surface sediments from the Great Kwa river estuary, Calabar, Southeast Nigeria. J. Nig. Environ. Soci., 1: 47-256.

EPA (1999). Sediment Q,uality Guidelines developed for the national status and trends program. Report No. 6/12/99. http://www.epa.gov/waterscience/cs/pubs.htm_(Accessed in May 2004).

Gomez AJL, Giráldez I, Sánchez-rodas D, Morales E (2000). Comparison of the feasibility of three extraction procedures for trace metal partitioning in sediments from south west Spain. Sci. Total. Environ. 246: 271-283.

Hammer MJ, Hammer MJ (2004). Water Quality. In: Water and Waste-water Technology. 5th Edn. New Jersey: Prentice-Hall, pp. 139-159.

Hati SS, Joseph CA, Ogugbuaja VO (2005). Comparative assessment of chromium discharge in Tannery wastewater in Kano State, Northern Nigeria. Proceed. Of the 28th Ann. Int. Confer. CSN 2 (1): 137-139.

Lamberson JO, Dewitt TH, Swartz RC (1992). Assessment of sediment toxicity to marine benthos. In: Burton GA (Ed): Sediment Toxicity Assessment, Lewis Pub, Boca Raton, FL, pp. 183-211.

Loizidou M, Haralambous, KJ, Sakellarides PO (1992). Environmental study of the Marins Part II. A study on the removal of metals from the marianas sediment. Environ. Tech. 3: 245-252.

Magalhaes C, Coasta J, Teixeira C, Bordalo AA (2007). Impact of trace metals on denitrification in estuarine sediments of the Douro River estuary, Portugal. Marine Chemistry, 107:332-341.

Marcovecchio JE, Botte SE, Freije RH (2007). Heavy Metals, Major Metals, Trace Elements. In: Handbook of Water Analysis. L.M. Nollet, (Ed.). 2nd Edn. London: CRC Press, pp.275-311.

Mendie U (2005). The Nature of Water. In: The Theory and Practice of Clean Water Production for Domestic and Industrial Use. Lagos: Lacto-Medals Publishers, p. 1-21.

Mouvet C, Morhain E, Sutter C, Counturiex U (1993). Aquatic mosses for detection and follow up of accidental discharge in surface water. Water, Air Soil Pollut., 67: 333-347.

Newman MC, Mcintosh AW (1991). Metal Ecotoxicology:Concepts and Applications. Lewis Publishing, Michigan. p.399.

Ottaway JH (1978) the Biochemistry of Pollution. Come Lot Press London. p. 231.

Ozdamar K (1991) Biostatistics with SPSS. Kann press Eskisehir, pp 2-23.

Pizzaro I, Gomez M, Camara C, Palacios MA (2003). Arsenic speciation in environmental and biological samples –Extraction and stability studies. Anal. Chim. Acta 495: 85-98.

Radojevic M, Bashkin VN (1999). Practical Environmental Analysis. The Royal Society of Chemistry, Cambridge, p. 466.

Rodriguez VM, Jimenez-capdeville ME, Giordano M (2003). The effects of arsenic exposure on the nervous system. Toxicol. Lett. 145: 1-18.

Ramos L, Fernaudez MA, Gonzalez MJ, Hernandez LM (1999). Heavy metal pollution in water, sediments and earthworms from the Ebro River, Spain. Bull. Environ. Contam. Toxicol., 63: 305-311.

Sanders MJ, Preez HH, Van vuren, JHJ (1999). Monitoring cadmium and zinc contamination in fresh water systems with the use of the freshwater crab, Potamanautius warrenii. Water SA. 25(1): 91-98. http://www.wrc.org.za/archives/watersa%20archive/1999/January/jan 99_p91.pdf

Sharma R, Pervez S (2003). Enrichment and exposure of particulate lead in a traffic environment in India. DOI: 10.1023/A: 1024520522083 Environ. Geochem. Health, 25: 297-306.

Shuhaimi MO (2008). Metals concentration in the sediments of Richard Lake, Sudbury, Canada and sediment Toxicity in an Ampipod Hyalella azteca. J. Environ. Sci. Technol., 1: 34-41.

Singh M, Ansari AA, Muller G, Singh IB (1997). Heavy metals in freshly deposited sediments of Gomti River (a tributary of the Ganga River): Effects of human activities. Environ. Geol. 29 (3-4): 246-252.

Stephen SR, Alloway BJ, Carter JE, Parker E (2001). Towards the characterization of heavy metals in dredged canal sediments and an appreciation of availability. Environ pollt., 113: 395-401.

Turnland JR (1988). Copper nutrition, Bioavailability and influence of dietary factors. J. Am. Dietetic Assoc., 1: 303-308.

WHO (2004). Guidelines for Drinking Water Quality. 3rd Edn. World Health Organization, ISBN: 92- 4-154638-7. p. 516.

WHO (2007). Water for Pharmaceutical Use. In: Quality Assurance of Pharmaceuticals: A Compendium of Guidelines and Related Materials. 2nd Updated Edn. World Health Organisation, Geneva, 2: 170-187.

18

Concentrations of heavy metals in some pharmaceutical effluents in Lagos, Nigeria

Chimezie Anyakora[1*], Kenneth Nwaeze[1], Olufunsho Awodele[2], Chinwe Nwadike[1], Mohsen Arbabi[3], and Herbert Coker[1]

[1]Department of Pharmaceutical Chemistry, University of Lagos, Nigeria.
[2]Department of Pharmacology, University of Lagos, Nigeria.
[3]Department of Environmental Health Engineering, School of Health, Shahrekord University of Medical Sciences, Iran.

The concentrations of some heavy metals in the effluents of nine pharmaceutical companies operating in Lagos, Nigeria were determined using atomic absorption spectrophotometer. The heavy metals analyzed in this study included Cadmium, Chromium, Lead, Nickel, Zinc and Copper. Most of the samples were found to contain the metals in varying concentrations. The highest concentration of heavy metal detected was Zinc with concentration of 1.437 mg/L. Mostly, the concentrations were above the WHO recommended maximum contaminant concentration level. The highest concentrations were found to be 0.132 mg/L for Nickel, 0.644 mg/L for Lead, 0.337 mg/L for Copper, 0.280 mg/L for Cadmium, 1.437 mg/L for Zinc, and 0.491 mg/L for Chromium. This study reveals the need for enforcing adequate effluent treatment methods before their discharge to surface water to reduce their potential environmental hazards.

Key words: Heavy metals, pharmaceutical effluents, ecotoxicity, surface water.

INTRODUCTION

Various devastating ecological effects and human disasters in the last 40 years have arisen majorly from industrial wastes causing environmental degradation (Abdel-Shafy and Abdel-Basir, 1991; Sridhar et al., 2000). The discharges from these industries constitute biohazard to man and other living organisms in the environment because they contain toxic substances detrimental to health (Adebisi et al., 2007; Adriano, 2001; Bakare et al., 2003). Recently, there has been an alarming and worrisome increase in organic pollutants (Nadal et al., 2004). Since many effluents are not treated properly, these products are discharged on the ground or in the water bodies (Odiete, 1999), and most of these discharges to water bodies accumulate in the system through food chain (Odiete, 1999).

Pharmaceutical effluents are wastes generated by pharmaceutical industries during the process of drugs manufacturing. Their risk to human health and environmental species cannot be overemphasized. In Nigeria, the increase in demand for pharmaceuticals has resulted in a consequent increase in pharmaceutical manufacturing companies in the country and hence increased pharmaceutical waste which most times contain substantial amount of heavy metals. These effluents are usually discharged into the environment and when improperly handled and disposed, they affect both human health and the environment (Osaigbovo and Orhue, 2006; Ayodele et al., 1996; Anetor et al., 1999).The uncontrollable growing use of pharmaceutical products now constitutes a new challenge. Most pharmaceutical effluents are known to contain varying concentrations of organic compounds and total solids including heavy metals. Heavy metals such as Lead, Mercury, Cadmium, Nickel, Chromium and other toxic organic chemicals or phenolic compounds discharged from pharmaceutical industries are known to affect the surface and ground waters (Foess and Ericson, 1980). Due to mutagenic and carcinogenic properties of heavy metals, much attention has been paid to them since they have direct exposures to humans and other organisms

*Corresponding author. E-mail: canyakora@gmail.com.

(Momodu and Anyakora, 2010).

Heavy metals are natural components of the earth crust. These metals enter into living organisms through food or proximity to emission sources. They tend to bioaccumulate and are stored faster than excreted (Lenntech, 2006; Daniel et al., 1997; Davies et al., 2006). Industrial exposure accounts for a common route of contact in adults and ingestion for children (Roberts, 1999).

This study was aimed at determining the presence of six heavy metals, namely Lead, Chromium, Cadmium, Zinc, Copper and Nickel in the effluents of nine selected pharmaceutical companies in Lagos, Nigeria. The results obtained may form the basis for intervention by encouraging the pharmaceutical companies to effectively treat their effluent before being discharged into the environment.

MATERIALS AND METHODS

Chemicals and reagents

All chemicals and reagents were of analytical grade and were obtained from BDH Chemicals Ltd, UK. Conc. HNO_3 was used for the digestion of the samples while corresponding metal salts [namely: $CdCl_2.H_2O$, $Cu(SO4)_2$, $Zn(SO_4)_2$, $Pb(NO_3)_2$, $NiCl_2.6H_2O$ and $Cr(NO_3)_3$,] were used as standards.

Instrumentation

AAS instrument (PERKIN ELMER A. Analyst 200; Germany) consisting of a hollow cathode lamp, slit width of 0.7 nm and an air-acetylene flame was used for this work. The samples were analyzed for six heavy metals namely, Chromium which was analyzed at wavelength of 357.87 nm, Nickel at 232.00 nm, Cadmium at 228.80 nm, Zinc at 213.86 nm, Lead at 283.31 nm and Copper at 324.75 nm.

Sampling

Nine companies were identified due to their intense production; these companies are represented here with the letters A to I. Samples was taken at three different times between March and September, 2008. The pharmaceutical effluent samples were collected using thoroughly cleaned 250 ml Pyrex bottles each at the point where the effluents leaves the companies to the environment and stored in a refrigerator at 2 - 8°C before analysis. A total of 27 samples were collected at various peak production periods.

Sample preparation

To ensure removal of organic impurities and prevent interference during analysis, each of 50 ml volume sample was digested using 10 ml conc. HNO_3 in a 250-ml conical flask placed on a fume cupboard (Momodu and Anyakora, 2010). The samples were covered properly with aluminum foil to avoid spillage and heated on a hot plate until the solution reduced to 10 ml. This was allowed to cool and made up to mark with distilled water before filtering into a 50-ml standard flask, labeled and ready for analysis. The blank constituted 5% HNO_3.

Standard preparation

This served as the stock solution equivalent to 1000 ppm. Subsequently lower concentrations of 2 ppm, 4 ppm and 6 ppm were prepared from the stock by serial dilution. The same method was adopted for Cr, Ni, Pb, Cu and Zn.

RESULTS AND DISCUSSION

Calibration curves were obtained using a series of varying concentrations of the standards for all six metals. Calibration curve for all the metals were linear with a correlation coefficient of approximately one. The effluents analyzed contain some heavy metals in varying concentrations. Tables 1 to 3 give the summary of the results obtained in this study while Table 4 summarizes guidelines by different international agencies.

Nickel concentrations appear to be higher than the normal acceptable contaminant level according to WHO standard of 0.02 mg/L with company E showing the highest value of 0.132 mg/L. Companies A, C, B and D follow in a decreasing order of above normal concentration. However, companies G, H and I have no detectable Nickel concentration in their effluents, which could be as a result of the absence of the metal from the raw material in use at the time of sample collection. Nickel toxicity can cause a devastating histological change on plants and animal tissues (Moore, 1991).

Lead on the other hand gave an alarming result of 0.644 mg/L in company E followed by A with 0.241 mg/L. The WHO maximum contamination level is 0.01 mg/L. Only companies C and D gave a non-detectable result for lead. This may be due to its absence from the raw materials used at the time of sample collection. Lead poisoning could cause abdominal pain, loss of appetite, insomnia and constipation. Severe kidney as well as brain damage has been reported on long term exposure (Momodu and Anyakora, 2010). It is much more worrisome as it can substitute for calcium in bone causing skeletal anomalies especially in children (Bottcher and Hamman, 1986). Lead toxicity has been reported in some plant parts such as root and leaf most of which are either used as food or as medicine (Hrsak et al., 2000). The presence of lead in the air as a result of incomplete combustion from vehicles and generating sets in Nigeria especially at urban centers also contribute to the build-up of the metal. This is more worrisome as 80% of 140 million estimated populations of Nigerians live or migrate to the urban cities either for a short or long stay. Out of this population, over 15 million people reside in Lagos State - south west of Nigeria, where these effluents are discharged. The heavy metals in the effluents may end up in drinking water and food chain.

The Copper concentrations in batch 1 are all within the maximum acceptable concentration of 2.0 mg/L but companies H and I had no detectable copper in the effluents. Zinc concentrations in all samples fall below

Table 1. Heavy Metal concentration in the first batch effluent samples.

Company	Ni^{2+} (mg/L)	Pb^{2+} (mg/L)	Cu^{2+} (mg/L)	Cd^{2+} (mg/L)	Zn^{2+} (mg/L)	Cr^{2+} (mg/L)
A	0.044	0.241	0.070	0.015	ND	ND
B	0.029	0.079	0.097	0.007	ND	ND
C	0.030	ND	0.046	0.028	ND	ND
D	0.002	ND	0.003	0.023	ND	ND
E	0.132	0.644	0.337	0.280	0.206	ND
F	0.014	0.031	0.072	0.054	1.437	ND
G	ND	0.068	0.040	0.020	0.858	ND
H	ND	0.043	ND	0.006	ND	ND
I	ND	0.038	ND	0.016	ND	ND

Table 2. Heavy metal concentration in the second batch effluent samples.

Company	Ni^{2+} (mg/L)	Pb^{2+} (mg/L)	Cu^{2+} (mg/L)	Cd^{2+} (mg/L)	Zn^{2+} (mg/L)	Cr^{2+} (mg/L)
A	ND	0.278	0.056	0.065	0.190	0.174
B	ND	0.089	0.069	ND	0.294	0.343
C	ND	ND	0.038	0.206	0.241	ND
D	ND	0.090	0.102	0.098	0.248	0.206
E	ND	0.152	0.054	ND	0.217	0.278
F	ND	ND	0.028	ND	0.055	0.513
G	ND	0.304	0.150	0.040	0.050	ND
H	ND	ND	0.162	ND	0.356	0.491
I	ND	0.094	0.071	ND	0.198	0.381

Table 3. Heavy metal concentration in the third batch effluent samples.

Company	Ni^{2+} (mg/L)	Pb^{2+} (mg/L)	Cu^{2+} (mg/L)	Cd^{2+} (mg/L)	Zn^{2+} (mg/L)	Cr^{2+} (mg/L)
A	ND	ND	0.162	ND	0.693	0.461
B	ND	ND	0.102	ND	0.345	0 0.489
C	ND	ND	0.281	0.050	0.306	0.171
D	ND	ND	0.100	ND	0.820	0.387
E	ND	ND	0.145	ND	1.227	0.479
F	ND	ND	0.206	0.129	1.245	0.294
G	0.058	0.0368	0.102	0.064	0.084	ND
H	ND	ND	ND	ND	ND	0.415
I	ND	ND	0.175	ND	0.874	0.457

Table 4. Comparison between International drinking water and FDA bottled water guidelines for the parameters analyzed in the study.

Parameter	USEPA MAC (mg/L)	Canada MAC (mg/L)	EU MAC (mg/L)	Japan MAC (mg/L)	WHO guideline (mg/L)	Bottled water US Federal drug administration level (mg/L)
Chromium	0.1	0.05	0.05	0.05	0.05	0.1
Cadmium	0.005	0.005	0.005	0.1	0.003	0.005
Copper	1.3	1.0	2.0	1.0	1-2	1.0
Lead	0.015	0.01	0.01	0.05	0.01	0.005
Nickel	0.1	-	0.02	0.01	0.02	-
Zinc	5.0	5.0	NS	1.0	3-5	-

USEPA – United State Environmental Protection Agency; MAC – Maximum Allowable Concentration; EU – European Union; WHO – World Health Organization.

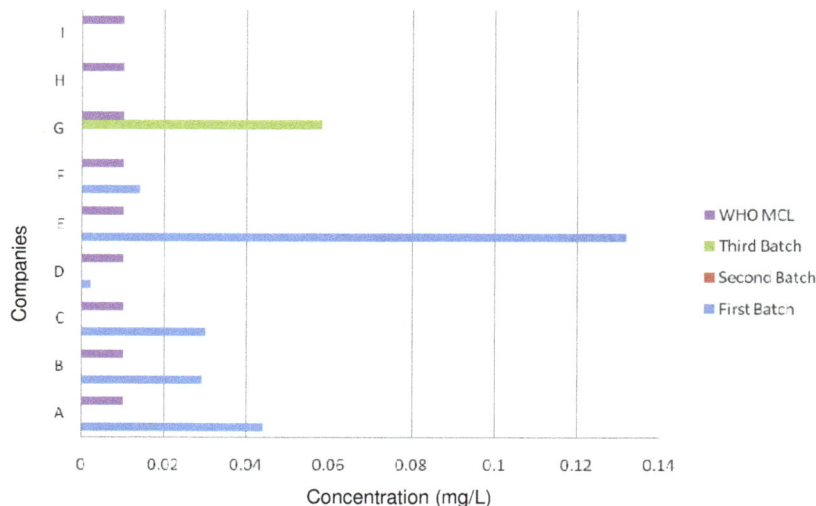

Figure 1. Nickel concentration in the samples.

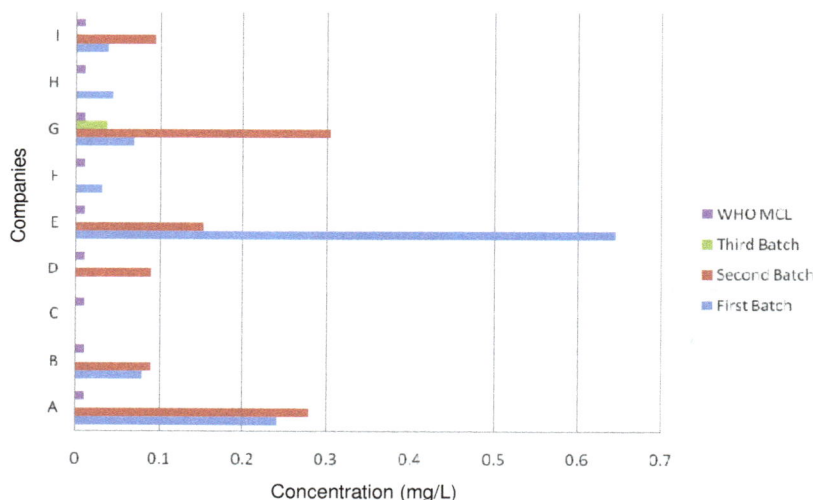

Figure 2. Lead concentration in the samples.

the WHO acceptable maximum of 5.0mg/L. Chromium was not detected in the samples. Cadmium concentrations were found to be higher than the WHO acceptable maximum of 0.003 mg/L. Company E effluent had cadmium concentration of 0.280 mg/L. Cadmium toxicity has been reported to cause food poisoning, mutation, hypertension, and cancer among others (Bottcher and Hamman, 1986). Long term exposure to cadmium has been found to cause serious damage to kidney, liver, bone and blood.

In subsequent batches, the concentrations of the different metals in the effluents fluctuated significantly giving credence to the fact that the metals are due to particular product being manufactured. For instance in the second and third sets of samples, Nickel concentration was only detected in one company's

effluent. This variation occurred for all metals. In batch 2, 70% of the samples contain high lead and chromium concentrations and 80% of the samples had detectable chromium concentration in the third batch. In the same batch, 90% of the samples had no detectable lead concentration. Figures 1 to 6 show the distribution of the metals in different samples, and they illustrate these variations well.

Conclusion

The results obtained in this study revealed a high presence of Nickel, Lead, Cadmium and Chromium in the effluents from the pharmaceutical companies analyzed. This may pose adverse consequences on health and

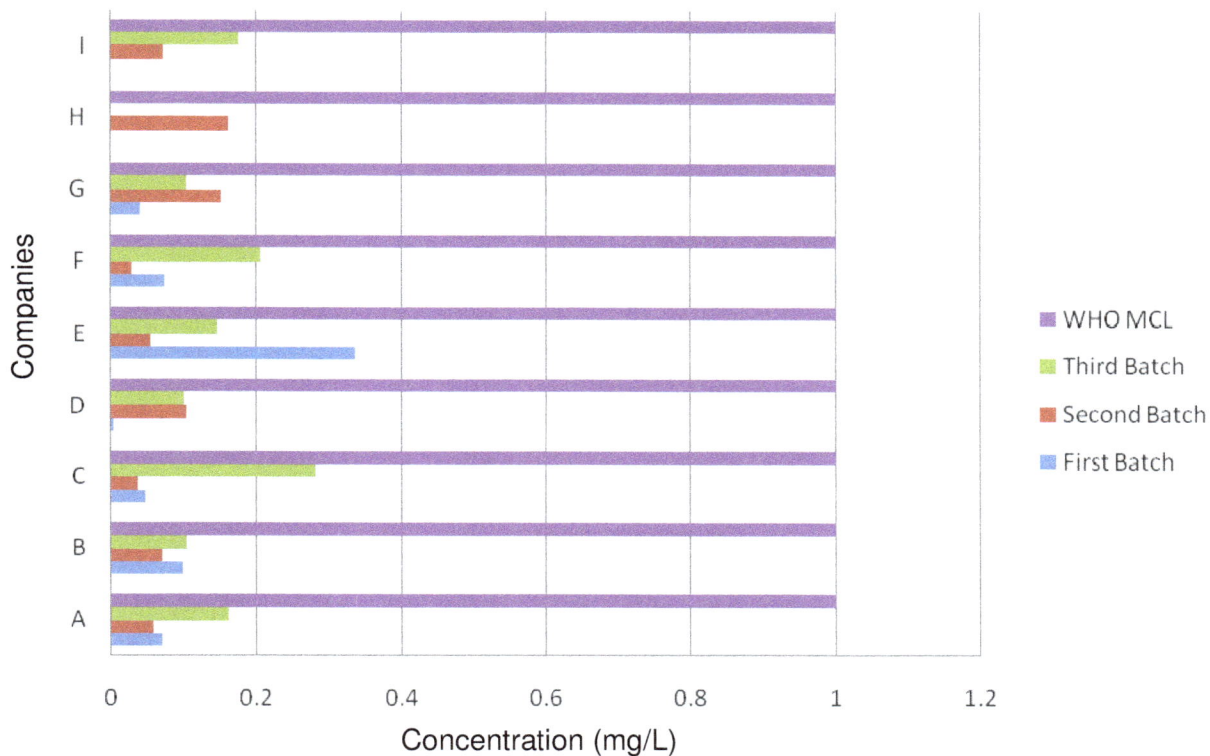

Figure 3. Copper concentration in the samples.

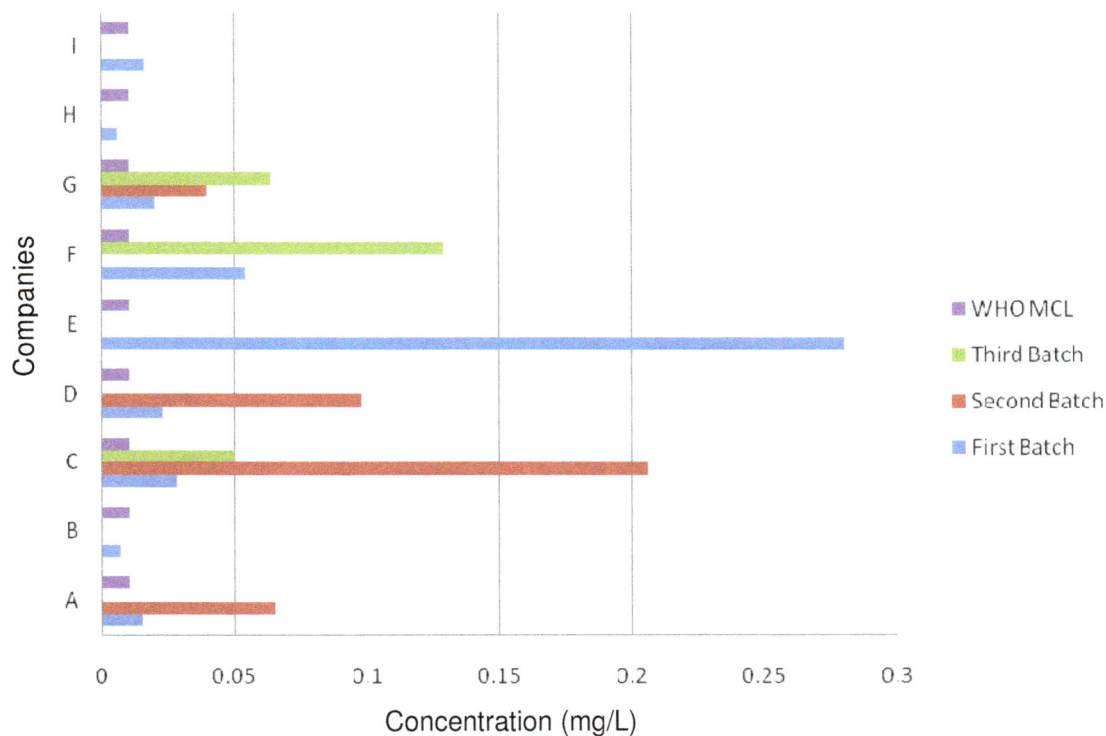

Figure 4. Cadmium concentration in the samples.

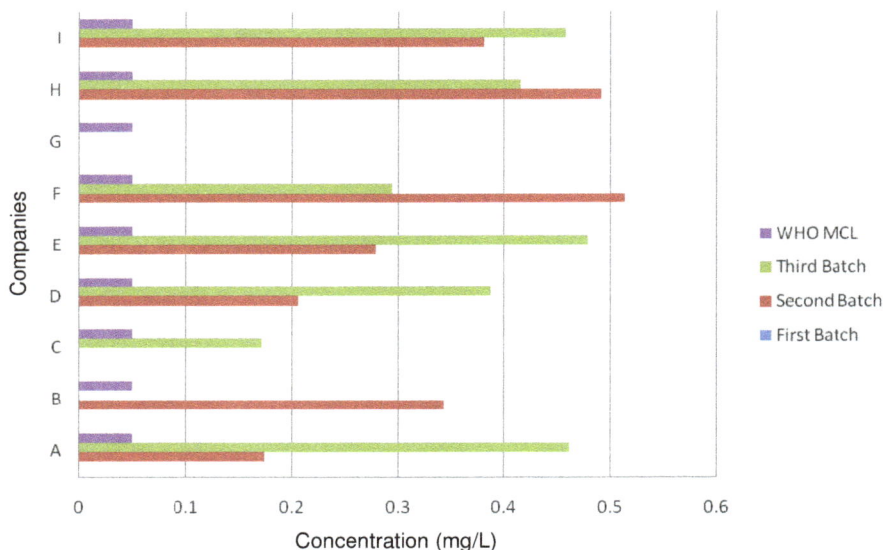

Figure 5. Chromium concentration in the samples.

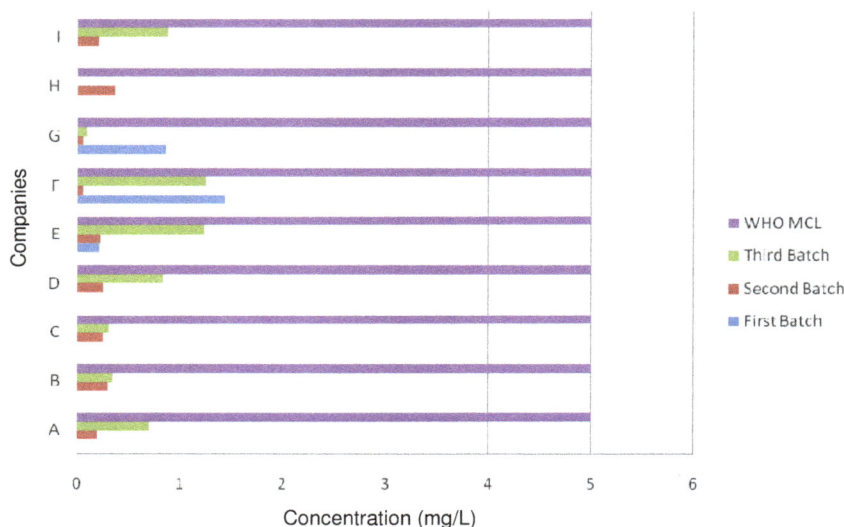

Figure 6. Zinc concentration in the samples.

environment. Therefore, there is an urgent need to enforce effluent treatment to reduce such environmental and health risks.

REFERENCES

Adriano D C (2001). Trace Elements in Terrestrial Environments: Biochemistry, bioavailability and risks of metals, Springer Verlag, p. 867.

Abdel-Shafy HI, Abdel-Basir SE (1991). Chemical treatment of industrial wastewater. Environ. Manage. Health, 2: 19-23.

Adebisi SA, Ipinromiti KO, Amoo IA (2007). Heavy Metals Contents of Effluents and Receiving Waters from Various Industrial Groups and their Environs. J. Appl. Sci., 2(4): 345-348.

Anetor JI, Adeniyi FA, Taylor GO (1999). Biochemical indicators of metabolic poisoning associated with lead based occupations in nutritionally disadvantaged communities. Afr. J. Med. Med. Sci., 28: 9-12.

Ayodele JT, Momoh RU, Amin M (1996). Determination of heavy metals in Sharada Industrial effluents, in Water Quality Monitoring and Environmental Status in Nigeria, Proceedings of the National Seminar on Water Quality Monitoring and Status in Nigeria, powered by Federal Environmental Protection Agency and National Water Resources Institute. Federal Environmental Protection Agency, October 16-18, 1991: 158-166.

Bottcher DB, Haman DZ (1986). Home Water Quality and Safety. Florida Co-operative Extension Services. Circular, p. 703.

Bakare AA, Lateef A, Amuda OS, Afolabi RO (2003). The aquatic toxicity and characterization of chemical and microbiological constituents of water samples from Oba River, Odo-oba, Nigeria.

Asian J. Microbiol. Biotechnol. Environ. Sci., 5: 11-17.

Daniel WE, Stockbridge HL, Labbe RF, Woods JS, Anderson KE, Bissell DM, Bloomer JR, Ellefson RD, Moore MR, Pierach CA, Schreiber WE, Tefferi A, Franklin GM (1997). Environmental chemical exposure and disturbances of hemesynthesis. Environ Health Perspect., 105 (Suppl 1): 37-53.

Davies OA, Allison ME, Uyi HS (2006). Bioaccumulation of heavy metals in water, sediments and periwinkle (Tympanotonus fuscatus var radula) from the Elechi Creek, Niger Delta: Afr. J. Biotechnol., 5: 10.

Foestner U, Wittmann GTW (1979). Metal Pollution in the Aquatic Environment. Springer, Heidelberg, p. 486.

Foess GW, Ericson WA (1980). Toxic Control. The trend of the future. Water Wastes Eng., pp. 21-27.

Ge Y, MacDonald D, Suave S, Hendershot W (2005). Modeling of Cd & Pb Speciation in soil solutions by WinHumicV and NICA-Donnan Model. Environmental Modeling and Software, 20: 353-359.

Ge Y, Murray P, Hendershot WH (2000). Trace metal speciation and bioavailability in urban soils. Environ. Pollut., 107: 1.

Goyer RA (1989). Mechanism of lead and cadmium nephrotoxicity.Department of pathology, University of Western Ontario, London N6A 5CI, Canada. Toxicol. Lett., 46 (1-3): 153-162.

Hrsak J, Fugass M, Vadjic V (2000). Soil Contamination by Pb, Zn, and Cd from a Lead smeltry. Environ. Monit. Assess., 60: 359-366.

Lenntech, "Water treatment and air purification" (2006). In: http://www.Lenntech.com/heavy-metals.htm.

McGrath SP, Smith S (1990). Chromium and Nickel, In: ALLOWAY .B.J. (Ed.). Blackie Glasgow. Heavy Metals, pp.125-150.

Momodu M, Anyakora C (2010). Heavy Metal Contamination of Groundwater: The Surulere Case Study. Res. J. Environ. Earth Sci., 2(1): 39-43.

Moore JW (1991). Inorganic Contaminants in surface water. Sprenger Verlag, New York, pp. 21-27.

Nadal M, Schuhmacher M, Domingo DL (2004). Metal pollution of soils and vegetation in an area with petrochemical industries. Sci. Total Environ., 321(1-3): 59-69.

Odiete WO (1999). Impacts associated with water pollution in Environmental Physiology of Animals and Pollution. Diversified Resources Ltd, Lagos 1st edition, pp. 187-219.

Osaigbovo, AE, Orhue ER (2006). Influence of pharmaceutical effluents on some soil chemical properties and early growth of maize (Zea mays L). Afr. J. Biotechnol., 5: 18.

Roberts JR (1999). Metal Toxicology in Children. In Training Manual on Pediatric Environmental Health: Putting it into practice Emeryville, C. A! Children's Environmental Health Network, p. 27.

Sawyer CN, McCarty PI, Parkin GF (1994). Chemistry for Environmental Engineering. McGraw-Hill, New York, pp. 634-635.

Sridhar MKC, Olawuyi JF, Adogame, LA, Okekearu OCO, Linda A (2000). Lead in the Nigerian environment: problems and prospects, in 11th Annual International Conference on Heavy Metals in the Environment. University of Michigan, School of Public Health, p. 862.

19

Heavy metal impact on growth and leaf asymmetry of seagrass, *Halophila ovalis*

Rohani Ambo-Rappe[1]*, Dmitry L. Lajus[2] and Maria J. Schreider[3]

[1]Department of Marine Sciences, Faculty of Marine Sciences and Fisheries, Hasanuddin University, Tamalanrea Km. 10 Makassar, 90245, Indonesia.
[2]Department of Ichthyology and Hydrobiology, Faculty of Biology and Soil Sciences, St. Petersburg State University 16 line V.O. 29, 199178, St. Petersburg, Russia.
[3]School of Environmental and Life Sciences, Ourimbah Campus, Newcastle University, Brush Road, Ourimbah, NSW, 2258, Australia.

A major threat to the seagrass ecosystem worldwide, due to the growth of human population along the coastal environment, is pollution or contamination resulting from industrial and urban development. Although seagrass appears to be rather resistant to heavy metal contaminants, these substances may possibly harm some components of the seagrass and such responses have not been examined to a significant extent. Lead (Pb) and copper (Cu) was tested on seagrass, *Halophila ovalis*, to see whether the metals are environmental stressor on the seagrass. Reduced growth rate of the seagrass was observed both in Pb and Cu treatments. Leaf size of the plant also reduced as the metal concentrations increased and when the plants were exposed to the heavy metal for longer duration. An increased leaf asymmetry was more apparent at the 2 mg/L Cu treatment and no significant increases in fluctuating asymmetry were found in Pb treatment or in low levels of Cu treatment. Further discussion were made in view of selecting non-costly bioindicators of heavy metal contamination.

Key words: Bioindicators, fluctuating asymmetry, *Halophila ovalis*, heavy metals, seagrass.

INTRODUCTION

Seagrass habitats are subjected to stronger anthropogenic pressure than many other marine communities. This is particularly related to the close proximity of this habitat to human activities (Walker et al., 2001; Duarte, 2002). Heavy metals are significant environmental contaminants of seagrass systems (Pergent-Martini and Pergent, 2000). Heavy metals can be incorporated into seagrass leaves and vascular tissue from either water column or sediments. In locations where elevated concentration of metals was suspected, seagrass leaves also contained an elevated concentration of metals (Ambo Rappe et al., 2007).

The presence of heavy metals in both water and sediment has been demonstrated to inhibit the growth of seagrass (Ward, 1989). Moreover, toxic concentrations of metals inhibited metabolic activity and interfered with vital biochemical pathways, such as photosynthesis (Ralph and Burchett, 1998). Therefore, it is essential to understand whether metal can kill, permanently damage or merely cause stress to the seagrass.

Morphological traits of seagrass, in particularly leaf dimension, have been investigated for use as indicators of environmental quality. It was found that narrower leaves were developed in more stressful conditions (McMillan, 1978; McMillan and Phillips, 1979; Phillips, 1980). Another morphological trait that can be used for assessing stress is fluctuating asymmetry (FA). FA represents the random deviations from perfect symmetry and usually increases under stressful conditions (Tracy et al., 1995; Kozlov et al., 1996; Anne et al., 1998; Hosken et al., 2000; Mal et al., 2002; Tan-Kristanto et al., 2003).

*Corresponding author. E-mail: rohani_amborappe@yahoo.com.

By its nature, FA represents random component of phenotypic variance standing on equal footing with genotypic and environmental components (Lajus et al., 2003).Fluctuating asymmetry has been proposed as a tool for monitoring the quality of the environment and is being considered as a sensitive monitor of stress (Tracy et al., 1995; Anne et al., 1998; Lajus and Zhang, 2003). It has been claimed to be impacted at concentrations less than those required to impact life history features (Anne et al., 1998; Hoffmann and Woods, 2003).

Moreover, this technique has been recommended because it is biologically relevant, non-destructive, and time- and cost-effective (Tracy et al., 1995). Thus, the main goal of this study was to observe the heavy metals effect on growth and leaf asymmetry of seagrass in a controlled environment.

MATERIALS AND METHODS

Plant material

Halophila ovalis was the seagrass species chosen for this study due to the ability of this species to grow faster than other seagrasses (Butler and Jernakoff, 1999), and its convenient size for the laboratory condition. *H. ovalis* were collected from uncontaminated area of Lake Macquarie (Fennel Bay; 151°66'E, 33°08'S) and transported to the laboratory free of sediment in the container filled with lake water, following McMillan (1980). The uncontaminated site were selected based on result from Ambo Rappe et al. (2007).

In the laboratory, the plants were sorted based on the size and the number of leaves. Individual with 3 pairs of mature leaves and 1 growing leaf were selected. The first pair of mature leaves was harvested and placed it in labeled jars for further analysis.The plants were grown in culture tubs (11.5 × 5 × 17 cm) filled with terrestrial sandy loam sediment. *In situ* sediment was not used as it may contain an unknown amount of contaminants, following Ralph and Burchett (1998).

Each culture tub contained 5 individual plants, which were arranged in the same direction of growth (that is, grown plant with older leaves was in the right section of the tub followed by the younger part to the left side). Twelve 15 L glass tanks were filled with filtered (200 μm) seawater and two culture tubs of *H. ovalis* were placed in each tank under 200 μmol quanta m^{-2} s^{-1} with a photoperiod of 16 h light and 8 h dark. The tank has a filter installed to aerate and filter the water during the entire experiment.

Heavy metal experimentation

A fully randomized experimental design with six treatments in two replicates totaling 12 tanks was used. The treatments were lead (10 and 50 mg/L), copper (0.5, 2, and 4 mg/L), and control (no addition of heavy metals). Lead and copper were selected for this experiment because these metals represent non-essential and essential metal, respectively, for seagrass growth. The metal concentrations were used in this study based on previous findings that exposure to copper concentrations more than 4 mg/L have a lethal effect on *H. ovalis*, whereas only limited effects of lead were shown to concentrations up to 10 mg/L (Ralph and Burchett, 1998).

Lead (Pb) and copper (Cu) solutions were produced by dissolving lead (II) nitrate ((Pb(NO$_3$)$_2$) and copper chloride salt (CuCl$_2$), respectively, with distilled water. 10 000 mg/L Pb and 1000 mg/L Cu stock solutions were made. Since the volume of the tank

was 15 L, addition of 15 and 75 ml of 10 000 mg/L Pb stock solution produced the treatment concentration of 10 and 50 mg/L Pb, respectively. Addition of 7.5, 30, and 60 ml of 10 000 mg/L Cu stock solution produced the treatment concentration of 0.5, 2, and 4 mg/L Cu, respectively. To remove any treatment effect caused by adding different amounts of stock solution, 60, 67.5, 45, 15, and 75 ml of distilled water was added to the 10 mg/L Pb treatment, 0.5 mg/L Cu treatment, 2 mg/L Cu treatment, 4 mg/L Cu treatment, and the control, respectively. Tanks were arranged in line in a random order to remove possible differences of environmental conditions in laboratory.

Water quality parameters such as salinity, temperature, pH, dissolved oxygen, and turbidity were measured weekly using Yeo-Kal (Model 611) water quality analyzer. Salinity was maintained by addition of tap water to compensate for losses due to evaporation, following McMillan and Moseley (1967) and McMillan (1976). The water quality parameters remained steady and constant between control and heavy metal dosed tanks with salinity values in the range of 33.88 to 34.08 ppt, temperature 26.65 to 26.79°C, pH 8.00 to 8.03, DO 5.05 to 5.27 ppm and turbidity 0.56 to 0.89 ntu. Therefore, the water quality parameters among the tanks do not appear to have had an effect on the overall stress response in this study.

Growth measurement

Images of the plants inside the tanks were taken using a digital camera at the beginning of experiment and then at 2, 5, 10, 18, 34, 42 and 51 days after the treatment. The measurements of the images were taken using 'Image Tool' software (developed at the University of Texas Health Science Centre at San Antonio, Texas). The overall growth of the *H. ovalis* during the experiment was estimated by counting the number of nodes and measuring the length of the rhizome. Growth rate per day was estimated from each treatment at two times, Time 0 day (T1) to time 18 days (T2) and time 18 days (T1) to time 51 days (T2), using the following formula:

(Length of the rhizome at T2 - Length of the rhizome at T1) / number of days

Fluctuating asymmetry (FA) and leaf dimension measurement

For the purpose of FA and leaf dimension data, a pair of leaves was harvested at the start of the experiment (before metals treatment), 18 days after the treatment, and 51 days (the end of the experiment). Time of harvesting was selected based on the information from the previous pilot study that *H. ovalis* produced a mature leaf at about 16 days in the tank. However, some of the plants in the copper treatments did not grow a mature leaf as quickly as other treatments; therefore, time 18 days was chosen to give sufficient time for these copper treatments. The final data of FA was taken at 51 days after observation that there was no further signs of new leaf growth, especially from copper treatment, thus the experiment was finalized.

Eight traits were analysed for leaf asymmetry of *H. ovalis* including number of intersections of lateral veins with the peripheral vein (Trait 1), number of intersections of lateral veins with the central vein (Trait 2), length of peripheral vein (Trait 3), length of central vein (Trait 4), length of first lateral vein (Trait 5), leaf perimeter (Trait 6), leaf length (Trait 7), and leaf width (Trait 8). Traits 1 to 5 were bilateral characters that measured leaf asymmetry within a leaf (Figure 1), whereas traits 6 to 8 was bilateral characters that measured leaf asymmetry between a pair of leaves. Length of veins was measured as sum of distances between intersections of lateral veins with peripheral or central

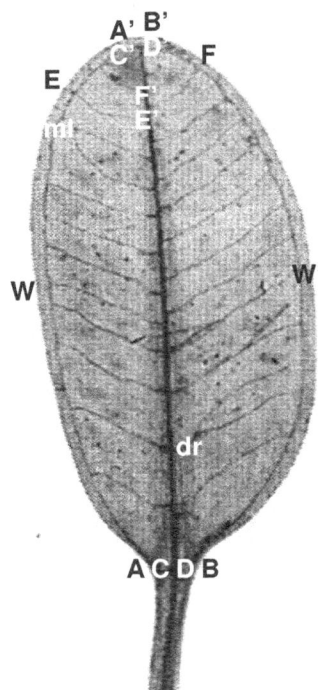

Trait 1, number of lateral vein intersections to the peripheral vein; A-A'(left) & B-B' (right)

Trait 2, number of lateral vein intersections to the central vein; C-C'(left) & D-D' (right)

Trait 3, length of peripheral vein: A-A' (left) & B-B' (right)

Trait 4, length of central vein: C-C' (left) & D-D' (right)

Trait 5, length of first lateral vein: E-E' (left) & F-F' (right)

Leaf-width (W-W') and leaf-length (D-D').
Additional points mark location of diverging or merging of veins using symbol
ml (merging on the left side),
dl (diverging on the left side),
mr (merging on the right side),
dr (diverging on the right side).

Figure 1. Measurement of fluctuating asymmetry (Traits 1 to 5) on the leaf of *H. ovalis*.

veins. Leaf perimeter was measured as a sum of distances between intersections of lateral veins with peripheral veins on both the left and the right side of the leaf. The length of mid-vein was measured as a leaf length and the width of leaf was measured from one side of the leaf margin to the other side at its widest section.

The initial and repeat measurements of each trait were evaluated by conducting two-way mixed ANOVA with side and individual as random factors. The test revealed that measurement error was responsible for 17.03, 17.28, 9.32, 168.08, 16.74, 11.77, 8.48 and 15.63% of traits 1 to 8, respectively. Due to the high measurement error on trait 4, this trait was excluded from further analyses. Other traits have low measurement error (< 20%) and could thus be used for the next step of analysis.

The two-way ANOVA also revealed the non-significant difference in factor "Side" in traits 1, 2, 3 and trait 8, indicating absence of directional asymmetry (DA). Other traits (Traits 4 to 7) showed directional asymmetry. Palmer and Strobeck (2003) suggested avoiding using traits that exhibit DA because it would complicate the FA analyses. Therefore, traits 5 to 7 were also excluded from further FA analysis.

There was also a significance of the interaction factors "Individual x Side" for all traits indicating the presence of non-directional asymmetry, which may relate to FA. However, it is important to know if another type of non-directional asymmetry, antisymmetry, contributes in the non-directional asymmetry observed in this data. To test for antisymmetry, the distribution of left minus right within each trait selected was observed visually and checked for the departure from normality using Kolmogorov-Smirnov test (Sokal and Rohlf, 1995). The data was also checked for statistical outlier that is a common source of skew or leptokurtosis in studies of fluctuating asymmetry (Palmer, 1994).

Kolmogorov-Smirnov test revealed that (L-R) distributions of trait 8 did not deviate from the normal distribution, but the distribution

deviated from zero for traits 1 to 3. Significant positive kurtosis was detected for (L-R) distribution of traits 1 to 3 ($p<0.001$) indicating leptokurtic distribution (Sokal and Rohlf, 1995). Leptokurtosis can be caused by number of reasons such as outliers, measurement errors, and the mixture of ideal FA and antisymmetry (Palmer and Strobeck, 2003). The other two reasons (outliers and measurement errors) were controlled in this study, however no methods have yet been suggested to statistically correct for the presence of antisymmetry, therefore care should be taken to show that inferred differences in developmental stability among samples are not confounded by this factor (Palmer, 1994).

From the two-way ANOVA test, significant variation was also found dealing with the overall trait size or shape among individuals. A dependence of asymmetry on trait size can influence inferences made in studies of developmental stability, therefore, a size-correction was conducted by standardizing the asymmetry data with leaf length. The FA index was then calculated based on unsigned left minus right characters $|L - R| / (R + L)$ (Palmer, 1994).

Statistical analysis

Analysis of variance (ANOVA) (GMAV 5, University of Sydney 2000) was conducted to compare all the variables measured between metal treatment (Pb and Cu) and control.

The length of plant and number of nodes were analyzed using 2 fixed orthogonal factors ANOVA: Time (8 levels) and treatment (6 levels), with $n = 20$ per combination of factors.

FA of bilateral characters within a leaf (Traits 1, 2, and 3) and leaf dimension characters (leaf length and leaf width) were first analyzed prior to the metal treatment (Time 0 day) using four factors ANOVA: Treatment (6 levels, fixed orthogonal), tank (2 levels, nested in treatment), tub (2 levels, nested in tank), plant (5

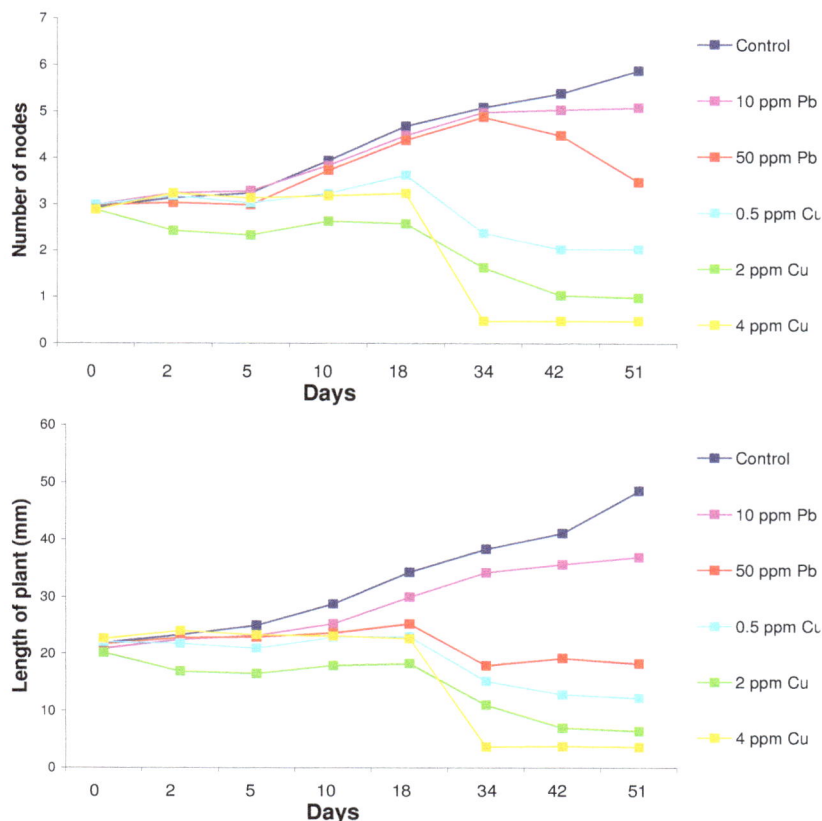

Figure 2. Mean number of nodes ($n = 20$) and length of plants ($n = 20$) at different metals treatments and at different duration of the experiment

levels, nested in tub), with $n = 2$ per combination of factors. Whereas, FA of bilateral character within the plant (Trait 8) was analyzed at time-0 day using three factors ANOVA: Treatment (6 levels, fixed orthogonal), tank (2 levels, nested in treatment), tub (2 levels, nested in tank), with $n = 5$ per combination of factors. These analyses were performed to see whether there were other factors rather than metal treatment confounding the result.

For the final result, data on FA and leaf dimension characters were pooled from individual plants, tubs, and tanks due to the limited number of data available, especially from the metal treatments. Therefore, analysis was only performed on the more important factors, in this case, treatment and time. Thus, two fixed factors ANOVA: Time (3 levels) and treatment (6 levels) was used. Prior to the analysis of variance, all data were tested for heterogeneity of variance using Cochran's test. When the Cochran's test was significant, the data were transformed. For all ANOVA analyses, significance was determined at the $\alpha = 0.05$. Moreover, Student-Newman-Keuls (SNK) tests were used if a significant effect was found to determine which level of the factor differed significantly.

RESULTS

Growth responses

Growth rate of *H. ovalis* was influenced by heavy metals (lead and copper) ($F_{5,114} = 10.17$, $p < 0.001$) and SNK tests indicated a difference between control and all metal treatments, with exception for 10 mg/L Pb. Plants in the control treatment grew constantly at a rate of 0.69 mm/day on the first 18 days and continued to grow at a slower rate (0.43 mm/day) during the period of 18 to 51 days. Although plants in the 10 mg/L Pb had a similar pattern of growth to the control, but the growth was lower (0.5 mm/day on the first 18 days and 0.21 mm/day at 18 to 51 days). Plants exposed to 50 mg/L Pb had a slower growth rate than plants in 10 mg/L Pb ($p < 0.05$) and the difference with control was even greater ($p < 0.01$).

The growth of the plant in the 50 mg/L lead treatment was only observed for up to 18 days of the experiment (Figure 2). Plants treated with the relatively low concentrations of Cu (0.5 mg/L) showed a significantly slower growth rate than the control ($p < 0.01$), and the growth was also observed for up to 18 days only. The toxic effects of copper on the growth of *H. ovalis* were more apparent at the higher concentrations (2 to 4 mg/L), since increasing inhibition of rhizome elongation was observed with increasing Cu concentration (Figure 2). The rhizomes of the plants (measured for this parameter) were still intact but some appeared decayed over the time. Therefore, the measured length was also becoming smaller.

Table 1. Summary of ANOVA for leaf length, leaf width, leaf width to length ratio of *H. ovalis* after the metal treatments.

Source of variation	df	Leaf length (LL)		Leaf width (LW)		Ratio LW/LL	
		MS	F	MS	F	MS	F
Time	2	64.03	244.83 ***	14.23	250.02 ***	0.004	17.86 ***
Treatment	5	5.80	22.19 ***	0.76	13.3 ***	0.002	8.55 ***
Time × Treatment	10	2.29	8.75 ***	0.43	7.57 ***	0.001	4.56 ***
Residual	18	0.26		0.06		0.000	

Cochran's tests were not significant, data were not transformed.*** denoted significant at $p<0.001$.df, degrees of freedom, MS, mean square, F, Fisher-test.

There was a positive correlation between length of plants (measured from the length of rhizome) with the number of nodes (Spearman correlation, $p<0.001$). The plants become longer as the number of nodes increases (Figure 2). In relation to the inhibition of growth, the length of the plant (measured from the length of rhizome) and the number of nodes were also influenced by the heavy metals ($F_{5.912} = 40.99$, $p<0.001$ and $F_{5.912} = 58.54$, $p<0.001$, respectively). SNK tests indicated a difference between control and all metal treatments. The plants exposed to 50 mg/L Pb and to different level of copper treatments were shorter with fewer nodes than control ($p<0.01$). The plants exposed to 10 mg/L Pb were shorter compared to the control ($p<0.05$), but there was no difference to control in the number of nodes.

The duration of metal exposure influenced the length of the plants and the node number ($F_{7.912} = 2.04$, $p<0.05$ and $F_{7.912} = 4.15$, $p<0.001$). Moreover, the significant effect of the interaction between the metal treatment and the metal exposure indicated that exposure time also influenced the effects of the metals in determining the plant characters ($F_{35.912} = 5.23$, $p<0.05$ for length of plant and $F_{35,912} = 7.33$, $p<0.001$ for number of nodes). SNK test revealed a difference between 0 day and 34 to 51 days of the metal exposure. Plants exposed to 10 mg/L Pb show a decrease in length and number of nodes after 51 days of exposure, whereas plants exposed to higher Pb concentration (50 mg/l Pb) show a decrease in length after 34 days of exposure and a decrease in number of nodes after 42 days of exposure. Plants treated with different levels of copper decreased the length and number of nodes after 34 days of copper exposure (Figure 2).

Leaf dimension characters

The initial leaf length and leaf width, before adding metal treatments, did not differ among control and the various treatment levels. There was also no difference in these parameters among the tubs used as a media culture. The only difference was observed among the individual plants within the tub, but this factor was pooled in the final analysis. After the treatment, leaf dimension characters (e.g. leaf length, leaf width and leaf width to length ratio) were significantly different between treatments and control and also between times (Table 1).

In this analysis, data were pooled from tubs and individual plants due to the limited number of leaves which survived. The mortality of leaves was especially high in copper treatments where more than 50% of the leaves became senescent and detached from the rhizome only after 2 days of the copper treatment. SNK tests indicated a difference between the control and copper treatments, but no such difference was observed between control and lead treatments. Plants exposed to lead had a similar leaf length and width to control, but plants treated with copper had a smaller leaves than the control (Figure 3). SNK tests also indicated an effect of the metal treatment when comparing leaf length and width in two intervals: From the beginning of the experiment up to 18 days after treatment, and for days 18 to 51. There was a general pattern across the treatments that new leaves growing in the tank were smaller in size (in term of reduced leaf length and width) with time. This general pattern may be due to the effect of the culture tank that had no nutrition addition during the experimentation. This phenomenon was also previously observed in the pilot study.

The interaction of metal treatment and exposure time of the treatment was also analyzed using SNK tests and the tests indicated plants in control and lead treatments only reduced their leaf length and width at 18 days and the leaves that grew after this period had the same or even larger size. Plants exposed to copper treatments, however, continued to have smaller leaves throughout the time with the effect becoming more apparent at the higher Cu concentrations (Figure 3).

High metal concentrations (50 mg/L Pb and 2 to 4 mg/L Cu) caused also change of shape of leaves measured as width to length ratio as indicated by SNK tests. Plants exposed to these higher metal concentrations had larger width to length ratio (that is, wider leaves) than control. The SNK tests also revealed a difference in width to length ratio of the plant between 0 and 51 days, but the pattern was not consistent among the treatments.

Leaf length (LL)

☐ 0 day ■ 18 days ☐ 51 days

Leaf width (LW)

☐ 0 day ■18 days ☐ 51 days

Ratio LW/LL

☐ 0day ■18 days ☐ 51 days

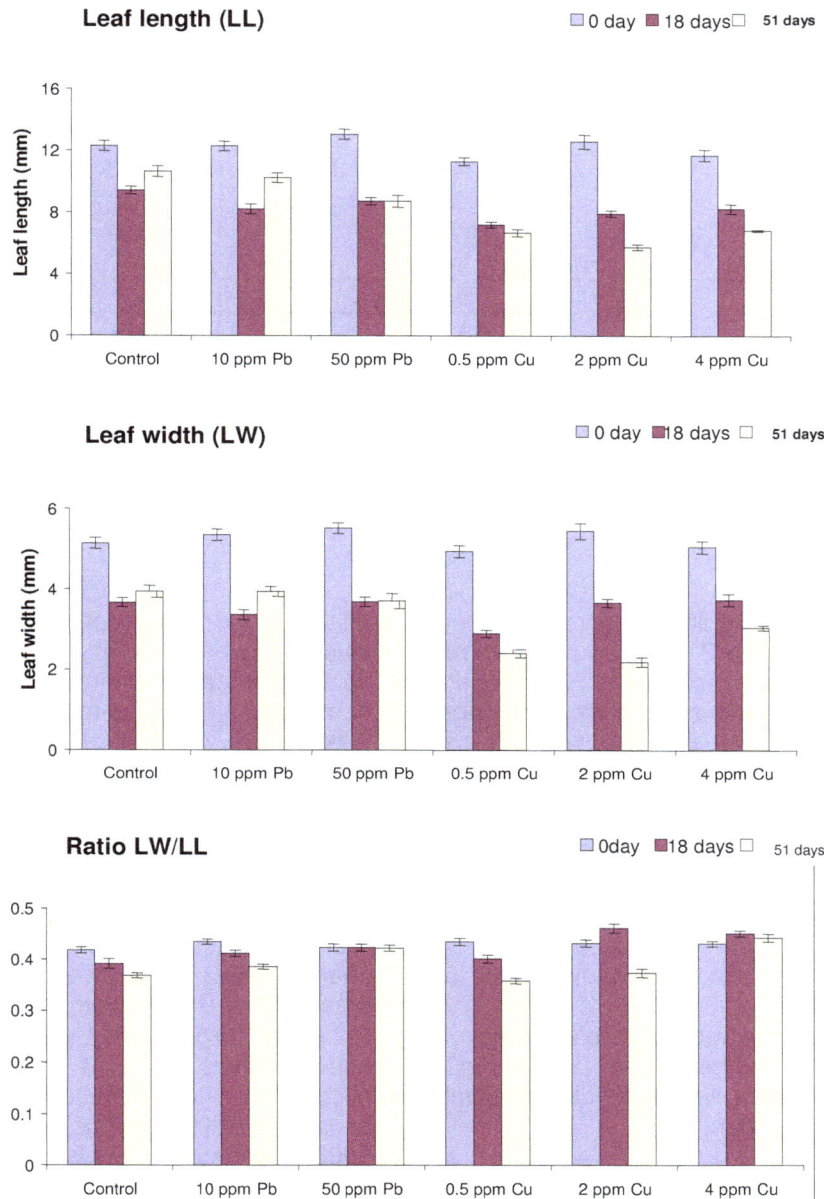

Figure 3. Mean (±SE, n = 40) length, width, and width to length ratio of *H. ovalis* leaves at different treatment levels and at different duration of the experiment.

Fluctuating asymmetry (FA)

The fluctuating asymmetry characters (Traits 1, 2, 3 and 8) before adding the metals did not differ among the control and different treatments. There was also no difference in fluctuating asymmetry between leaves from different tanks and from different individual plants. After the treatment, however, there was an increased fluctuating asymmetry (showing from all traits measured) between treatments (Table 2) with SNK tests indicating a difference between control and 2 mg/L Cu, which exhibited an increased FA. Although other metals treatments show a higher FA than control, the differences were not significant (Figure 4).

The difference in time exposure of the metals on the leaf asymmetry was significant for traits 1 and 2 (Table 2) with SNK tests indicated a significant increased in leaf asymmetry after 51 days of the metal exposure. Traits 1 and 2, which are meristic characters dealing with number of veins, show a similar effect in relation to the treatment. These traits were correlated each other with the coefficient of correlation (r = 0.83). These traits are in concordance, thus using one of the traits will be able to reflect the FA in the same way as using both traits. These

Table 2. Summary of ANOVA for FA characters of *H. ovalis* leaf after the metal treatments.

Source of variation	df	Trait 1		Trait 2	
		MS	F	MS	F
Time	2	0.51	5.64 *	0.46	5.21 *
Treatment	5	0.34	3.76 *	0.38	4.33 **
Time × Treatment	10	0.19	2.13 ns	0.12	1.39 ns
Residual	18	0.09		0.09	

Source of variation	df	Trait 3		Trait 8	
		MS	F	MS	F
Time	2	0.00	1.39 *ns*	0.00	0.93 *ns*
Treatment	5	0.00	4.94 **	0.00	4.88 **
Time × Treatment	10	0.00	2.79 *	0.00	2.24 *ns*
Residual	18	0.00		0.00	

Traits 3 and 8 were not transformed (Cochran's test >0.05), Ln(X) transformation for traits 1 and 2. *ns* - not significant, * - significant at $p<0.05$, ** - significant at $p<0.01$. df, degrees of freedom, MS, mean square, F, Fisher-test.

traits consistently showed a significant increase in leaf asymmetry of plants exposed to 2 mg/L Cu compared to control. Moreover, the leaf asymmetry significantly increased in all copper treatments after 51 days of copper exposure. Although plants exposed to 50 mg/L Pb also show an increased FA after 51 days of the treatment, the pattern was not significant (Figure 4).

Fluctuating asymmetry of the length of peripheral vein (Trait 3), on the other hand, showed more complicated pattern and this trait only indicated the significant increase in leaf asymmetry of the plants treated with 2 mg/L of copper. Similar to traits 1 and 2, this trait also exhibited significantly higher FA than control after 51 days of the copper exposure.

Trait 8, which comparing the width of a pair of leaves, has greater FA in high copper treatments (2 to 4 mg/L) than control, but the difference was significant in plants exposed to 2 mg/L Cu only (Figure 4).

DISCUSSION

Lead and copper inhibited the growth of seagrass, *H. ovalis*. The effect was more pronounced at higher concentration of heavy metals and at longer exposure. For example, 0.5 mg/L Cu had reduced growth of *H. ovalis* in comparison with control, but the plant still grew very slowly up to 18 days, and no growth was observed after this period. Moreover, the plants exposed to 2 to 4 mg/L Cu had progressively reduced the extension of rhizome resulting in shorter plants with a reduced number of nodes. In that higher copper treatment, the plant rhizome became decayed shortly after the treatment. Thus, there was no growth observed in the plants treated with those higher copper concentrations.

Copper, although an essential nutrient for plants, when absorbed in excess amounts can cause deleterious effects at morphological, physiological and ultrastructural levels (Ouzounidou et al., 1992). Specifically, excess copper can cause chlorosis, inhibition of root growth and damage to plasma membrane permeability, leading to ion leakage (De Vos et al., 1991; Ouzounidou et al., 1992). A reduction of root extension at the concentration of 30 to 160 μM of copper have also been demonstrated in a terrestrial plant *Alyssum montanum*, where Cu^{2+} attack sulphydryl groups causing damage to permeable layers and allowing the diffusion of ions into the chloroplast (Ouzounidou, 1994).

Leaf senescence was also occurred in *H. ovalis* 2 days after exposing the plant to copper treatments, with more than 50% of leaves prematurely senesced and detached from the rhizome after exposure to the higher copper concentrations (2 to 4 mg/L). The effect of copper exposure on leaf senescence found in this study was consistent with previous studies (Ralph and Burchett, 1998; Prange and Dennison, 2000). This phenomenon is suggested to be associated with the stimulation of phytochrome activity, leading to increased abscisic acid and ethylene production, compounds that are precursors of leaf abscission and loosening of cell walls (Malea, 1994).

Similar to copper, lead also reduced the *H. ovalis* growth rate and the effect was more apparent at the greater concentration and longer exposure period. Thus, 10 mg/L Pb had reduced the growth rate of the plant to about 28% up to 18 days and the growth rate reduced to 51% thereafter until the end of the experiment (a further 33 days). The more extreme effect was observed at higher concentrations (50 mg/L), where the plants had reduced growth rate of about 71% up to 18 days, and no growth was observed after this period. Comparing with the copper effect, however, the effect of lead on *H. ovalis* growth was weaker. Lead (a non-essential element) displayed only limited effect on the growth of *H. ovalis*,

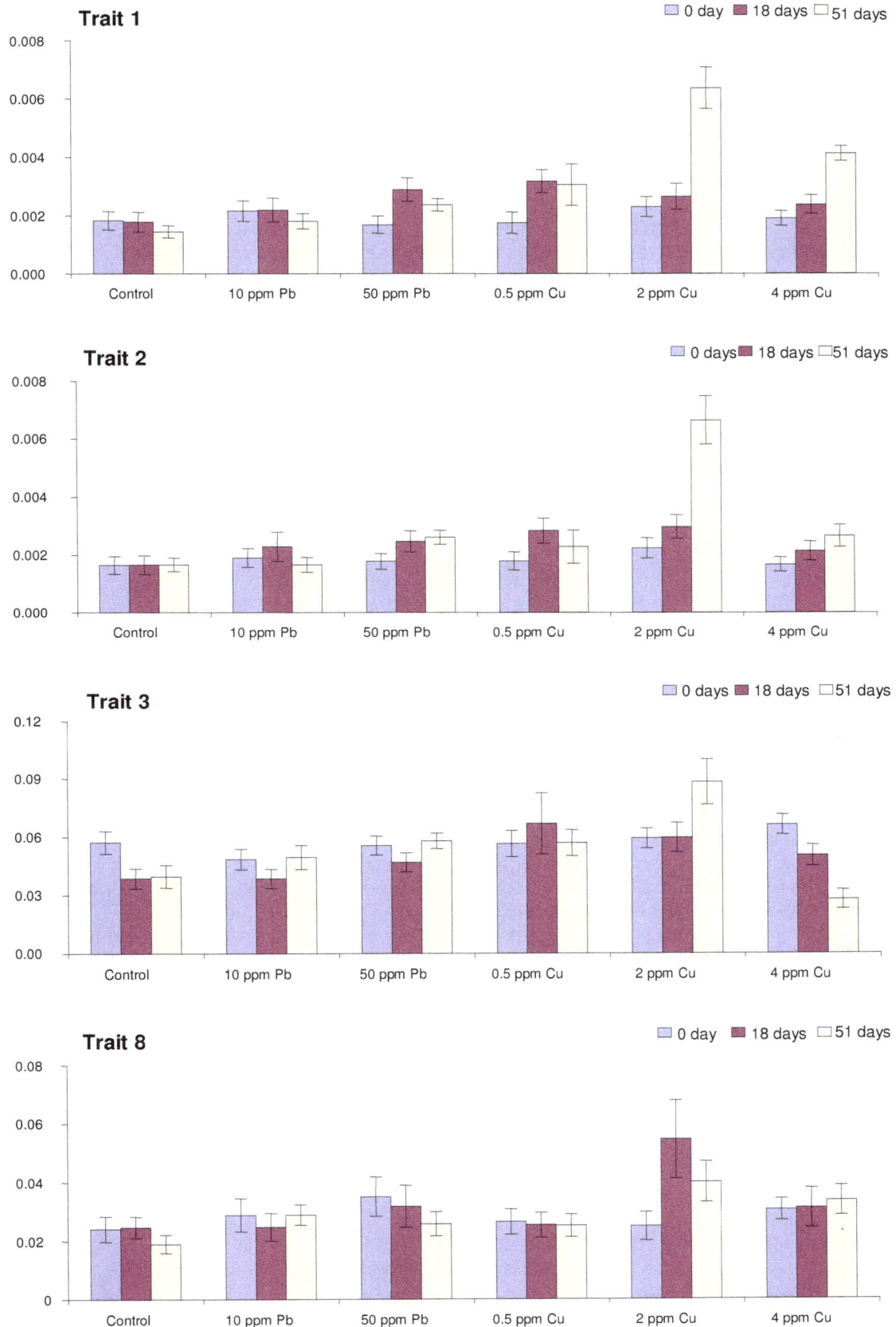

Figure 4. FA of *H. ovalis* leaves at different treatment levels and at different duration of the experiment.

whereas copper (an essential element), even at lower concentrations, had significantly greater effect. The evident negative effect of essential micronutrient heavy metal on growth could be associated with uptake and exclusion mechanisms.

Since copper is required for metabolic processes, the plant actively takes up copper; however, when exposed to increased concentrations, uptake may exceed metabolic requirements and result in a toxic impact. On the other hand, lead is not required for growth or development; the plant may actively exclude this element to minimize the toxic impact by isolating this element in storage tissues, where it does not interfere with the metabolic activity. Ward (1989) postulated that seagrasses tolerate high concentrations of heavy metals by sequestering them into structural components of the leaf tissue, therefore preventing them from affecting the more sensitive metabolic processes. Similar mechanisms have been reported in algae (Pinto et al., 2003). Moreover, MacFarlane and Burchett (2000) reported the limited translocation of Pb in the grey mangrove Avicennia marina. They found Pb concentrated in root tissue, with only minimal transport to the shoot. This evidence also suggested that Pb has a limited effect on plants growth and development due to the limited translocation of this element through the vascular system to the leaves/shoot, where the metabolic activity is mostly occurred.

The results of the effect of Cu and Pb on the growth of the seagrass, H. ovalis, presented in this study are in good agreement with other studies. For example, it has been reported that 1 mg/L copper had an effect on photosynthetic activity of H. ovalis, and the concentrations of 5 to 10 mg/L Cu had a lethal effect on this plant, whereas concentrations of 1 to 10 mg/L Pb had only limited effect (Ralph and Burchett, 1998). It has also been suggested that Cu can be toxic to microalgae at concentrations of 0.19 to 0.3 µg/L, which is only marginally above those found in oceanic waters, while Pb reduces growth of this microscopic plant at concentration above 20 µg/L (Langston, 1990). Lead treatment did not significantly affects the leaf dimension characters, but copper treatment did affect these characters by a decrease in leaf size (measured by leaf length and width) and the effect became more apparent as the copper concentrations increased and also with a longer time exposure.

On the other hand, width to length ratio of the H. ovalis leaves was greater in higher metal levels for both lead and copper. Reduced leaf size and an increased width to length ratio may be a mechanism of the plant to survive under environmental pressures, in this case under heavy metal stress. It was suggested that toxic concentrations of heavy metals inhibit metabolic activity, and plants that survive do so with decrease growth and slower development (Clijsters and Van Assche, 1985). Seagrass, Zostera capricorni, from unpolluted locations

decreased leaf length when growing in polluted sediment (Conroy et al., 1991). In contrast, it was found from the field study that seagrass, H. ovalis, grew in the metal polluted location had considerably longer and wider leaves compared to the ones in unpolluted locations suggesting the effects of uncontrolled factors such as nutrients, which may potentially compensate or even surpass the effect of heavy metals on the seagrass (Ambo Rappe et al., 2008).

The present experiment also demonstrates that heavy metals stress could increase FA in H. ovalis. FA as an ecological indicator may thus have potential as sensitive tools for monitoring the quality of the environment (Tracy et al., 1995; Anne et al., 1998; Hoffmann and Woods, 2003; Lajus and Zhang, 2003; Leung et al., 2003). Reduced growth rate and also leaf size of H. ovalis have been documented, especially with increased copper treatment and increasing of the duration of exposure to the metal. Increased of FA at this metal treatment was observed as well. The experiment thereby succeeded in discriminating between non-stressful and stressful conditions due to heavy metals. Meristic traits (Traits 1 and 2) dealing with the number of veins seems to be suitable, especially for leaf asymmetry with a bilateral character. Metric traits (Traits 3 and 8) dealing with the perimeter of the leaf is also a potential trait to be used, even though they still need to be examined further due to the inconsistency of the results on some occasions. It has been suggested that different organs of plants exhibit multiple forms of symmetry.

Consequently, many different estimates of FA can be made (Freeman et al., 2003). Moreover, different estimates of FA can display different sensitivities (Freeman et al., 2005). Different characters may also exhibit different patterns of fluctuating asymmetry depending on their size (Lajus, 2001). Selecting certain traits to be used as estimates of asymmetry clearly depends on the knowledge of the biology of the species being studied (Freeman et al., 2005). Meristic traits that have proven to have a good sensitivity to heavy metal stresses in H. ovalis, is cross veins that join and interconnect a midvein with two lateral 'intra-marginal veins'. These branched cross veins is positioned on either side of the midrib (Kuo and den Hartog, 2001; Coles et al., 2004; Kuo and den Hartog, 2006).

Like other veins, these branched cross veins have an important function in the vascular system of plants, which is related to the transport of solution within the plant (Kuo and den Hartog, 2006). It might be possible that the heavy metals pressure causes the imbalance of development of this structure on the right and the left side of the midrib that is normally in balance number (Freeman et al., 1993), leading to increased developmental instability (measured as FA). Plants exposed to 2 mg/L Cu exhibited significantly higher leaf asymmetry, while no significant leaf asymmetries were observed in plants treated with lead or with low levels of

Cu (0.5 mg/L).

Conclusion

This experimental study showed that copper (Cu) could lead to reduced growth rate and increased leaf asymmetry in *H. ovalis*. Leaf size of the plant also reduced as Cu level increased and when the plants were exposed to the heavy metal for longer duration. Copper, being an essential metal, resulted in a more pronounced effects on the plant characteristics than lead (non-essential metal), especially at higher concentrations. Thus, the experimental data showed quite a notable effect of heavy metals on various characteristics of seagrass, *H. ovalis*. It is, therefore, important to emphasize that the effect is observed at concentrations, which do occur in the wild and thus can be used for interpretation of the field data.

REFERENCES

Ambo Rappe R, Lajus DL, Schreider MJ (2007). Translational fluctuating asymmetry and leaf dimension in seagrass, *Zostera capricorni* Aschers in a gradient of heavy metals. Environ. Bioindic., 2: 99-116.

Ambo Rappe R, Lajus DL, Schreider MJ (2008). Increased heavy metal and nutrient contamination does not increase fluctuating asymmetry in seagrass *Halophila ovalis*. Ecol. Indic., 8: 100-103.

Anne P, Mawri F, Gladstone S, Freeman DC (1998). Is fluctuating asymmetry a reliable biomonitor of stress? a test using life history parameters in soybean. Int. J. Plant Sci., 159: 559-565.

Butler A, Jernakoff P (1999). Seagrass in Australia: Strategic Review and Development of an R & D Plan. CSIRO, Collingwood, Australia, p. 209.

Clijsters H, Van Assche F (1985). Inhibition of photosynthesis by heavy metals. Photosyn. Res., 7: 31-40.

Coles RG, McKenzie LJ, Campbell S, Mellors JE, Waycott M, Goggin L (2004). Seagrasses in Queensland Waters. CRC Reef Research Centre Ltd., Townsville, Queensland, p. 5.

Conroy BA, Lake P, Buchhorn N, McDouall-Hill J, Hughes L (1991). Studies on The Effects of Heavy Metals on Seagrass in Lake Macquarie. In: Whitehead JH, Kidd RW, Bridgman HA (eds.) Lake Macquarie: An Environmental Reappraisal. University of Newcastle, Department of Community Programmes, NSW, Australia, pp. 55-65.

De Vos CHR, Schat H, De Waal MAM, Vooijs R, Ernst WHO (1991). Increased resistance to copper-induced damage of the root cell plasmalemma in copper tolerant *Silene cucubalus*. Physiologia Plantarum, 82: 523-528.

Duarte CM (2002). The future of seagrass meadows. Environ. Conserv., 29: 192-206.

Freeman DC, Graham JH, Emlen JM (1993). Developmental stability in plants: symmetries, stress, and epigenesis. Genetica, 89: 97-119.

Freeman DC, Graham JH, Emlen JM, Tracy M, Hough RA, Alados CL, Escos J (2003). Plant Developmental Instability: New Measures, Applications, and Regulations. In: Polak M (ed.) Developmental Instability: Causes and Consequenses. Oxford University Press, New York, pp. 367-386.

Freeman DC, Brown ML, Duda JJ, Graham JH, Emlen JM, Krzysik AJ, Balbach H, Kovacic DA, Zak JC (2005). Leaf fluctuating asymmetry, soil disturbance and plant stress: A multiple year comparison using two herbs, *Ipomoea pandurata* and *Cnidoscolus stimulosus*. Ecol. Indic., 5: 85-95.

Hoffmann AA, Woods RE (2003). Associating Environmental Stress with Developmental Stability: Problems and Patterns. In: Polak M (ed.) Developmental Instability: Causes and Consequences. Oxford

University Press, New York, pp. 387-401.

Hosken DJ, Blanckenhorn WU, Ward PI (2000). Developmental stability in yellow dung flies (*Scathophaga stercoraria*): Fluctuating asymmetry, heterozygosity and environmental stress. J. Evol. Biol., 13: 919-926.

Kozlov MV, Wilsey BJ, Koricheva J, Haukioja E (1996). Fluctuating asymmetry of birch leaves increases under pollution impact. J. Appl. Ecol., 33: 1489-1495.

Kuo J, den Hartog C (2001). Seagrass Taxonomy and Identification Key. In: Short FT, Coles RG, Short CA (eds.) Global Seagrass Research Methods. Elsevier, Amsterdam, pp. 31-58.

Kuo J, den Hartog C (2006). Seagrass Morphology, Anatomy, and Ultrastructure. In: Larkum AWD, Orth RJ, Duarte CM (eds.) Seagrasses: Biology, Ecology and Conservation. Springer, The Netherlands, pp. 51-87.

Lajus DL (2001). Variation patterns of bilateral characters: variation among characters and among populations in the White Sea herring (*Clupea pallasi marisalbi*). Biol. J. Linn. Soc., 74:237-253.

Lajus DL, Graham JH, Kozhara AV (2003). Developmental Instability and The Stochastic Component of Total Phenotypic Variance. In: Polak, M. (Ed.), Developmental Instability: Causes and Consequences. Oxford University Press, New York, pp. 343-363.

Lajus DL, Zhang X (2003). Application of fluctuating asymmetry in studying effects of stress on aquatic organism. Adv. Mar. Sci., 21: 236-242.

Langston WJ (1990). Toxic Effects of Metals and The Incidence of Metal Pollution in Marine Ecosystems. In: Furness RW, Rainbow PS (eds.) Heavy Metals in The Marine Environment. CRC Press, Boca Raton, Florida, pp. 101-122.

Leung B, Knopper L, Mineau P (2003). A Critical Assessment of the Utility of Fluctuating Asymmetry as a Biomarker of Anthropogenic Stress. In: Polak M (ed.) Developmental Instability: Causes and Consequences. Oxford University Press, New York, pp. 415-426

MacFarlane GR, Burchett MD (2000). Cellular distribution of copper, lead and zinc in the grey mangrove, *Avicennia marina* (Forsk.) Vierh. Aquat. Bot., 68: 45-59.

Mal TK, Uveges JL, Turk KW (2002). Fluctuating asymmetry as an ecological indicator of heavy metal stress in *Lythrum salicaria*. Ecol. Indic., 1: 189-195.

Malea P (1994). Uptake of cadmium and the effects on viability of leaf cells in the seagrass *Halophila stipulacea* (Forsk.) Aschers. Botanica Marina, 37: 67-73.

McMillan C (1976). Experimental studies on flowering and reproduction in seagrasses. Aquat. Bot., 2: 87-92.

McMillan C (1978). Morphogeographic variation under controlled conditions in five seagrasses, *Thalassia testudinum*, *Halodule wrightii*, *Syringodium filiforme*, *Halophila engelmannii*, and *Zostera marina*. Aquat. Bot., 4: 169-189.

McMillan C (1980). Culture Methods. In: Phillips RC, McRoy CP (eds.) Handbook of Seagrass Biology: An Ecosystem Perspective. Garland STPM Press, New York, pp. 57-69.

McMillan C, Moseley FN (1967). Salinity tolerances of five marine spermatophytes of Redfish Bay, Texas. Ecology, 48: 503-506.

McMillan C, Phillips RC (1979). Differentiation in habitat response among populations of New World seagrass. Aquat. Bot., 7: 185-196.

Ouzounidou G (1994). Copper-induced changes on growth, metal content and photosynthetic function of *Alyssum montanum* L. plants. Environ. Exp. Bot., 34: 165-172.

Ouzounidou G, Eleftheriou EP, Karataglis S (1992). Ecophysiological and ultrastructural effects of copper in *Thlaspi ochroleucum* (Cruciferae). Can. J. Bot., 70: 947-957.

Palmer AR (1994). Fluctuating Asymmetry Analyses: A Primer. In: Markow TA (ed.) Developmental Instability: Its Origins and Evolutionary Implications. Kluwer, Dordrecht, Netherlands, pp. 335-364.

Palmer AR, Strobeck C (2003). Fluctuating Asymmetry Analyses Revisited. In: Polak M (ed.) Developmental Instability: Causes and Consequences. Oxford University Press, New York, pp. 279-319.

Pergent-Martini C, Pergent G (2000). Marine phanerograms as a tool in the evaluation of marine trace-metal contamination: An example from the Mediterranean. Int. J. Environ. Pollut., 13: 126-147.

Phillips RC (1980). Phenology and Taxonomy of Seagrasses. In:

Phillips RC, McRoy CP (eds.) Handbook of Seagrass Biology: An Ecosystem Perspective. Garland STPM Press, New York, pp. 29-40.

Pinto E, Sigaud-Kutner TCS, Leitão MAS, Okamoto OK, Morse D, Colepicolo P (2003). Review Heavy metal–induced oxidative stress in algae. J. Phycol., 39: 1008-1018.

Prange JA, Dennison WC (2000). Physiological responses of five seagrass species to trace metals. Mar. Pollut. Bull., 41: 327-336.

Ralph PJ, Burchett MD (1998). Photosynthetic response of *Halophila ovalis* to heavy metal stress. Environ. Pollut., 103: 91-101.

Sokal RR, Rohlf FJ (1995). Biometry: The Principles and Practice of Statistics in Biological Research. W.H. Freeman and Company, New York, p. 887.

Tan-Kristanto A, Hoffmann A, Woods R, Batterham P, Cobbett C, Sinclair C (2003). Translational asymmetry as a sensitive indicator of cadmium stress in plants: A laboratory test with wild-type and mutant *Arabidopsis thaliana*. New Phytol., 159: 471-477.

Tracy M, Freeman DC, Emlen JM, Graham JH, Hough RA (1995). Developmental instability as a biomonitor of environmental stress. In: Butterworth FM, Corkum LD, Guzman-Rincon J (eds.) Biomonitors and Biomarkers as Indicators of Environmental Change. Plenum Press, New York, pp. 314-337.

Walker DI, Hillman KA, Kendrick GA, Lavery P (2001). Ecological significance of seagrasses: Assessment for management of environmental impact in Western Australia. Ecol. Eng., 16: 323-330.

Ward TJ (1989). The accumulation and effects of metals in seagrass habitats. In: Larkum AWD, McComb AJ, Shepherd SA (eds.) Biology of seagrasses: A treatise on the biology of seagrasses with special reference to the Australian region. Elsevier, New York, pp. 797-820.

20

Modification and validation of a microwave-assisted digestion method for subsequent ICP-MS determination of selected heavy metals in sediment and fish samples in Agusan River, Philippines

Elnor C. Roa[1,2]*, Mario B. Capangpangan[1] and Madeleine Schultz[3]

[1]Department of Chemistry, College of Science and Mathematics, Mindanao State University-Iligan Institute of Technology (MSU-IIT), 9200 Iligan City, Philippines.
[2]Institute of Fisheries Research and Development, Mindanao State University-Naawan, 9023 Naawan, Misamis Oriental, Philippines.
[3]School of Physical and Chemical Sciences, Queensland University of Technology, Brisbane, Queensland 4001, Australia.

This study investigated, validated, and applied the optimum conditions for a modified microwave-assisted digestion method for subsequent ICP-MS determination of mercury, cadmium, and lead in two matrices relevant to water quality, that is, sediment and fish. Three different combinations of power, pressure, and time conditions for microwave-assisted digestion were tested, using two certified reference materials representing the two matrices, to determine the optimum set of conditions. Validation of the optimized method indicated better recovery of the studied metals compared to standard methods. The validated method was applied to sediment and fish samples collected from Agusan River and one of its tributaries, located in Eastern Mindanao, Philippines. The metal concentrations in sediment ranged from 2.85 to 341.06 mg/kg for Hg, 0.05 to 44.46 mg/kg for Cd and 2.20 to 1256.16 mg/kg for Pb. The results indicate that the concentrations of these metals in the sediments rapidly decrease with distance downstream from sites of contamination. In the selected fish species, the metals were detected but at levels that are considered safe for human consumption, with concentrations of 2.14 to 6.82 µg/kg for Hg, 0.035 to 0.068 µg/kg for Cd, and 0.019 to 0.529 µg/kg for Pb.

Key words: Mercury, cadmium, lead, CRM, pollution, dissolution, analysis.

INTRODUCTION

Although, standard methods of chemical analyses are available from many recognized sources, there are occasions when they cannot be used or when their use is inconvenient, e.g., when no standard method exists for a particular constituent or characteristic, such as extremely low heavy metal concentrations. In most analytical procedures the sample is physically destroyed by dissolution, calcinations or by other digestion methods (Quevauviller, 1993). This stage of analysis can possibly cause losses of the analytes and correspondingly affect the accuracy and precision of the final results obtained. The accuracy can be determined by extracting a relevant certified reference material (CRM) and calculating the percentage recoveries relative to the certified values. The use of CRMs is also a good way to identify losses due to the method used. Several sample dissolution methods are commonly used for heavy metal analyses, including open vessel and microwave-assisted digestions. In the past few years, the latter has proven to be a useful tool in sample preparation for both organic and inorganic analytes.

This technique is advantageous compared to other

*Corresponding author. E-mail: bebottcroa@yahoo.co.uk.

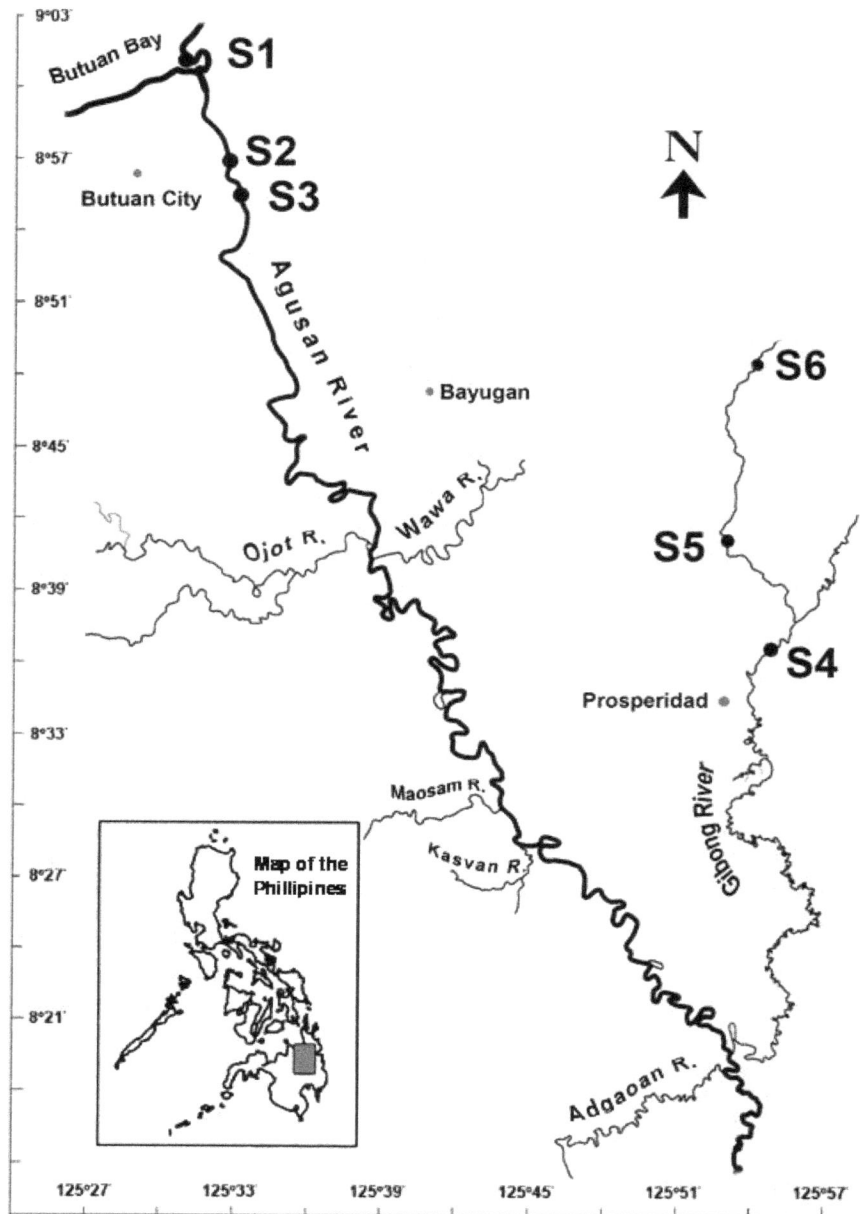

Figure 1. Location of study sites within the Agusan (S1 to S3) and Gibong (S4 to S6) Rivers, in northeastern Mindanao, Philippines.

methods as it provides rapid sample preparation and reduces the risk of contamination. The group of Shalini Ashoka (2009) presented a comparison of six acid digestion methods involving nitric acid in conjunction with other reagents to digest three certified marine biological samples (DOLT-3, DORM-3, IAEA-407) and a fish bone homogenate (prepared from *Merluccius australis*) for subsequent ICP-MS determination of 40   elements. Results of their study showed that microwave digestion with nitric acid and hydrogen peroxide (Method III) gave the   most   acceptable   results   as   determined   by

comparison with the certified values. Deshpande et al. (2008) used nitric acid and perchloric acid in digesting fish samples (including mackerel, pomfret, king fish and Indian salmon collected from Mumbai Docks) on a hot plate and further reconstituted the digest using HCl for subsequent analysis using a voltametric analyzer.

The U.S. EPA Method 3051 is a microwave-assisted digestion method and is known to be a regulatory alternative to the earlier microwave-assisted Method 3050 (Lorentzen and Kingston, 1996; U.S. EPA, 1995) that provides rapid, safe, efficient digestion that is invulnerable to losses of volatile metallic analytes (Hewitt and Reynolds, 1990). A modification of this method, U.S. EPA Method 3051A, is also a microwave-assisted digestion method for sediments, sludges, soils, and oils for trace metal analysis. However, this method has some conditions that cannot be applied using some types of commercial laboratory microwave digestors. For example, this method specifies 175°C temperature that corresponds to approximately 14 - 17 atm or 205 - 250 psi (U.S. EPA, 1997) depending on the type of sample used. This pressure range exceeds the 200 psi maximum allowable pressure for CEM Microwave Oven (CEM Corporation, Matthews, NC, USA). U.S. EPA Method 3052 is another microwave-assisted digestion designed for digesting siliceous and organically based matrices. In this method, hydrofluoric acid (HF) is added at varying concentrations depending on the silicon dioxide (silica, SiO_2) concentration in the sample. Obviously, it would be attractive to use a method that does not require the use of dangerous HF. Open vessel digestion methods can also be applied to analysis of heavy metals such as cadmium (Cd) and lead (Pb) but are not applicable to the volatile analyte mercury (Hg). Thus, closed digestion systems are preferred to minimize possible contamination of the digest and avoid losses of volatile elements.

Several analytical instruments have been used in the final determination of the analyte concentrations in different environmental matrices. For instance, flame and electrothermal atomic absorption spectrometry (ET-AAS) are not used for the analysis of Hg in sediments because of very poor detection limits but can be used when coupled with pre-concentration steps (Ciceri, 2008). Later, it was reported that ET-AAS has improved its limits of detection for As, Se, Pb, Te, Cd, Sn, Sb, Bi, and Au, as well as Hg by employing chemical vapor generation and in-atomizer trapping techniques (Lampugnani et al., 2003; Moscoso-Pérez et al., 2003). Then, ICP-OES with fast multi-element technique was also used, as it has dynamic linear range, although with low to moderate limits (Sturgeon, 2000). The problem of low to moderate detection limits was overcome by using ICP-MS. The ICP-MS offers high sensitivity and pre-concentration steps are not even mandatory (Ciceri et al., 2008).

The aims of this study were: (1) to modify a microwave-assisted digestion technique for subsequent ICP-MS determination of Hg, Cd, and Pb from various environmental matrices; (2) to validate the method using CRMs for sediment and fish samples; and (3) to apply the validated method in the analysis of sediment and fish samples from Agusan River and Gibong River, a tributary of the Agusan River, in Agusan del Sur, eastern Mindanao, Philippines (Figure 1).

The geographical locations of both rivers in Figure 1 suggest that they are carriers and passageways of effluents draining from surrounding areas, which are involved in either or both agricultural and industrial activities, predominantly artisanal gold mining. Gibong River merges with Agusan River, which finally empties to Butuan Bay. Appleton et al. (2006) reported that Hg, Cd, and Pb concentrations of 27 to 124, <1-5, and 14 to 186 mg/kg, respectively, were observed in the suspended solids, and 17 to 33, < 1, and 17 to 19 mg/kg in bottom sediment samples collected from the upstream region of Agusan River below Naboc River whose headwater was the focus of the mining activity in Mt. Diwalwal (Appleton et al., 2006). However, the impact of similar activities on other tributaries of Agusan River has not been well documented. Thus, the validated microwave-assisted digestion method of this study was applied to sediments and fish samples collected from several locations along both Agusan River and Gibong River to determine the effects of human activities in these areas.

MATERIALS AND METHODS

Certified reference materials

The CRM for sediments in this study, coded as NCSDC 73372, contains certified values of 60 elements including Hg, Cd, and Pb, obtained from the mean of 14 independent laboratories, provided by the Institute of Geophysical and Geochemical Exploration (China) to China National Analysis Center for Iron and Steel, and was distributed by Graham B. Jackson PTY LTD, Dandenong 3175 Victoria. This CRM is a lake deposit in powdered form with size less than 0.074 mm.

The fish CRM, coded as LUTS-1, is a non-defatted lobster hepatopancreas reference material for 16 trace metals including Cd and Pb (National Research Council, Canada). The certification process of this CRM was verified by 5 other external expert laboratories.

Standards and reagents

The standards for ICP-MS were prepared from stock solutions of Hg, Cd, and Pb at 1000 mg/L concentrations obtained from Sigma-Aldrich, Australia, and labelled as Fluka TraceCert Ultra. Spiked solutions were prepared from the stock as necessary. Ultrapure nitric acid (HNO_3) and hydrochloric acid (HCl) reagents were obtained from J.T. Baker Inc. All other reagents and solvents used in this study were analytical grade obtained from Sigma-Aldrich, Australia. All water used in washing laboratory glassware, other apparatus, and in the preparation of sample and standard solutions were deionized (resistance < 18 mΩ, Academic Milli-Q Ultra Pure Water System, Australia).

Instruments and apparatus

All microwave-assisted digestions were performed using CEM MDS 2000 630 WATT system (CEM Corporation, Matthews, NC, USA). This system was designed to hold 12 polytetrafluoroethylene (PTFE) digestion vessels. The digestion vessels had rupture membranes for safe operation under 200 psi. The elemental analyses were conducted using an Agilent 7500ce inductively coupled plasma mass spectrometer (ICP/MS) (AgilentTechnologies, USA). Optimization of the instrument was done as outlined in the Agilent 7500ce user's manual. The operating conditions of the instrument are shown in Table 1. This type of ICP-MS is fitted with an octopole reaction-collision cell, operated in collision mode that will minimize the isobaric interferences from polyatomic ions such as oxides (MO^+) that may occur from either the matrix or the plasma

Table 1. Operating conditons of the Agilent 7500ce ICP-MS (AgilentTechnologies, USA).

Operating conditions	Value
Plasma gas flow rate (L/min)	13
Auxiliary gas flow rate (L/min)	1.5
Carrier gas flow rate (L/min)	0.85
Makeup gas flow rate (mL/min)	0.28
Collision gas/flow rate (mL/min)	He 4.2
RF power (W)	1550
Nebuliser	Micro mist
Torch injector internal diameter (mm)	2.5
Sample depth (mm)	8.0
Interface cone	Pt sampler cone, Ni skimmer
CeO^+/Ce^+ (%)	0.5

Table 2. Microwave-assisted digestion conditions used in the study.

Stage	Power (%)	Pressure (psi)	Total time (min)	Time at that pressure(min)
		Set I[a]		
1	75	75	35	30
2	100	75	20	15
3	60	20	35	30
		Set II[b]		
1	40	20	10	5
2	40	40	10	5
3	40	85	10	5
4	40	135	10	5
5	40	175	10	5
		Set III[c]		
1	70	20	10	5
2	100	40	10	5
3	100	85	10	5
4	100	135	10	5
5	90	175	10	5

[a]Set I has 3 stages with variable percent power (75%, 100%, 60%) and pressure (20 to 75 psi) for a total digestion time of 90 minutes. [b]Set II is designed for digesting 2 sample vessels and has 5 stages with constant percent power (40%) but different pressures (20 to 175 psi) for a total digestion time of 50 min. This was based from Cem-Microwave sample preparation note for digestion of fish tissue in a closed vessel using pressure controlled microwave heating for the determination of metals by spectroscopic methods. [c]Set III has also 5 stages with variable percent power (70%, 100%, 100%, 100%, 90%) and variable pressure (20 to 175 psi).

through kinetic energy bias. Additionally, this instrument is extremely sensitive, with a very large dynamic range that can measure analytes from the very low ng/L (ppt) level to the upper mg/L range (1000 mg/L) in one run or a total range of 8 to 9 orders of magnitude.

Modification and optimization of the microwave-assisted digestion method

In this study, three different sets or combinations of power, pressure, and time conditions for microwave-assisted digestion (Table 2) were tested using the two CRMs representing sediment

and fish to determine the optimum set of conditions. Set I was based from the work of Lim et al. (2007) which determined the effects of fuel characteristics and engine operating conditions on elemental composition of emissions from twelve heavy duty diesel buses in Brisbane, Australia; Set II was based on the method outlined for CEM Microwave Sample Preparation System (Microwave sample preparation note for digestion of fish tissue in a closed vessel using pressure controlled microwave heating for the determination of metals by spectroscopic methods) (CEM Corporation, Matthew, NC, USA). This method was described for use with only 2 vessels (5% power was added for every increase of sample vessel), so Set III is a modification of Set II (by this study) using 12 microwave vessels simultaneously.

These three sets of microwave-assisted digestion conditions involved the digestion of 0.5 g of the CRM with 10 mL HNO_3 in a CEM microwave oven. After digestion, the vessels were allowed to cool until the pressure of the vessel was reduced to below 50 psi. The caps of each vessel were then carefully removed and the contents were filtered using 0.45 μm microfibre filter (Banksia Scientific, Australia), diluted to 20 mL in a volumetric flask using deionized water, and stored in polyethylene vial prior to the final determination of the trace metals' concentration.

Validation of the optimized microwave-assisted digestion method

From the three sets of conditions tested, the one that gave the highest recovery (the optimum set) was validated using three validation techniques. First, the three CRMs were digested and the percentage recoveries were compared to the certified values of the CRM. In the case of LUTS-1, which does not include certified value for mercury, 400 μL of a 1000 mg/L (or 400 μg) Hg standard was spiked to the CRM prior to digestion. The second and third techniques for validating the microwave methods were by comparing the recoveries to recoveries obtained when the same CRMs were analyzed using two standard methods: AOAC Method 990.08 for Cd and Pb and EPA Method 200.7 for Hg, Cd, and Pb, respectively. These two standard methods are both open vessel digestion procedures with the temperature maintained at 95°C. In the validation relative to EPA Method 200.7, the three CRMs were spiked with 400 μL of 1000 mg/L (or 400 μg) each of Hg, Cd, and Pb standards. Spiking was done to elevate the analyte concentration in the CRMs and particularly to supply Hg to LUTS-1.

Detection limits

The minimum detection (MDL) and instrument detection limits (IDL) were calculated for the three elements (Hg, Cd, and Pb) as 3 times the standard deviation of the concentrations of the seven blanks, and sum of the MDL and mean metal concentration of the blank, respectively (Agilent Technology, 2005).

Application of the validated microwave-assisted digestion method

Study sites

Sediment samples were collected from three sites each in Agusan River and in Gibong River (Figure 1) on April 16, 2009. Agusan River is located in the eastern part of Mindanao Island in the Philippines. It has several creeks and river tributaries and has become the recipient of waste deposits carried by Naboc River which receives wastes from mine tailings of Mt. Diwalwal, the biggest and most controversial gold mining area in the Philippines and a significant global gold producer (Appleton et al., 2006). Gibong River is one of the major river tributaries of Agusan River and carries both agricultural and industrial effluents and wastes from Agusan del Sur. The basis for selecting the sites was the absence of data comparing the selected trace metals' (Hg, Cd, Pb) concentrations in Agusan River and its tributary, the Gibong River.

Types of sample and sampling protocol

Approximately 2 kg sediment samples were collected at a depth of approximately 2 ft. from the designated sampling sites using a modified corer made of PVC pipe (2 ft. × 6 in i.d.) and wet-sieved *in situ* using a screen with an opening of $^1/_8$ inch. This was done to remove the twigs, leaves, rocks and other materials that were incorporated and collected together with the sediment samples in the corer. The collected sediments were allowed to stand for 1 h in a basin and were then sieved to separate from the water. The sieved sediments were kept in polyethylene bags, stored in an ice chest, and transported to the MSU-IIT Chemistry Laboratory. In the laboratory, the sediments were freeze-dried, ground, and sieved to obtain a particle size of < 75 μm. Fish samples were bought from the fishermen fishing near Study Site 1 (SS1) during the time when sediment sampling was done. The samples included 5 pieces of *Arius maculatus* (sea catfish), 10 pieces *Ambassis commersonii* (glassfish), 10 pieces *Johnieops vogleri* (croaker fish), and 10 pieces *Sillago sihama* (common whitings). These fish species are all eaten in the local region and form an important part of the diet. In the laboratory, the meat was separated from the fishbone, then homogenized, freeze dried, transferred to polyethylene bags, and kept in desiccators until the analyses.

Analyses of mercury, cadmium, and lead

The operating conditions of the ICP-MS instrument were optimized before the analysis was performed. In particular, the nebulizer gas flow rate, the ion lens voltage(s) and the ICP RF power were adjusted to yield the highest signal intensities possible while maintaining low levels of oxides and doubly-charged ion production (both should be less than ~ 3%). After the instrument had been optimized, appropriate calibration standards were then measured. As part of the quality assurance protocol, at least six-point calibrations of different ranges (0.10 to 15.0 mg/L and 0.05 to 2000 μg/L) were carried out for Hg, Cd, and Pb to generate a calibration curve with correlation coefficient of 0.999 or better. The sample solution concentrations were then determined from the corresponding calibration curve. Calibration standards were also analyzed at regular intervals during analytical runs for the ICP-MS to ensure that the instrument continued to meet acceptable sensitivity and linearity criteria.

Data treatment and statistical analysis

Basic statistical parameters such as mean value, standard deviation, and relative standard deviation were calculated using EXCEL 2007 (Microsoft Inc., Redmond, USA) while the statistical analysis was carried out using the Statistical Package for Social Sciences (SPSS) (Version 12.0) software (SPSS Inc., Chicago, USA).

RESULTS AND DISCUSSION

Performance evaluation of the microwave-assisted digestion method under different conditions

Figure 2 shows the percent recoveries of the three target analytes obtained during the performance evaluation using the three sets of microwave-assisted digestion conditions. It is noticeable in Figure 2 that higher percent recoveries of all target analytes from sediment CRMs were obtained using Set I conditions as compared to those obtained using Set II. On the other hand, the percent recovery of Hg, Cd, and Pb from fish CRM obtained using Set II was higher compared to those obtained using Set I. Note that, neither set of conditions

Sediment CRM

Fish CRM

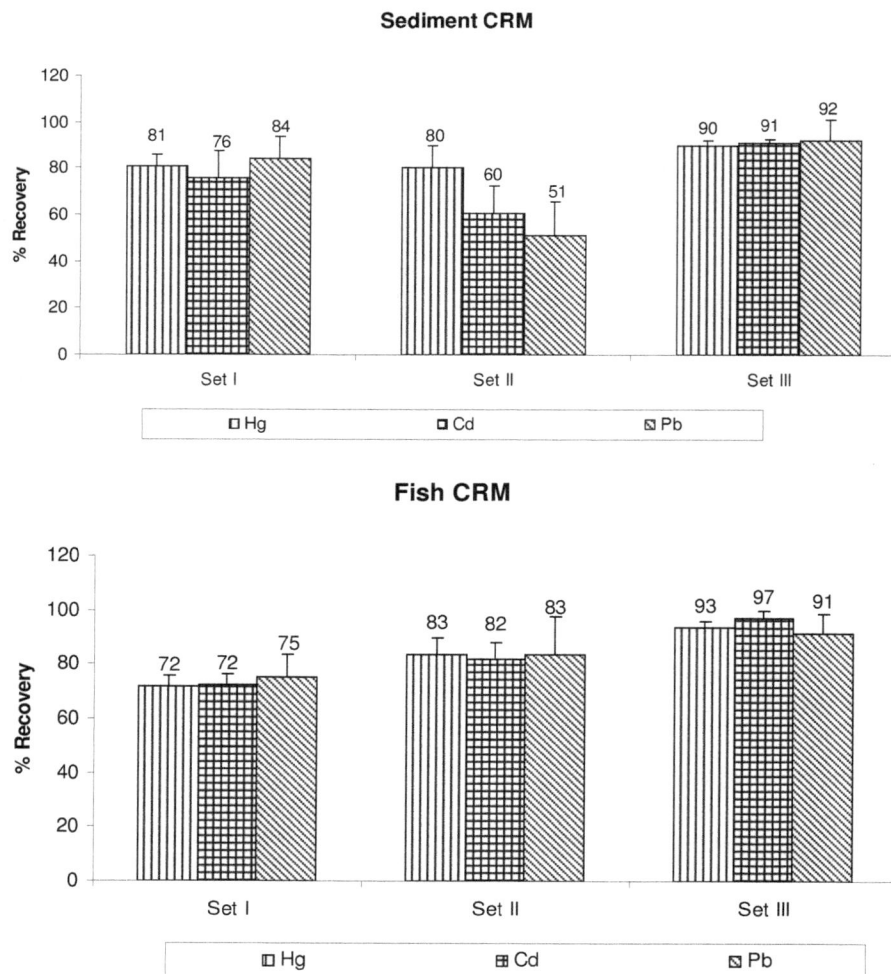

Figure 2. Percent recoveries of Hg, Cd, and Pb from sediment and fish CRMs using Sets I, II, and III microwave-assisted digestion conditions.

gave consistent and acceptable recoveries of all of the target analytes from the two different matrices. However, Set III, which was a modification (by this study) of Set II, gave improved and acceptable percent recoveries (90 to 93% Hg, 91 to 97% Cd, and 91 to 92% Pb) of all the target analytes from the two matrices.

Validation against certified reference materials: First validation technique

Table 3 shows the validation results of the modified and optimized microwave-assisted digestion method (Set III) using two different CRMs (sediment CRM, fish CRM). As shown in Table 3, the recoveries (in mg/kg) obtained using the method is in good agreement with the certified values of Hg, Cd, and Pb in the certified reference materials. For example, Cd in LUTS-1 CRM was recovered at a concentration of 2.06 ± 0.15 mg/kg while

its certified value is 2.12 ± 0.15 mg/kg. This translates to 97% accuracy at a precision of 2.6% RSD. Thus, the method also showed very good accuracy and precision as indicated by the values of the percent recovery (Column 4) and percent RSD of less than 10% (Column 5).

Validation against EPA Method 200.7 and AOAC Method 990.08: Second and third validation techniques, respectively

The accuracy of the method was further validated by comparing the percent recoveries from analysis of CRMs obtained using the modified and optimized microwave-assisted digestion method to those obtained using two standard methods - EPA Method 200.7 and AOAC Method 990.08 (Figure 3).

The results shown in Figure 3 indicate that the modified

Table 3. Validation results showing the percent recoveries[a+] (accuracy) and percent RSD (precision) of Hg, Cd, and Pb from sediment and fish CRMs as obtained using the modified and optimized microwave-assisted digestion methods.

Analyte	Certified value (mg/kg)	Microwave digestion validation results (Using the First Validation Technique)[b]		
		A (mg/kg±SD)	B (%)	C (%)
Sediment CRM: NCSDC 73372 (Lake deposits)				
Hg	0.030	0.027±0.001	90	1.9
Cd	0.10	0.091±0.002	91	1.8
Pb	25	23.024±2.212	92	9.6
Fish CRM: LUTS-1 (non-defatted lobster hepatopancreas)[c]				
Hg[c]	1000	934.600±26.888	93	2.9
Cd	2.12±0.15	2.057±0.054	97	2.6
Pb	0.010±0.002	0.009±0.001	91	8.2

[a]Relative to the certified values. [b]A - Recovered Concentration Values; B - Recovery (Accuracy); C - Relative Standard Deviation (Precision).
[c]LUTS-1 does not have certified values for Hg. Thus, 400 µL of 1000 mg/l Hg (or 400 µg) was spiked to the CRM instead.

and optimized microwave-assisted digestion method achieved significantly better recovery values (91 to 93% Hg, 90 to 96% Cd, and 92% Pb) of all target analytes from the two different CRMs (sediment CRM, fish CRM) compared to the other methods (EPA Method 200.7: 89% Hg, 88 to 90% Cd, 86 to 90% Pb; AOAC Method 990.08: 85 to 90% Cd, 87 to 95% Pb;) ($p < 0.05$). For example, the Pb in CRM-NCSDC 73372 (lake deposits) was only 86 and 87% recovered using EPA Method 200.7 and AOAC Method 990.08, respectively, compared to 92% using the method being validated. Furthermore, the new method also showed very good precision and repeatability with % RSD values ranging from 1.8 to 9.6%.

In general, based from the results obtained such as percent recovery and precision (Table 3), the modified microwave-assisted digestion method (Set III conditions) has been validated to be more accurate and precise compared to the other two microwave methods (Set I and Set II) and the two open digestion methods (EPA Method 200.7 and AOAC Method 990.08) considered in the study. Thus, it can be used in the analysis of Hg, Cd, and Pb in different matrix types such as sediments and fish tissues using a single set of microwave digestion conditions. It should also be noted that the microwave method is quicker, cleaner, and easier to perform if a microwave is available, compared with the open digestion methods. Furthermore, this modified and vali-dated microwave-assisted digestion method combined with ICP-MS is able to extract and detect with very good precision and accuracy extremely low levels of heavy metal concentrations in sediment and fish matrices.

Detection limits

The minimum detection limits or method detection limits (MDL) for Hg, Cd, and Pb using the modified and validated

microwave assisted digestion were 1.409 µg/L for Hg, 0.001 µg/L for Cd, and 0.002 µg/L for Pb, while the instrument detection limits (IDL) were 1.750 µg/L for Hg, 0.002 µg/L for Cd, and 0.003 µg/L for Pb.

Application of the validated microwave-assisted digestion method

The newly validated microwave-assisted digestion method (Set III conditions) was subsequently applied in the analyses of the selected analytes (Hg, Cd, and Pb) in the sediment and fish samples collected from Agusan and Gibong Rivers. Tables 4 and 5 show the mean concentrations of Hg, Cd, and Pb observed in sediments and fish samples collected from these study sites, respectively.

Concentrations of mercury in sediments

An extremely high mean Hg concentration (341.06 mg/kg) was observed in Study Site 6 (SS6) ($p < 0.05$) which is located approximately 5 m from the drainage of the previously operated plant for gold extraction by amalgamation followed by SS5 (69.78 mg/kg Hg) which is approximately 40 km from SS6 (Figure 1 and Table 4). The Hg concentration dropped significantly from SS6 downstream towards the mouth of the Agusan River; the three sites near the mouth of the river have similar, much lower values. A similar trend has been reported by Appleton et al. (1999), indicating that in the Agusan River basin, elevated sedimentary Hg concentrations occur close to point sources.

Note that this is the highest Hg concentration observed in this study and is much higher than the previously reported 27 to 124 mg/kg Hg in bottom sediment samples collected in Agusan River (Appleton et al., 2006).

Sediment CRM

Fish CRM

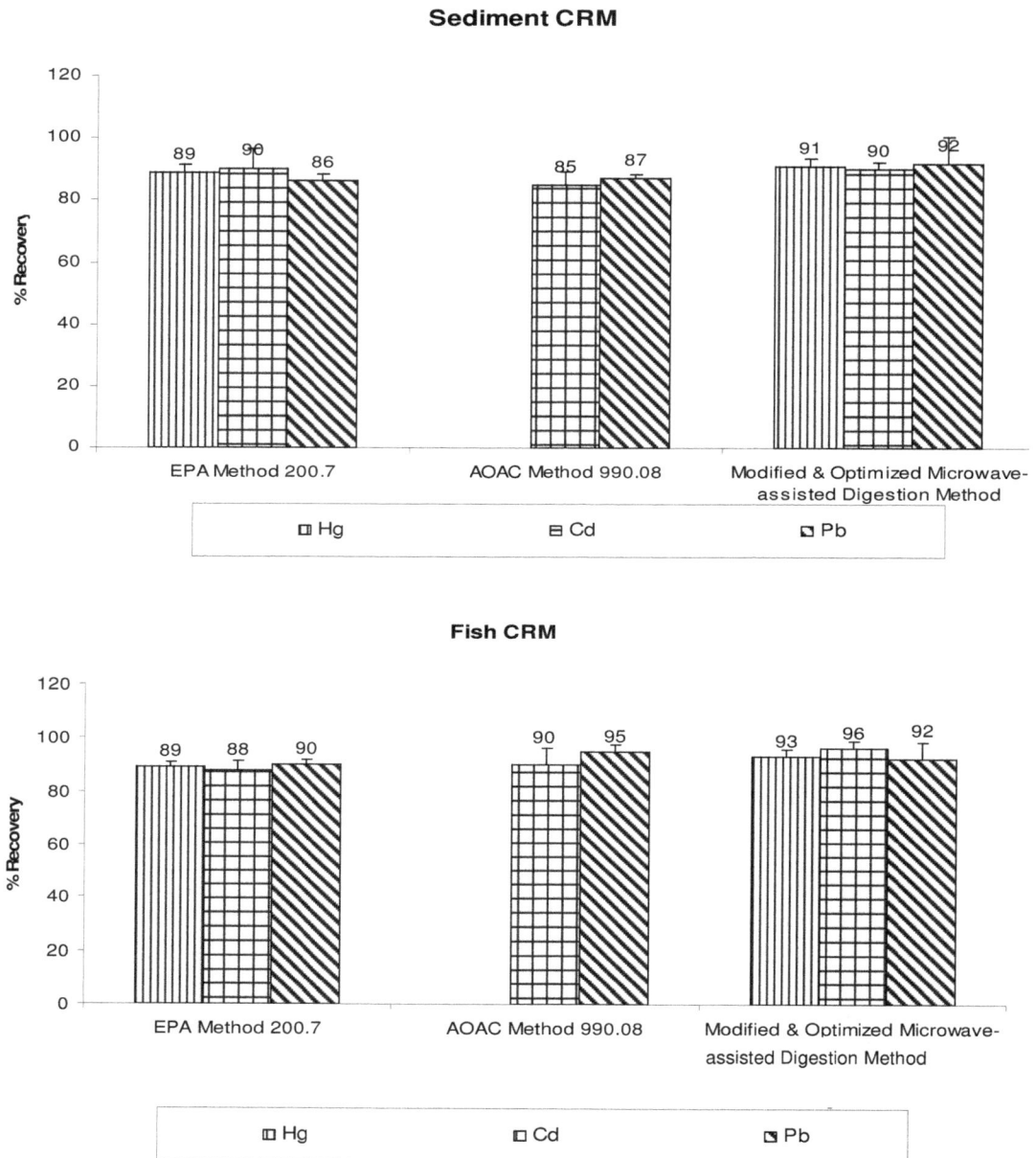

Figure 3. Comparison of percent recoveries (accuracy) of Hg, Cd, and Pb from certified reference materials (sediment and fish) using EPA Method 200.7, AOAC Method 990.08, and the modified and optimized microwave-assisted digestion method. Note that in AOAC Method 990.08, Hg is not included in the analysis.

However, that part of Agusan River, as mentioned in the introduction, is further upstream than the stretch of the river included in this study. Of significance, the Hg concentrations found in sediments at all six sites in the Agusan and Gibong Rivers exceed the permissible threshold effect level of 1 mg/kg Hg set by Environment Canada 1999 (Haines et al., 1994; CCME, 1999), Federal Water Quality Administration Criterion for dredge material, the maximum acceptable concentration in the United States (Pavlou et al., 1987), and Sediment Quality Guidelines in Australia and New Zealand (ANZECC, 1998).

Concentrations of cadmium and lead in sediments

In this study, the Cd and Pb concentrations found in sediments range from 0.05 to 44.46 mg/kg and 2.20 to 1256.16 mg/kg, respectively. Again, the highest concentrations of these metals were found at S6; Cd and

Table 4. Mean (± SD) and percent RSD of Hg, Cd, and Pb concentrations (mg/kg dry weight) in sediments from selected sites in Agusan River and Gibong River as obtained using the validated method[*].

Study site (Location)	Hg (mg/kg±SD)	%RSD	Cd (mg/kg±SD)	%RSD	Pb (g/kg±SD)	%RSD
SS1 (N 9.01548º) (E125.51810º)	$13.92^b \pm 0.39$	2.81	$0.22^a \pm 0.003$	1.33	$7.25^a \pm 0.42$	5.80
SS2 (N 8.94703º) (E125.54669º)	$5.69^a \pm 0.30$	5.32	$0.05^a \pm 0.001$	2.30	$2.20^a \pm 0.09$	4.27
SS3 (N 8.92985º) (E125.55300º)	$2.85^a \pm 0.12$	4.39	$0.37^a \pm 0.018$	4.85	$11.51^a \pm 0.49$	4.28
SS4 (N 08.61035º)(E125.91387º)	$8.86^{ab} \pm 0.29$	3.30	$0.07^a \pm 0.001$	0.75	$3.53^a \pm 0.10$	2.68
SS5 (N 08.68178º)(E125.66371º)	$69.78^c \pm 1.84$	2.63	$0.66^a \pm 0.020$	2.96	$17.23^a \pm 1.07$	6.22
SS6 (N 08.81667º)(E125.90000º)	$341.06^d \pm 6.76$	1.96	$44.46^b \pm 1.579$	3.55	$1256.16^b \pm 50.07$	3.99

[*]Note: Lower case superscript indicates *anova* result at α=0.05. (The same superscripts written in the same column indicate that the corresponding values have no significant differences)

Table 5. Mean (± SD) and percent RSD of Hg, Cd, and Pb concentrations (µg/kg dry weight) in fish samples from Agusan River (near SS1) as obtained using the validated method.[*]

Sample	Hg, µg/kg±SD	% RSD	Cd, µg/kg±SD	% RSD	Pb, µg/kg±SD	% RSD
Arius maculatus	$6.82^a \pm 0.26$	3.79	$0.440^a \pm 0.030$	6.86	$0.529^a \pm 0.044$	8.38
Ambassis commersonii	$3.51^b \pm 0.18$	5.02	$0.035^b \pm 0.003$	8.39	0.151 ± 0.008	5.44
Sillago sihama	$2.22^c \pm 0.06$	2.56	$0.068^b \pm 0.002$	2.47	$0.029^c \pm 0.003$	8.83
Johnius vogleri	$2.14^c \pm 0.13$	6.24	$0.037^b \pm 0.001$	3.69	$0.019^b \pm 0.001$	1.66

[*]Note: Lower case superscript indicates *anova* result at α = 0.05. (The same superscripts written in the same column indicate that the corresponding values have no significant differences).

Pb concentrations were significantly lower in the other study sites (Table 4). There is no significant trend observed for sediment data from SS5 going towards SS1, which is situated at the mouth of Agusan River. Thus, the proximity of SS6 to the point source is presumably the major factor responsible for the highest concentrations of all metals observed in this site. Birch and Taylor (1999) reported a Pb concentration as high as 1000 µg/g (1000 mg/kg) and a moderate concentration of >2 µg/g (2 mg/kg) for Cd in superficial sediments collected from Port Jackson Estuary, Australia. The distribution of these metals exhibits a strong declining trend in concentration from the upper reaches of the estuary towards the estuary mouth (Birch and Taylor, 1999).

IMPLICATIONS FOR THE ENVIRONMENT

The sediment samples collected from both Agusan and Gibong Rivers contain detectable concentrations of the selected hazardous heavy metals Hg, Cd, and Pb. The distribution of these metals showed a strong decreasing pattern in concentration from the upper reaches of the river towards the river mouth. The heavy metal pollution in both rivers drops off with distance from a polluting source, confirming the old adage that "the solution to pollution is dilution". It also shows an observable close relationship between the land use in the catchment area and the concentration of the heavy metal. The highest concentration of all metals was observed in SS6 which is situated in the headwater of Gibong River, and which is the location of one of the mining establishments in the area. Thus, waste discharged from mining industries is a major point source of these metal pollutants suggesting that their presence is anthropogenic in nature.

Concentrations of mercury, cadmium, and lead in selected fish species

The Hg concentrations in the four fish species considered in this study (Table 5) were found to be 2.14 to 6.82 µg/kg. This range of Hg concentrations is significantly lower than the 500 µg/kg maximum recommended limit for Hg (UNEP, 2002) commonly allowed in fish for human consumption in most countries. It is also lower than the findings of Maramba et al. (2006) for Hg concentrations in marine biota (1.65 to 1152.01 µg/kg) from three communities near abandoned gold mines in Zamboanga del Norte, Philippines. As shown in Table 5, Cd values were found to be in the range of 0.035 to 0.44 µg/kg. EU and Turkish Limits for Cd are the same value (0.1 mg/kg) while the World Health Organization specifies a 0.5 mg/kg maximum permissible Cd content in fish (EU, 2001; FAO/WHO, 1987). Note that the Cd content in A. maculatus is significantly higher than in the other three fish species, but still much lower than these allowed limits.

The Pb concentration detected in the four fish species ranged from 0.019 to 0.53 µg/kg. This range of Pb concentrations in fish is significantly lower than the maximum permissible and tolerable values of 0.1 mg/kg Pb and 0.5 mg/kg Pb, set by EU 2001 and FAO/WHO 1987, respectively, and does not demonstrate risk if utilized for human consumption. Note that the highest concentration was exhibited in the same fish species (A. maculatus) whose Cd and Hg concentrations were also the highest. A. maculatus is carnivorous and a bottom-dweller. The adults of A. maculatus feed on insect larvae, shrimps, worms, fish and whatever organic debris that occur on the bottom (Conlu, 1986). Two of the other species tested, J. vogleri and S. sihama are also bottom dwellers but do not feed on detritus. Presumably, ingestion of detritus is increasing the levels of heavy metal in the tissues of this fish.

The Hg, Cd, and Pb values in the fish species in this study (2.14 to 6.82 µg/kg Hg, 0.035 to 0.44 µg/kg Cd, 0.019 to 0.53 µg/kg Pb) are much lower than the reported Hg, Cd, and Pb levels in Tilapia spp. (0.125 to 0.277 µg/g Hg, 0.13 µg/g Cd, and 0.16 µg/g Pb) and Taiwan clams (0.315 to 0.869 µg/g Hg, 0.12 to 0.16 µg/g Cd, and 0.22 to 0.24 µg/g Pb) collected in Naboc and Agusan Rivers (Appleton et al., 2006) upstream of where this study was completed. Thus, based on the Hg, Cd, and Pb levels determined in the four fish species considered in this study, it can be concluded that all fish species in the area are safe for human consumption. These results showed that fortunately, in spite of industrial activities along the two rivers, fish that is being consumed has tolerable heavy metal concentrations that should not affect the health of the human consumers.

Metal intake related to fish consumption

Fish is one of the most indicative factors in freshwater system for estimating heavy metal pollution and risk potential to human health (Papagiannis et al., 2004). From the results of this study, a consumption of 1 kg of fish per week from Agusan River is equivalent to intakes of 2.14 to 6.82 µg Hg, 0.035 to 0.068 µg Cd, and 0.019 to 0.53 µg Pb per week (Table 5). These Hg, Cd, and Pb intakes per week are several folds lower than the Provisional Tolerable Weekly Intake (PTWI) of 300 µg Hg (FAO/WHO, 2003), 420 µg Cd (WHO, 2006), and 1470 µg Pb (WHO, 2000) recommended by the Joint FAO/WHO Expert Committee on Food Additives. Also, the reported weekly Hg intake of 277 µg Hg by a 60 kg person eating 250 g of fish four times in a week from Naboc River (Appleton, 2006) is enormously greater than the weekly Hg intake obtained here.

ACKNOWLEDGEMENTS

The researchers would like to acknowledge the Philippines Commission on Higher Education (CHED)-Higher

Education Development Project-Faculty Development Program for financial support, the Mindanao State University at Naawan, together with the local government unit of Naawan (in Misamis Oriental, Philippines), for the assistance during the collection of samples for the method application phase, and Queensland University of Technology (in Brisbane, Queensland, Australia) particularly the School of Physical and Chemical Sciences, for technical assistance in the use of the laboratories and instruments needed in the conduct of the study.

REFERENCES

Agilent Technologies (2005). ICP-MS: Inductively coupled plasma mass spectrometry. A primer. Agilent Technologies: USA. p.80

Appleton JD, Weeks JM, Calvez JPS (2006). Impacts of mercury contaminated mining waste on soil quality, crops, bivalves, and fish in Naboc River are, Mindanao, Philippines. Sci. Total Environ. 354:198-211.

Appleton JD, Williams TM, Breward N, Apostol A, Miguel J, Miranda C (1999). Mercury contamination associated with artisanal gold mining on the island of Mindanao, the Philippines. Sci. Total Environ. 228:95-109.

ANZECC (1998). Australian and New Zealand Environmental and Conservation Council. Draft ANZECC guidelines for water quality in fresh and marine waters. Canberra, Australia. 8:1-30

Ashoka S, Paeke BM, Bremner G, Hageman KJ, Reid MR (2009). Comparison of digestion methods for ICP-MS determination of trace elements in fish tissues. Analytica Chimica Acta 653:191–199.

Birch G, Taylor S (1999). Source of heavy metals in sediments of the Port Jackson estuary, Australia. Sci. Total Environ. 227:123-138.

CCME (1999). Canadian Council of Ministers of the Environment. Canadian Environmental Quality Guidelines. Winnipeg, Manitoba. p.61

Ciceri E, Recchia S, Dossi C, Yang L, Sturgeon RE (2008). Validation of an isotope dilution, ICP-MS method based on internal mass bias correction for the determination of trace concentrations of Hg in sediment cores. Talanta. 74:642-647.

Conlu PV (1986). Guide to Philippine flora and fauna. Fishes. Natural Resources Management Center, Quezon City. 9:495.

Deshpande A, Bhendigeri S, Shirsekar T, Dhaware D, Khandekar RN (2008). Analysis of heavy metals in marine fish from Mumbai Docks. Environ. Monit. Assess. DOI 10.1007/s10661-008-0645-3. 159:493-500

EU (2001). Commission Regulation as regards heavy metals, Directive 2001/22/EC, No: 466.

FAO/WHO (1987). Priciples of the Safety Assessment of Food Additives and Contaminants in Food Environmental Health Criteria, Geneva, No: 70

FAO/WHO (2003). Summary and conclusions of the sixty-first meeting of the Joint FAO/WHO Expert Committee on Food Additives (JECFA), Rome, JECFA/61/SC.

Haines ML, Brydges K, MacDonald MJ, Smith SL, MacDonald DD (1994). A Review of environmental quality criteria and guidelines for priority substances in the Fraser River Basin. Supporting Documentation DOE FRAP 1994-31. Vancouver: Environment Canada.

Hewitt AD, Reynolds CM (1990). Dissolution of metals from soils and sediments with a microwave-nitric acid digestion technique. Atomic Spectrosc. 11:187-192.

Lampugnani L, Salvetti C, Tsalev DL (2003). Hydride generation atomic absorption spectrometry with different flow systems and in-atomizer trapping for determination of cadmium in water and urine—overviews of existing data on cadmium vapour generation and evaluation of critical parameters. Talanta. 61(5):683-698.

Lim MC, Ayoko GA, Morawska L, Ristovski Z, Jayaratne R (2007). The effects of fuel characteristics and engine operating conditions on the elemental composition of emissions from heavy duty diesel buses. Fuel, 86:1831-1839.

Lorentzen EML, Kingston HM (1996). Comparison of microwave-assisted and conventional leaching using EPA Method 3050B. Anal. Chem. 68:4316-4320.

Maramba NPC, Reyes JP, Francisco-Rivera AT, Crisanta L, Panganiban R, Dioquino C, Dando N, Timbang R, Akagi H, Castillo MT, Quitoriano C, Afuang M, Matsuyama A, Eguchi T, Fuchigami Y (2006). Environmental and human exposure assessment monitoring of communities near abandoned mercury mine in the Philippines: A toxic legacy. J. Environ. Manage.81:135-145.

Moscoso-Perez C, Moreda-Piñeiro J, Lopez-Mahia P, Muniatequi-Lorenzo, Fernadez-Fernandez E, Prada-Rodriquez D (2003). Bismuth determination in environmental samples by hydride generation-electrothermal atomic absorption spectrometry. Talanta. 61(5):633-642.

Papagiannis I, Kagalou I, Leonardos J, Petridis D, Kalfakaou V (2004). Copper and zinc in four freshwater fish species from Lake Pamvotis (Greece). Environ. Int., 30: 357–362.

Pavlou S, Kadeg R,Turner A, Marchlik M (1987). Sediment quality criteria methodology validation: Uncertainty analysis of sediment normalization theory for nonpolar organic contaminants. Envirosphere Company. Bellevue, Washington. Submitted to Washington Environmental Program Office. U.S. Environmental Protection Agency.1987: Washington, District of Columbia. p. 105.

Quevauviller P, Imbert JL, Olle M (1993). Evaluation of the use of microwave oven systems for the digestion of environmental samples. Microchim. Acta, 112:147-154.

Sturgeon RE (2000). Current practice and recent developments in analytical methodology for trace metal analysis of soils, plants and water. Commun. Soil Sci. Plant Anal., 31:1512-1530.

UNEP (2002). Global Mercury Assessment. Geneva: UNEP Chemicals. p.270

U.S. EPA (1995). Test methods for evaluating solid waste. Vol. IA: Laboratory manual physical/chemical methods. SW 846.3rd ed.U.S. Gov. Print. Office, Washington, DC. p.3500

U.S. EPA (1997). Method 3051a: Microwave assisted acid dissolution of sediments, sludges, soils, and oils. 2nd ed. U.S. Gov. Print. Office, Washington, DC. p.30

WHO, World Health Organization (2000). Evaluation of Certain Food Additives and Contaminants (Fifty-third Report of the Joint FAO/WHO Expert Committee on Food Additives). WHO Technical Report Series, No. 896, Geneva.

WHO, World Health Organization (2006). Evaluation of Certain Food Additives and Contaminants (Sixty-fourth Report of the Joint FAO/WHO Expert Committee on Food Additives). WHO Technical Report Series, No. 930, Geneva.

21

Ethylenediaminetetraacetate (EDTA)-Assisted phytoremediation of heavy metal contaminated soil by *Eleusine indica* L. Gearth

Garba, Shuaibu Tela[1]*, Osemeahon, Akuewanbhor Sunday[2], Maina, Humphrey Manji[2] and Barminas, Jeffry Tsaware[2]

[1]Department of Chemistry, P. M. B. 1069. University of Maiduguri, Borno State.Nigeria.
[2]Department of Chemistry, P. M. B. 2076. Federal University of Technology Yola (FUTY), Adamawa State, Nigeria.

This study was designed to assess the natural and chemically enhanced phytoextraction ability of *Eleusine indica* (grass). Three sets of laboratory pot experiment were conducted. Viable seeds of the grass were seeded into one kilogram of the experimental soil placed in each plastic pot. The shoot, root and the experimental soil around root were analyzed for the preliminary levels of the heavy metals: Copper (Cu), Cadmium (Cd), Chromium (Cr), Cobalt (Co) and Lead (Pb). The preliminary levels of Cu, Cd, Cr, Co and Pb in soil, root and shoot of the grass are: soils: 104.5, 5.1, 36.4, 13.3, 14.4 µg/g; root: 164.2, 4.3, 153.9, 11.5 and 24.7 µg/g and shoot of the grass are: 111.5, 2.9, 51.2, 11.1, and 60.7 µg/g respectively. The phytoextraction ability was assessed in terms of its metal transfer factors; Enrichment Coefficient (EC) and Translocation Factor (TF). Copper, Chromium and Lead had the highest EC of 1.07, 1.41 and 4.22 respectively. The levels of the elements in the roots and shoots of the grass at the end of the laboratory experiment shows that more than the bioavailable pool of Cu, Cd, Cr Co and Pb were taken up in the roots with slow translocation of Pb to the shoot: t_1Cu 236.0 to 108.2 μgg^{-1} root-shoot; t_2Cu 137.5 to 316.8 μgg^{-1} root to shoot; t_1Cr 228 to 84.3 μgg^{-1} root-shoot; t_2 Cr 242.6 to 94.2 μgg^{-1} root to shoot; t_1Pb 54.8 to 176.2 μgg^{-1} root to shoot and t_2 Pb 96.0 to 326.0 μgg^{-1}root-shoot. Inductively Coupled Plasma to Optical Emission Spectroscopy - ICP-OES (for Pb determination) and X-ray fluorescence (XRF) (for Cu, Cr, Cd and Co determination) were used for heavy metals determination in this study. The grass showed relatively good response to EDTA application and the higher levels of Cu and Cr concentration in the root suggested that the grass may be a good metal excluder with the possibility of extracting Pb from contaminated soils.

Key words: Phytoextraction, phytostabilization, pollution, soil, grass, cadmium, cobalt, copper, lead and chromium.

INTRODUCTION

Our environment has always been under natural stresses but its degradation was not as severe as it is today. The importance of the study of environmental hazards and their impact on living beings needs no emphasis. The human use of soil can lead to its deterioration by the degradation of soil organic matter and the lowering of its fertility due to erosion and the introduction of various polluting substances including heavy metals.

Environmental pollution by heavy metals is now a global issue that requires considerable attention. Soils contaminated with heavy metals usually lack established vegetation cover either due to the toxic effects of the heavy metal or to the incessant physical disturbances such as erosion (Salt et al., 1995). Most heavy metals are emitted from anthropogenic sources such as industries and transportation. Manure and herbicides as well as, sewage silt used in agriculture are also sources of heavy metals in the environment (Fargasova, 1999). Persistence of these heavy metals in soils and

---

*Corresponding author. E-mail: ms.tg13@yahoo.com.

continuous exposure to them can directly or indirectly lead to their accumulation in plants, animals and subsequently humans.

Trace amount of some heavy metals such as Cu, Zn, Fe, and Co are required by living organisms, however, any excess amount of these metals can be detrimental (Berti and Jacobs, 1996). Non-essential heavy metals include arsenic, antimony, cadmium, chromium, mercury, lead, etc; these metals are of particular concern because they cause air, soil and water pollution (Kennish, 1992). Decontamination of such soils has therefore, become imperative for the safety of animals and humans. A number of techniques have been developed to remove metals from contaminated soils. However, many sites remain contaminated because of economic and environmental costs of the available technologies. Techniques such as excavation and disposal of contaminated soils in landfills are not environmentally friendly and may serve as secondary pollution sources. Therefore, new environmentally friendly and less expensive techniques are required.

Phytoremediation of heavy metal contaminated soils is an emerging technology that extracts or inactivates metals in soils. It is defined as the engineered use of green plants (including grasses, shrubs and woody species) to remove, concentrate, or render harmless such environmental contaminants as heavy metals, trace elements, organic compounds, and radioactive compounds in soil or water (Hinchman et al., 1996). It is environmentally friendly, of low cost, in situ applicable technique for the clean-up of sites contaminated with toxic metals or organic pollutants. Depending on the degree of contamination and the size and volume of the polluted area, different technologies can be used to achieve the desired goals (Henry, 2000; McGrath, 1998; Salt et al., 1998). Phytoextraction seems to be the most promising technique and has received increasing attention from researchers since it was proposed by Chaney (1983) as a technology for reclaiming metal polluted soils. Several approaches have been used but the two basic strategies of phytoextraction, which have finally been developed are: Chelate or chemically assisted phytoextraction or induced phytoextraction, in which artificial chelates are added to increase the mobility and uptake of metal contaminant and continuous or natural phytoextraction which measures the natural ability of the plant to remediate soil. Only the number of plant growth repetitions is therefore controlled (Salt et al., 1995, 1998).

In view of the fact that the rate of bioremediation is directly proportional to the plant growth rate and the total amount of bioremediation is correlated with plant's total biomass, the integration of specially selected high biomass crops with improved plant husbandry and innovative soil management practices is a promising alternative strategy towards achieving high biomass and metal accumulation rates from contaminated soil

(Nowack et al., 2006; Evangelou et al., 2007). Many chemical amendments, such as ethylenediaminetetraacetic acid (EDTA), Hydroxyethylene-diaminetriacetic acid (HEDTA), Nitrilotriacetic acid (NTA) and organic acids have been used in pot and field experiments to enhance extraction rates of heavy metals and to achieve higher phytoextraction efficiency (Blaylock et al., 1997; Wu et al., 2006). There is much evidence confirming that EDTA is one of the most efficient chelating agents in enhancing Pb phytoavailability in soil and subsequent uptake and translocation in shoots (Chen and Cutright, 2001; Shen et al., 2002).

Majority of studies on phytoremediation was based on pot experiments and hydroponic culture, and only a few reports evaluated the phytoextraction potential of hyperaccumulators or high biomass crops under field conditions (McGrath et al., 2006; Zhuang et al., 2007). Only a few attempts have been made to evaluate the possibility of metal removal in response to modifications of agronomic practices (Marchiol et al., 2007). Some weeds of the grass family have been experimented to be suitable for phytoremediation because of their multiple ramified root systems.

In this study, we assessed the natural and chelated phytoextraction potential of the native tropical grass: *Eleusine indica*, and when chemically enhanced with EDTA, to evaluate the ability of the grass to remediate soils contaminated with multiple heavy metals.

MATERIALS AND METHODS

Sampling

Four samples of soil and grass were collected from a refuse dumping site along Gombe road at the outskirt of Maiduguri metropolis (Figure 1). Fresh plant samples were collected in the morning by pulling carefully from the soil to avoid damage to the roots and washed with tap water. They were then separated into shoots and roots. Soil samples were collected from the surface to subsurface portion of the soil around the plant roots (Rotkittikhum et al., 2006) at a range interval of 20 to 30 square meter apart.

Sample preparation and analysis

Both the soil and plant samples collected were dried at 60°C to a constant weight, grounded into fine powder, sieved with 2 mm wire mesh and analyzed for the preliminary levels of the heavy metals: Cd, Cu, Co, and Cr were analyzed using X-ray fluorescence (XRF) while Pb was determined using inductively coupled plasma optical emission spectroscopy (ICP-OES) following aqua- regia digestion (McGrath and Cunliffe, 1985). The dried soil sample was also characterized for its physicochemical properties (Lombi et al., 2001). The concentration of Cd, Cu, Co, and Cr in the shoots and roots of the grass samples were also determined by X-ray fluorescence (XRF) while ICP-OES was used to determined the level of Pb. Using 0.5 g of the powdered sample, digested with HNO_3 and $HClO_4$ acid (Lombi et al., 2000).

Physicochemical analysis of soil samples

Soil texture was determined by the Bouyoucos hydrometer method.

Table 1. Physicochemical properties of experimental soil.

Soil parameter	Mean ±S.D
Clay (%)	25.90 ±1.80
Silt (%)	21.70 ±2.50
Sand (%)	50.40 ±2.80
pH	7.80 ±0.10
Organic matter (%)	4.15 ±0.05
Nitrogen (%)	0.05 ±0.02
C EC (mol/ 100 g soil)	11.27 ±0.76
EC (mS/cm)	464.00 ±0.10
Potassium (µg/g)	22.73 ±2.63
Moisture content (%)	34.00±1.80

Measurements are averages of three replicates ± S.D (Standard deviation); CEC: Cation exchange capacity; EC: Electrical conductivity.

The moisture content of soil was calculated by the weight difference before and after drying at 105°C to a constant weight. The pH and electrical conductivity (EC) were measured after 20 min of vigorous mixed samples at 1: 2.5. Solid: deionized water ratio using digital meters (Elico, Model LI-120) with a combination pH electrode and a 1-cm platinum conductivity cell respectively. Total nitrogen was determined according to the standard methods of the American Public Health Association (1998). Cation exchange capacity (CEC) was determined after extraction with ammonium acetate at pH 7.0 and the organic carbon was determined by using Walkley–Black method (Jackson, 1973).

Three sets of controlled and artificial laboratory experiment were conducted. Plastic pots were used for the experiment. 0.5 to 1.00 kg soils of known chemical composition were placed into each of the pots and viable seeds of grass were seeded to soil. Soils of known chemical concentration were contaminated with various grams of the metals; Cu, Co, Cr, Cd and Pb. The contaminated soil received the metals Cd as Cd $(NO_3)_2$; Co as $CoCl_2$; Cr as Chromic acid; Cu as $CuCl_2$ and Pb as Pb $(NO_3)_2$ at the concentration of 50, 150, 250, 250 and 150 mgkg⁻¹ respectively. EDTA was applied to another soil of known chemical composition, amended with the same level of the said elements. This was done at the rates of one gram per kilogram (2.7 mmolkg⁻¹ of soil), four weeks after germination of the grass.

Experiments were exposed to natural day and night temperatures. Since humidity is one of the factors ensuring the growth of plants and the necessary physiological processes, the experimental grass in the pots were watered every 5 days with 200 ml of deionized water (Lombi et al., 2001). To prevent loss of nutrients and trace elements out of the pots, plastic trays were placed under each pot and the leachates collected were put back in the respective pots. This was done for a period of three months. Four replicates for each pot of grass were planted for statistical data handling. The samples of grass collected at the end of the experiment, were separated into roots and shoots, dried at 60°C to a constant weight, grounded into fine powder, sieved with 2 mm wire mesh and analyzed using X-ray fluorescence (XRF) for the level of the metals; Cu, Co, Cr, Cd, while ICP-OES was used to determine the level of Pb.

Statistical analysis

All statistical analyses were performed using the SPSS 17 package. Differences in heavy metal concentrations among different parts of the grass were detected using One-way ANOVA, followed by multiple comparisons using Turkey tests. A significance level of (p < 0.05) was used throughout the study.

RESULTS

The taxonomic classification of the experimental soil (Table 1) was sandy loam with pH of 7.8, EC of 464 mS/cm. The high pH level of the soil is generally within the range for soil in the region; soil pH plays an important role in the sorption of heavy metals, it controls the solubility and hydrolysis of metal hydroxides, carbonates and phosphates and also influences ion-pair formation, solubility of organic matter, as well as surface charge of Fe, Mn and Al-oxides, organic matter and clay edges (Tokalioglu et al., 2006).

The preliminary concentration levels of Cr and Co observed in experimental soil are 36.4 and 13.30 µg/g respectively. Maiduguri metropolitan highway road networking has been characterized with high level of Cu (Garba et al., 2007). It is specifically adsorbed or fixed in soils, making it one of the trace metals (Baker and Senft, 1995). Hence, the level of Cu observed in experimental soil used in this study (104.50 µg/g) was the highest of all the five metals studied. The level of Pb in the soil was found to be 14.4 µg/g. Cadmium is considered to be mobile in soils but is present in much smaller concentrations (Zhu et al., 1999). This could explain why the level of Cd (5.10 µg/g) observed in the experimental soil used in this study was the lowest when compared to the other metals (Table 2). It has been reported that the level and impact of heavy metals on the environment is greatly dependent on their speciation in soil solution and solid phase which determine their environmental availability, geochemical transfer and mobility pathways (Pinto et al., 2004).

Uptake and accumulation of metals by the grass plant _E. indica_

Table 2 shows the preliminary naturally desorbed

Table 2. Preliminary concentrations (µg/g) of Cu, Cd, Cr, Co and Pb observed in the roots, shoots and the experimental soil samples from the sampling site.

Element \ Sample	Root Mean ±SD	Shoot Mean ±SD	Soil Mean ±SD
Cu	$164.20^k \pm 2.93$	$111.50^c \pm 1.61$	$104.50^d \pm 1.94$
Cd	$4.30^h \pm 0.88$	$2.90^b \pm 1.94$	$5.10^e \pm 1.03$
Cr	$153.90^g \pm 3.18$	$51.20^q \pm 2.16$	$36.4^{0f} \pm 2.68$
Co	$11.50^w \pm 2.87$	$11.10^b \pm 2.42$	$13.30^a \pm 2.36$
Pb	$24.70^n \pm 2.59$	$60.7^{0x} \pm 2.57$	$14.40^a \pm 2.09$

The mean differences of elements in the same column with same letters are not significant at ($p<0.05$). (n=4).

Table 3. Enrichment coefficient (EC) and Translocation factor (TF) of the metals by the grass.

Element \ Sample	Translocation factor (TF)	Enrichment coefficient (EC)
Cu	0.68	1.07
Cd	0.67	0.57
Cr	0.33	1.41
Co	0.97	0.83
Pb	2.46	4.22

TF is calculated by the relation: - ratio of concentration of metal in the shoot to the concentration of metal in the roots (Cui et al., 2007). EC is given by the relation: - The ratio of the concentration of metal in the shoots to the concentration of metal in the soil (Chen et al., 2004).

concentration of the metals observed in the grass root and shoot of this study. In the roots, the levels of Cu, Cr, Pb, Co and Cd observed are: 164.20; 153.90; 24.70, 11.50 and 4.30 µg/g respectively. And in the shoot the levels for Cu, Cr, Pb, Co and Cd are 111.5, 51.2, 60.7, 11.10 and 2.90 (µg/g) respectively. Most of the metals (Cu, Cr and Cd) were found at higher level greater in root than the shoot. It has been reported that most grass specie are known to concentrate heavy metals in the roots, with only very low translocation to the shoot (Speir et al., 2003; Bennett et al., 2003). Several studies have demonstrated that the concentration of metals in plant tissue is a function of the metal content in the growing environment (Grifferty and Barrington, 2000). The results indicated that accumulation of Pb, Cu, Cd, Co and Cr in the roots can be arranged in the order: Cu>Cr>Pb>Co>Cd. However, the levels of the elements in either the root or the shoot of the grass plant; E. indica cannot determine its hyperaccumulating potential. Soil-to-plant metal transfer ratio is an important component of phytoextraction, it determines which part of a plant, root or shoot that accumulate in terms of translocation factor (TF) and enrichment coefficient (EC) (Frissel, 1997).

Metal transfer coefficients

Table 3 shows the enrichment coefficient (EC) and translocation factor (TF) of the elements naturally absorbed by the grass. Translocation factor is a measure of the ability of plants to transfer accumulated metals from the roots to the shoots. It is given by the ratio of concentration of metal in the shoot to that in the roots (Cui et al., 2007; Li et al., 2007). The TF observed for Pb was 2.46, the only element that has TF greater than one. The enrichment coefficient (EC) was used to evaluate the ability of plant to accumulate heavy metals in the root. Enrichment coefficient was given by the ratio of the concentration of metal in the shoots to the concentration of metal in the soil (Chen et al., 2004). In this study, the EC of 1.07, 1.41 and 4.22 were observed for the elements: Cu, Cr and Pb respectively. Plants of high EC greater than one, accumulates metals in the root with less or poor translocation to the aerial parts (shoot), they mainly restrict metal in their roots.

Effect of EDTA application on metal uptake by the grass

Most metals in soils exist in unavailable forms, thus, soil conditions have to be altered to promote phytoextraction since the phenomenon, depends on a relatively abundant source of soluble metal for uptake and translocation to shoots. Table 4 gives the level of the metals desorbed when EDTA was applied. The observed level of Pb was found to increase higher in the shoots, 326.0 µg/g than the root compared to what was observed preliminarily.

Table 4. Mean concentration (μg/g) of the metals in roots and shoots of the grass from the laboratory pot experiment.

Element	Sample	Root Mean ±SD	Shoot Mean ±SD
Cu	t_1	236.00±3.72	108.20±2.12
	t_2	137.50±4.22	316.80±2.82
Cr	t_1	228.10±4.39	84.30±4.42
	t_2	242.60±2.57	94.20±2.57
Pb	t_1	54.80±3.57	176.20±1.75
	t_2	96:00±3.22	326.00±4.26

t_1=soil contaminated with heavy metal concentrations, t_2 =soil amended with EDTA and SD= Standard deviation. The mean differences of the elements were found significant at ($p<0.05$; n=4).

The concentration level of Cu on the other hand, was also found to increase in the shoot 316.8 μg/g. The application of EDTA to the experimental soil, increased the level of Cr in the root (242.6 μg/g) with less or poor translocation of the element to the shoot (94.2 μg/g). Plant uptake of metal in soil solution has been observed to depend on a number of factors: physical processes such as root intrusion, water and ion fluxes; biological parameters, including kinetics of membrane transport, ion interactions, and the ability of plants to adapt metabolically to changing metal stress in the environment (Cataldo and Wildung, 1978).

DISCUSSION

Uptake of contaminants from the soil by plants occurs primarily through the root system in which the principle mechanisms of preventing contaminant toxicity are found. The root system provides an enormous surface area that absorbs and accumulates the water and nutrients that are essential for growth, but also absorbs other non-essential contaminants (Arthur et al., 2005) such as Pb and Cd. Naturally the grass was found to accumulate most of the elements of interest in the root. The heavy metals: Cu, Cr and Cd were found at high levels in the root than the shoot with no sign of toxicity. Cadmium, for instance has been reported in many studies to be accumulated at higher concentrations in the roots than in the leaves (Boominathan and Doran, 2003). Pulford et al. (2001), in a study with temperate plants confirmed that Cr was poorly taken up into the aerial tissues but was held predominantly in the root. Similarly the grass E. indica in this study expressed high level of Cr in its roots. One of the mechanisms by which uptake of metal occurs in the roots may include binding of the positively charged toxic metal ions to negative charges in the cell wall (Gothberg et al., 2004) and the low transport of heavy metal to shoots may be due to saturation of root metal uptake, when internal metal concentrations are high.

Although, adverse effects of Cr on plant height and shoot growth has been reported (Rout et al., 1997); a significant reduction in plant height in Sinapsis alba when Cr was given at the rates of 200 or 400 mg kg[-1] soil has been reported (Hanus and Tomas, 1993). Wenger et al. (2003) reported that the critical toxicity level of Cu in the shoots of crop plants is greater than 20 to 30 mg kg[-1]. But no sign of toxicity at all was expressed by the grass E. indica in this study.

Ethylenediaminetetraacetate (EDTA) has been reported to be the most effective amendment in phytoextraction research. It has been successfully utilized for instance, to enhance phytoextraction of lead and other metals from contaminated soils (Cunningham and Ow, 1996; Chen et al., 2004). In this study, EDTA was found to enhance the bioavailability and to improve the uptake and translocation of Cu and Pb to the shoot. Huang et al. (1997) showed that EDTA was the most efficient chelator for inducing the hyperaccumulation of Pb in pea plants shoots. Vassil et al. (1998) also found that Indian mustard exposed to Pb and EDTA in nutrient solution accumulated 11,000 mg kg[-1] Pb in dry shoot tissue. The poor translocation of Cr to the shoots despite the addition of EDTA could be due to sequestration of most of the Cr in the vacuoles of the root cells to render it non-toxic which may be a natural toxicity response of the plant (Shanker et al., 2004). Phytoextraction is a long-term remediation practice. It requires many cropping cycles to decontaminate contaminated sites to an acceptable level favourable for human use. It has been reported that for phytoremediation, grasses are the most commonly evaluated plants (Ebbs and Kochian, 1998; Shu et al., 2002). The large surface area of their fibrous roots and their intensive penetration of soil reduces leaching, runoff, and erosion via stabilization of soil and offers advantages for phytoremediation.

Conclusion

This study therefore has proved the possibility of using

the grass *E. indica* for phytoremediation especially phytostabilization of Cu, Cr and possible phytoextraction of Pb. These techniques reduces leaching, runoff, and erosion via stabilization of soil and may decontaminate the soil of Pb contamination.

ACKNOWLEDGEMENT

Authors are sincerely grateful to Mr. Fine Akawu, the Laboratory Technologist who relentlessly assisted in the preparation of chemicals and the successful watering of the plants.

REFERENCES

American Public Health Association (APHA), (1998). Standard methods for the examination of water and waste water. 20th Edition, Washington DC, U.S.A.

Arthur EL, Rice PJ, Anderson TA, Baladi SM, Henderson KLD, Coats JR (2005). Phytoremediation - An overview. Crit. Rev. Plants Sci., 24:109-122.

Baker DE, Senft JP (1995). Copper. In: Heavy metals in soils, B.J. Alloway (ed.), (Second edition): Blackie Academic and Professional, London, 8: 179-205.

Bennett LE, Burkhead JL, Hale KL, Terry N, Pilon M, Pilon-Smits EA (2003). Analysis of transgenic Indian mustard plants for phytoremediation of metal-contaminated mine tailings. J. Environ. Qual., 32: 432-40.

Berti WR, Jacobs LW (1996). Chemistry and phytotoxicity of soil trace elements from repeated sewage sludge application. J. Environ. Qual., 25: 1025–32

Blaylock MJ, Dushenkov S, Zakharova O, Gussman C, Kapulnik Y, Ensley BD, Salt DE, Raskin I (1997). Enhanced accumulation of Pb in Indian mustard by soil-applied chelating agents. Environ. Sci. Technol., 31: 860-865.

Boominathan R, Doran PM (2003). Cadmium tolerance antioxidative defenses hyperaccumulator, *Thlaspi caerulescens*. Biotechnol. Bioeng., 83: 158-167.

Cataldo DA, Wildung RE (1978). Soil and plant factors influencing the accumulation of heavy metals by plants. Environ. Health Perspect., 27: 149-159

Chaney RL (1983). Plant uptake of inorganic waste constituents. In: Parr JF, Marsh PB, Kla JM (ed.): Land Treatment of Hazardous Wastes. Park Ridge Noyes Data Corp., London, pp. 50-76.

Chen H, Cutright T (2001). EDTA and HEDTA effects on Cd, Cr and Ni uptake by *Helianthus annus*. Chemosphere, 45: 21–28.

Chen Y, Shen Z, Li X (2004). The use of vetiver grass (*Vetiveria zizanioides*) in the phytoremediation of soils contaminated with heavy metals. Appl. Geo-chem., 19: 1553-1565.

Cui S, Zhou Q, Chao L (2007). Potential hyperaccumulation of Pb, Zn, Cu and Cd in endurant plants distributed in an old smeltery, northeast China. Environ. Geol., 51: 1043-1048.

Cunningham SD, Ow DN (1996). Promises and Prospect of phytoremediation. Plant Physiol., 110(5): 715-719.

Ebbs SD, Kochian LV (1998). Phytoextraction of zinc by oat (Avena sativa), barley (*Hordeum vulgare*), and Indian mustard (*Brassica juncea*). Environ. Sci. Technol., 32: 802-806.

Evangelou MHW, Nauer U, Ebel M, Schaeffer A (2007). The influence of EDDS and EDTA on the uptake of heavy metals of Cd and Cu from soil with tobacco *Nicotianna tabacum*. Chemosphere, 68: 345-353.

Fargasova A (1999). Root growth inhibition, photosynthetic pigments production, and metal accumulation in *Sinapis alba* as the parameters for trace metals of effect determination. Bull. Environ. Contam. Toxicol., 61: 762-769.

Frissel M (1997). Protocol for the experimental determination of the soil to plant transfer factors (concentration ratios) to be used in radiological assessment models. UIR Newslett., 28: 5-8.

Garba ST, Ogugbuaja VO, Samali A (2007). Public Health and The Trace Elements: Copper (Cu), Chromium (Cr) and Cobalt (Co) in Roadside Dust in Maiduguri Metropolis. J. Health Educ. Sport Sci. (JOHESS), 6(1): 152–157.

Gothberg A, Greger M, Holm K, Bengtsson BE (2004). Influence of nutrient levels on uptake and effects of mercury, cadmium, and lead in water spinach. J. Environ. Qual., 33: 1247-1255.

Grifferty A, Barrington S (2000). Zinc uptake by young wheat plants under two transpiration regimes. J. Environ. Qual., 29: 443-446.

Hanus J, Tomas J (1993). An investigation of chromium content and its uptake from soil in white mustard. Acta Fytotech., 48: 39–47.

Henry JR (2000). An Overview of Phytoremediation of Lead and Mercury. NNEMS Report, Washington, D.C., pp. 3-9.

Hinchman RR, Negri MC, Gatliff EG (1996). Phytoremediation: Using Green Plants to Clean Up Contaminated Soil, Groundwater and Wastewater. In Proceedings, International Topical Meeting on Nuclear and Hazardous Waste Management, Spectrum 96. Seattle. American Nuclear Society.

Huang JW, Chen JJ, Berti WR, Cunningham SD (1997). Phytoremediation of lead-contaminated soils: role of synthetic chelates in lead phytoextraction. Environ. Sci. Technol., 31: 800-880.

Jackson ML (1973). Soil chemical analysis. Prentice Hall Inc., Englewood Clifs, N.J. Library of Congress, USA.

Kennish MJ (1992). Ecology of Estuaries: Anthropogenic Effects. CRC Press, Inc., Boca Raton, FL., p. 494.

Li MS, Luo YP, Su ZY (2007). Heavy metal concentrations in soils and plant accumulation in a restored manganese mine land in Guangxi, South China. Environ. Pollut., 147: 168-175.

Lombi E, Wenzel WW, Adriano DC (2000). Arsenic-contaminated soils: II. remedial action. In: Wise DL, Tarantolo DJ, Inyang HI, Cichon EJ, eds. Remedial of hazardous waste contaminated soils. New York, NY, USA: Marcel Dekker Inc., pp. 739–758.

Lombi E, Zhao FJ, Dunham SJ, McGrath SP (2001). Phytoremediation of Heavy metal contaminated soils: natural hyperaccumulation versus Chemically enhanced phytoextraction. J. Environ. Qual., 30: 1919-1926.

Marchiol L, Fellet G, Perosa D, Zerbi G (2007). Removal of trace metals by *Sorghum bicolor* and *Helianthus annuus* in a site polluted by industrial wastes: A field experience. Plant Physiol. Biochem., 45(5): 379–387.

McGrath SP, Cunliffe CH (1985). A simplified method for the L181 (Annex A): 6–12. extraction of metals Fe, Zn, Cu, Ni, Cd, Pb, Cr, Co and Mn From Cooper, E.M., J.T. Sims, S.D. Cunningham, J.W. Huang, and W.R. soils and sewage sludge. J. Sci. Food Agric., 36: 794–798.

McGrath SP (1998). Phytoextraction for soil remediation. In: Plants that Hyperaccumulate Heavy Metals. (Ed.): R.R. Brooks. CAB International, Wallingford, UK, pp. 261-288.

McGrath SP, Lombi E, Gray CW, Caille N, Dunham SJ, Zhao FJ (2006). Field Evaluation of Cd and Zn phytoextraction potential by the hyperaccumulators *Thlaspi caerulescens* and *Arabidopsis halleri*. Environ. Pollut., 141(1): 115–125.

Nowack B, Schulin R, Robinson B (2006). Critical assessment of chelants-enhanced metal phytoextraction. Sci. Total Environ., 40(17): 5225-5232.

Pinto AP, Mota M, De Varennes A, Pinto FC (2004). Influence of organic matter on the uptake of cadmium, zinc, copper and iron by sorghum plants. Sci. Total Environ., 326: 239-247.

Pulford ID, Watson C, McGregor SD (2001). Uptake of chromium by trees: Prospects for phytoremediation. Environ. Geochem. Health, 23: 307–311.

Rout GR, Samantaray S, Das P (1997). Differential chromium tolerance among eight mung bean cultivars grown in nutrient culture. J. Plant Nutr., 20: 473–483.

Rotkittikhum P, Kroatrachue M, Chaiyarat R, Ngernsansaruay C, Pokethitiyook P, Paijitprapaporn A, Baker AJM (2006). Uptake and Accumulation of Lead by Plants from Ngam Lead Mine Area in Thailand. Environ. Pollut., 144: 681-688.

Salt DE, Blaylock M, Nanda Kumar PBA, Dushenkov V, Ensley BDI, Raskin I (1995). Phytoremediation: A novel strategy for the removal of toxic metals from the environment using plants. Biotechnology, 13: 468-474.

Salt DE, Smith RD, Raskin I (1998). Phytoremediation. Ann. Rev. Plant Physiol. Plant Mol. Biol., 49: 643–668.

Shanker AK, Djanaguiraman M, Sudhagar R, Chandrashekar CN, Pathmanabhan G (2004). Differential antioxidative response of ascorbate glutathione pathway enzymes and metabolites to chromium speciation stress in green gram (*Vigna radiata* (L) R Wilczek, cv CO 4) roots. Plant Sci., 166: 1035–1043.

Shen ZG, Li XD, Wang CC, Chen HM, Chua H (2002). Lead phytoextraction from contaminated soil with high-biomass plant species. J. Environ. Qual., 31: 1893-1900.

Shu WS, Xia HP, Zhang ZQ, Lan CY, Wong MH (2002). Use of vetiver and three other grasses for revegetation of Pb/Zn mine tailings: Field experiment. Int. J. Phytoremed., 4: 47-57.

Speir TW, Van Schaik AP, Percival, Close ME, Pang L (2003). Heavy metals in soil, plants and groundwater following high-rate sewage sludge application to land. Water Air Soil Pollut., 150: 349-358.

Tokalioglu S, Kavtal S, Gultekin A (2006). Investigation of heavy metal uptake by vegetables growing in contaminated soil using the modified BCR sequential extraction method. Int. J. Environ. Anal. Chem., 88(6): 417-430.

Vassil AD, Kapulnik Y, Raskin I, Salt DE (1998). The role of EDTA in lead transport and accumulation by Indian mustard. Plant Physiol., 117: 447-453.

Wenger K, Gupta SK, Furrer G, Schulin R (2003). The role of Nitrilotriacetate in Copper uptake by Tobacco. J. Environ. Qual., 32: 1669-1676.

Wu QT, Deng JC, Long XX, Morel JL, Schwartz C (2006). Selection of appropriate organic additives for enhancing Zn and Cd phytoextraction by hyperaccumulators. J. Environ. Sci., 18(6): 1113–1118.

Zhu Y L, Pilon-Smits EAH, Tarun AS, Weber SU, Jouanin L, Terry N (1999). Cadmium tolerance and accumulation in Indian mustard is enhanced by over expressing-glutamylcysteine synthase. Plant Physiol., 121: 1169–1177.

Zhuang XL, Chen J, Shim H, Bai Z (2007). New advances in plant growth promoting rhizobacteria for bioremediation. Environ. Int., 33: 406-413.

22

Removal of heavy metal ions in aqueous solutions using palm fruit fibre as adsorbent

IDERIAH, T. J. K.[1]*, DAVID, O. D.[2] and OGBONNA, D. N.[3]

[1]Institute of Pollution Studies, Rivers State University of Science and Technology, Rivers State, Nigeria.
[2]Department of Environmental and Applied Biology, Rivers State University of Science and Technology, Port Harcourt Rivers State, Nigeria.
[3]Institute of Geosciences and Space Technology, Rivers State University of Science and Technology, Rivers State, Nigeria.

The ability of the biomass of palm fruit fiber in removing Pb, Cu, Ni and Cr from aqueous solution was investigated as a function of concentration, contact time and pH variations. Palm fruit fiber from the study locations were washed with deionized water, air-dried and ground using electric grinder. The powdered fiber was sieved and treated with 0.3 M HNO_3 solution for 24 h, washed with deionized water until pH 7.2 and oven dried at 60°C. The biomass was added to 1 M stock metal ion solutions made from Copper sulphate, Potassium dichromate, Lead nitrate and Nickel sulphate. The concentrations, contact time and pH of each stock solution were varied. The mixtures were shaken, filtered and analyzed by GBC Avanta Atomic Absorption Spectrophotometer version 2.02. The results showed mean percentage recovery of 51.08% Pb, 54.75% Cu, 46.96% Ni and 44.91% Cr for concentrations, 96.96% Pb, 9.79% Cu, 49.21% Ni, and 7.63% Cr for contact time and 87.48% Pb, 82.86% Cu, 56.71% Ni and 37.68% Cr for pH. The application of the biomass to waste water showed percentage removal of 73% Pb, 78% Cr, 82% Cu and 87% Ni. The mean percentage removal value revealed Pb as the highest and Cr as the least adsorbed. The sorption capacity of the biomass decreased with increasing concentration of metal ion but increased with decreasing pH and increasing contact time. Chemical modification of the biomass enhanced its capacity. Thus the palm fruit fiber biomass is cost effective and has great potential for use as adsorbent in removing heavy metals from aqueous solutions.

Key words: Palm fruit fiber, heavy metal, biomass, ion, adsorbent, contact time, percent removal.

INTRODUCTION

Man's activities through industrialization, urbanization, technological development and agricultural activities discharge heavy metals into the environment; water, land and air which has become a matter of concern over the past few decades, due to the characteristics of metals to cause objectionable effects by impairing welfare and reducing the quality of life in the environment. The presence of metal ions in the environment is detrimental to many living species (Benhima et al., 2008). The removal of heavy metals is very important because they

are non-biodegradable. Heavy metals affect flora, fauna and other abiotic components of the ecosystem by irreversibly binding to the active sites of enzymes, resulting to metabolic alterations and changes that pose severe health injury and hazard (Rau and Amit, 2002).

Waste water collected from industries, municipalities and communities must ultimately return to the receiving water and land (Raji and Anirudhan, 1997). For example when waste water is released to land without treatment, it undergoes various processes such as diffusion, mechanical dispersion, chemical reaction and adsorption, the end product of which is the accumulation of metals in the soil for years. Furthermore, when untreated waste water contaminated with heavy metals is released into aquatic system, some microorganisms and plants in the

*Corresponding author. E-mail: itubonimi@yahoo.com.

system have the ability to bioaccumulate and biomagnify these heavy metals in the environment. The ingestion of these contaminants may affect not only the productivity and reproductive capacities of these organisms, but may also ultimately affect the health of man that depends on these organisms and plants as major source of protein (Davies et al., 2006).

Virtually all metals including the essential and non-essential metals are toxic to aquatic organisms as well as humans if exposure levels are sufficiently high (Law, 1993), and the presence of heavy metals has a potentially damaging effect on human physiological and biological systems when tolerance levels are exceeded (Fatoki et al., 2002).

A wide range of physical and chemical processes and techniques are available for the removal of heavy metal ions from waste water. These include reduction – precipitation, ion exchange, reverse osmosis, carbon sorption and electrolytic methods (Osinowo and Olayinka, 1998; Blanco et al., 1999; Babel and Kurniawan, 2003; Igwe and Abia, 2003; Mohan and Pittman Jr., 2007; Chiban et al., 2011a, b, c; Soudani et al., 2011 a, b). A survey of the advantage and disadvantage of these techniques reveals that; reduction-precipitation methods offer economic advantage but the draw back is the production of sludge in huge quantity and when different heavy metal ions are contained in same effluent, simultaneous treatment is difficult because of change in pH required for their precipitation (Okoye et al., 2001). Ion exchange and reverse osmosis are found to be effective and can remove 90-95% of most metals (Osinowo and Olayinka, 1998) but cost of operation is economically not appealing (Saifuddin and Kumaran, 2005). Electrolytic method is economically expensive; energy requirement is high and release other products that require disposal (Juang and Shiau, 2000). Adsorption using commercially activated carbon (CAC) can remove heavy metals from waste water but remains expensive material for most industries.

A review of these techniques and processes requires the need to explore a low cost technique and materials that are capable of being efficient and can compete favourably with other techniques; hence the principle of adsorption using biopolymer wastes products for the removal of metal ions in aqueous solution. The focus of recent researches is to use natural adsorbents which can be economic and cost effective (Benhima et al., 2008; Chiban et al., 2011a, b).

The use of palm oil milling wastes (fiber-biopolymers) is considered from the fact that natural biopolymers are capable of lowering transition metals ions to parts per billion (Saifuddin and Kumaran, 2005). Furthermore, certain wastes from agricultural operation may have the potential to be used as low cost adsorbents as they represent unused resources and are widely available and environment friendly (Deans and Dixon, 1992).

In Nigeria, oil palm (*Elaeis Guineenses*) is an important commercial crop and the expansion of this industry has generated enormous amount of vegetable wastes such as fiber, empty fruit bunches, palm fronds, palm kernel shells, trunks and effluents/sludge. The discharge or release of these wastes especially effluents and fibers without proper treatment may damage the environment both in the receiving water and foul smell to neighbourhood and communities. When these wastes are released into the receiving water, it increases the biochemical oxygen demand (BOD) of the water and at higher concentration, death of fishes and other living organisms occur due to proliferation of micro-organisms that consume the dissolved oxygen in water leading to depletion of oxygen in the system.

The use of processed palm oil wastes-fibers will add economic value; provide a potentially inexpensive material that will be used for the removal of metal ions from aqueous solution and help reduce the cost of waste disposal. Waste fibers are equally available at little or no cost; however there is the associated problem of disposal due to the enormous wastes generated in the industry that requires its treatment and disposal. Efforts are only made to convert empty fruit bunches to agricultural manure leaving the fibers and effluents to pollute the environment hence the need to carry out this study to explore the usefulness of this fibers. Thus this study is aimed at utilizing waste palm fruit fibers as adsorbents in the removal of copper, chromium, lead and nickel in aqueous solution by varying the experimental conditions such as contact time, metal ion concentration and pH which could be applied in reduction of heavy metal pollution in the environment.

MATERIALS AND METHODS

Study area

The study area Oshika, Elele-Alimini and Ubima are located in Ahoada West and Ikwerre Local Government Areas of River State, Nigeria. The industrial oil mill at Ubima is about 10,000 hectares of oil palm estate and extends up to Elele-Alimini. Oshika lies between longitude 6° 33′ 42.54″ E and latitude 5° 04′ 0.24″ N. Elele-Alimini lies between longitude 6° 43′ 31.26″ E and latitude 5° 03′.96″ N. Ubima lies between longitude 6° 52′ 57″ E and latitude 5° 8′ 14″ N. Between the Risonpalm and Elele Alimini are series of local palm oil processing mills releasing huge quantity of palm fruit fibers waste on land and into aquatic environments. The major activities in the area are farming of cash crops and processing of palm fruits. Figure 1 is the map of the study areas showing the sampling locations.

Preparation of biomass (Organic wastes)

Fresh palm fruit fiber collected from the study locations were washed properly with deionized water, air-dried and ground using an electric grinder. The powdered fiber material was sieved through a mesh size of 105 µm to obtain fine biomass. The finely sieved biomass was treated with 0.3 M HNO_3 solution for 24 h, followed by washing with deionized water until pH of 7.2 was achieved and oven dried at 60°C with constant mixing. The prepared biomass was stored in desiccators.

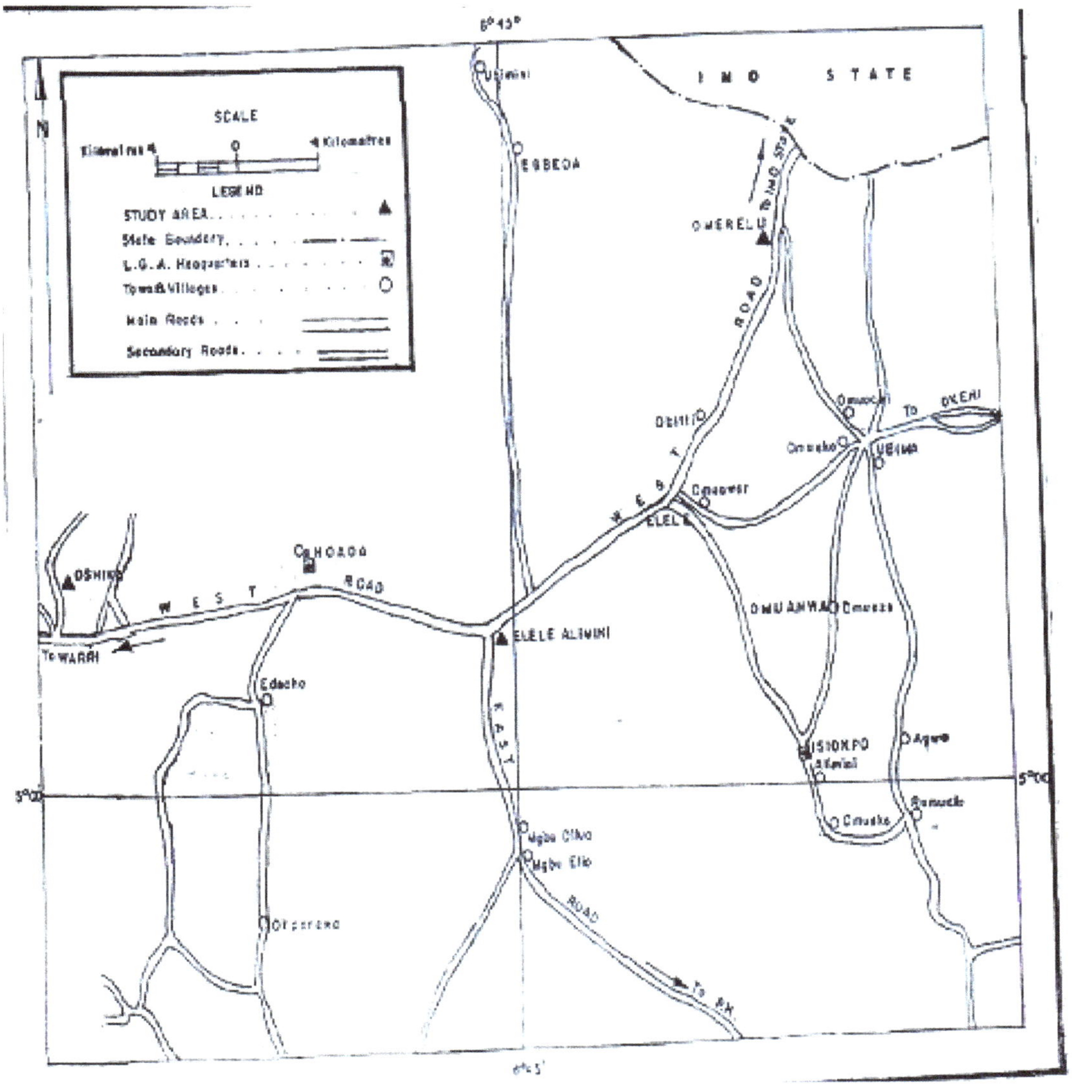

Figure 1. Map of the study areas showing the sampling locations.

Metal salts used and preparation of stock solution of metal ions

The following metal salts were used to prepare the stock solution of metal ions in aqueous form: Copper sulphate ($CuSO_4$. $5H_2O$), Potassium dichromate ($K_2Cr_2O_7$), Lead nitrate (Pb (NO_3)$_2$ and Nickel sulphate Ni (SO_4). $6H2O$.

For each of the metal ions, a stock solution of 1M was prepared separately by dissolving the appropriate weight of the metal salts in 1000 ml of distilled water. The solutions were prepared using a standard flask. The range of concentrations used was prepared by serial dilution of the stock solution with deionized water.

Batch sorption experiment

The sorption studies were carried out at room temperature using a conical flask, containing 50.0 ml of the test solution, with a known metal ion concentration.

Before mixing with adsorbent, the pH of the solution was adjusted to 7.0 with 1M HC1 and or NaOH. A known amount of dried pretreated fiber (2 g) referred as biomass was added into the conical flask containing metal ion solution and tightly covered with cellophane and shaken thoroughly, allowing sufficient time for adsorption equilibrium (also to be examined). The content of the flasks was filtered on Whatman filter paper of 0.42 µm porosity and centrifuged at 2800 rpm for 5 min and the supernatants were analyzed by GBC Avanta Flame Atomic Absorption Spectrophotometer version 2.02.

Determination of the effect of metal ion concentration

A 50.0 ml solution of 0.1, 0.2, 0.4, 0.6, 0.8 and 1.0 M Pb^{2+} concentrations was prepared and poured into six conical flasks labeled A, B, C, D, E and F respectively with the pH maintained at 7.0 for all. Then 2 g of the pretreated biomass was added to each of the standard metal ion solution of Pb^{2+} and the flasks tightly covered with cellophane and shaken vigorously for 2 h. The same process was repeated for various standard solutions of the other

Table 1. Concentrations and % Removal of metals in waste water during pre and post application of biomass.

Application	Pb	Cr	Ni	Cu
Pre ($\times 10^{-3}$ mg/g)	25.0	20.50	8.50	16.0
Post ($\times 10^{-3}$ mg/g)	6.75	4.51	1.11	2.88
% Removal	73	78	87	82

metal ions of Cu^{2+}, Cr^{2+} and Ni^{2+}.

Determination of the effect of pH

A 50.0 ml of the standard Pb^{2+} solution was poured into each of six conical flasks and the pH condition varied from 2 to 12 as 2, 4, 6, 8, 10, and 12 respectively by the addition of either dilute HC1 or NaOH using a pH meter. Then 2 g of the pretreated biomass was added to the standard metal ion solutions of Pb^{2+} and the flasks tightly covered with cellophane and shaken vigorously for 2 h. The suspensions were filtered and analyzed. The same process was repeated for standard solutions of the other ions of Cu^{2+}, Cr^{2+} and Ni^{2+}.

Determination of the effect of contact time

A 50.0 ml solution of the standard Pb^{2+} solution was poured into each of six conical flasks and the contact time varied from 20 to 120 min as 20, 40, 60, 80, 100, and 120 min respectively and maintaining the pH condition at 7.0. The same process was repeated for standard solutions of the other ions of Cu^{2+}, Cr^{2+} and Ni^{2+}.

Analysis of metal content

The metal ions Pb^{2+}, Cu^{2+}, Cr^{2+} and Ni^{2+} content in each experiment were determined with a GBC Avanta Flame Atomic Absorption Spectrophotometer (FAAS) version 2.02. Analytical grade standards were used to calibrate the instrument, which were checked periodically throughout the analysis for instrument's response.

Application of the experiment

Waste water was allowed to flow through a 4 inch plastic pipe containing palm fruit fiber biomass wrapped with white sieve cloth. The waste water samples were collected before and after passing through the biomass and analyzed for heavy metals, Pb, Cu, Cr and Ni. The statistical tools used to analyze data from the measurements include ANOVA, correlation and regression.

Calculation of metal ions uptake by biomass

The metal uptake was calculated by simple concentration difference method, while the adsorption capacity that is; amount of metal ions (mg/g) adsorbed per g (dry mass) of biomass was calculated using the equation below:

$$Q = \frac{V(C_i - C_r)}{W \times 1000}$$

Where; Q = mass (g) of metal ion adsorbed per g of sample (Biomass); C_i = initial metal ion concentration mg/l; C_r = residual metal ion concentration mg/l at equilibrium; W = weight of biomass in reaction mixture; V = volume of metal ions solution used (ml).

Percentage removal of metal ions can also be computed using the following equation:

$$\%R = \frac{C_i - C_f}{C_i} \times 100$$

Where; % R = Percentage removal, C_i = Initial metal ion concentration; C_f = Final metal ion concentration.

Furthermore, the apparent capacity of the biomass for metal ions was determined as follows:

$$C = \frac{\%R \times C_i}{100} \times \frac{V}{W}$$

Where; C_i = Initial metal ion concentration (M) (mg/kg); V = Volume of metal ion solution used (ml); W = Weight of biomass used (g); C = Apparent capacity.

RESULTS

The results of the use of palm fruit fiber biomass as adsorbent for the removal of metal ions in solutions are presented in Table 1 and Figures 2 to 7.

Table 1 shows the concentration and percentage (%) of metal ions removed from waste water before and after application of the biomass. The concentrations of metal ions before and after treatment indicate Pb as the highest and Ni as the least. The trend of the metal ion concentrations removed is of the order Pb > Cr > Cu > Ni. That is the percentage removal of Ni as the highest and Pb as the least.

The percentage of metal ions removed from the different concentrations of solutions varied between 1.17% (Pb in 0.8 M) and 92.32% (Pb in 0.1 M) (Figure 2). The % removal generally decreased with increasing concentration of the metal ion in solution for all the metals except Pb which suddenly increased at 1 M. The trend of

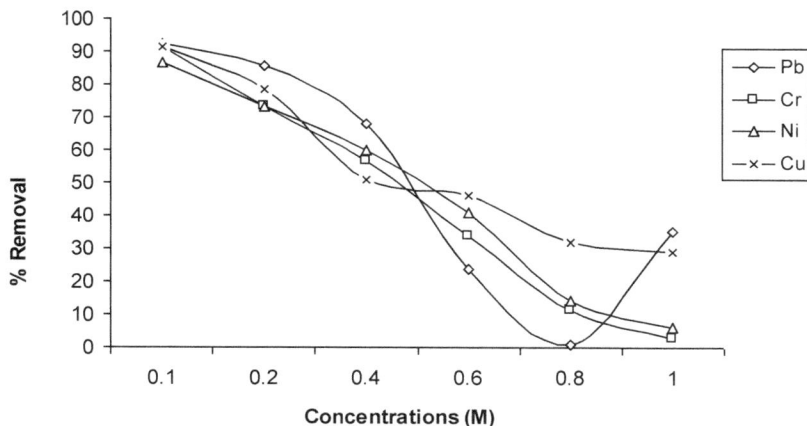

Figure 2. Variations of % removal of metal ion by palm fruit fibre biomass with concentration.

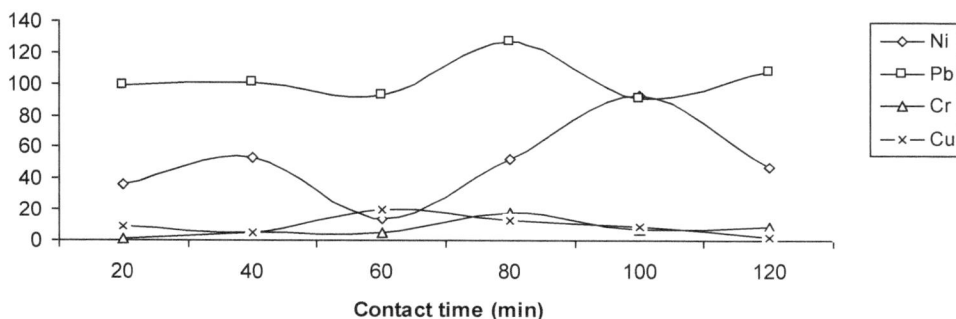

Figure 3. Variations of % removal of metal ion by palm fruit fiber biomass with contact time.

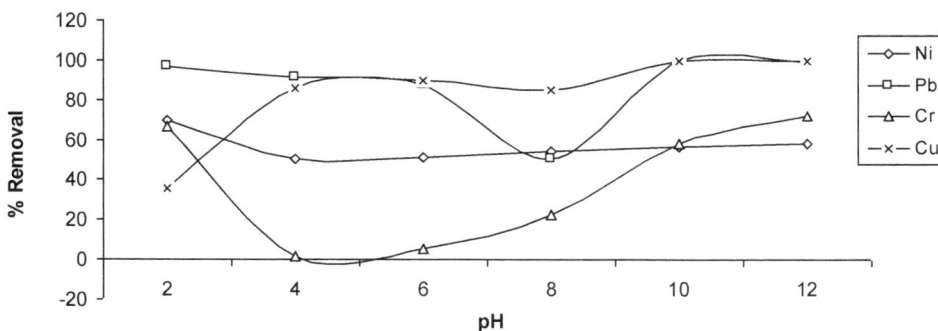

Figure 4. Variations of % removal of metal ion by palm fruit fibre biomass with pH.

percentage removal of metal ions was Pb > Cr > Cu > Ni for 0.1 M, Pb > Cu > Ni > Cr > for 0.2 M, Pb > Ni > Cr > Cu for 0.4 M, Cu > Ni > Cr > Pb for 0.6 M, Cu > Ni > Cr > Pb for 0.8 M and Pb > Cu > Ni > Cr for 1 M (Figure 2).

Figure 3 shows the percentage (%) of metal ions removed from the solutions resulting from varying the contact time of biomass from 20 to 120 min. The trend of percentage removal of metal ions was Pb > Ni > Cu > Cr for 20 min, Pb > Ni > Cr > Cu for 40 min, Pb > Cu > Ni >

Cr for 60 min, Pb > Ni > Cr > Cu for 80 min, Ni > Pb > Cu > Cr for 100 and Pb > Ni > Cr > Cu for 120 min. The results of percentage removal did not show any regular pattern with increasing contact time for Cr and Cu but increased after 60 min for Pb and Ni.

The percentage (%) removal of metal ions from the solution by the biomass due to varied pH is shown in Figure 4. The % removal ranged between 2.02% Cr at pH 4 and 100% Cu at pH 10 and 12 and 100% Pb at pH 10.

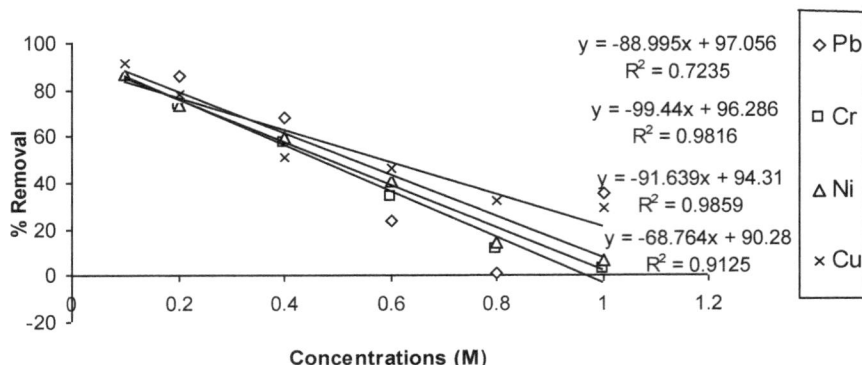

Figure 5. Variation in % removal of metal ion with concentrations of metal solutions.

The trend of percentage removal of metal ions was Pb > Ni > Cr > Cu for pH 2, Pb > Cu > Ni > Cr for pH 4, Cu > Pb > Ni > Cr for pH 6, Cu > Ni > Pb > Cr for pH 8, Pb > Cu > Cr > Ni for pH 10 and Cu > Pb > Cr > Ni for pH 12. Generally, the results showed that Cu removed increased with increasing pH up to pH 12 though with slight decrease at pH 8, Cr showed similar trend from pH 4. The percentage removal of Ni and Pb decreased with increasing pH up to pH 6 for Ni and pH 8 for Pb.

DISCUSSION

Effect of metal concentration

The effect of metal ion concentration showed that the percentage of metal ion removed by the fiber biomass decreased with increasing concentration of metal ion solution. This is because at lower concentrations, the biomass removed large amount of metal ion from the solution. However, as the concentration of metal ion increased the binding sites gradually became occupied and reduce the amount of metal ion being removed. The sorption capacity of the biomass decreased with increase in concentration. This observation agrees with the report of Ayawei et al. (2005) who studied the effect of concentration on the adsorption of metal ion by *Rhizophora mangle* waste biomass. The observations also indicate that surface saturation is dependent on the metal ion concentration and that the active sites took up the available metal more quickly at low concentrations. At higher concentrations, the surface saturation occurred on the biomass surface leading to reduction in biomass concentration as metal ion concentration increases. The differences in the metal ion removal could also be attributed to differences in the size (ionic radii) of the metals.

The relationship between sorption and concentration was fitted into linear regression equation:

$$y = mx + c$$

The slopes of the regression equations for the effect of concentrations were all negative but significant (Figure 5). The slopes of the regression equations for all the metals are negative but greater than unity. This implies higher variation of concentration than % removal of metal ions (low % removal at high concentration). The squared correlation coefficients R^2 of the linear model, (0.7235, 0.9816, 0.9859, 0.9125) for Pb, Cr, Ni, Cu respectively were significant. This implies that sorption is feasible and the metal uptake was high at low concentrations.

Analysis of variance (ANOVA) on the differences between the metal concentrations and uptake, adsorption and % removal by the biomass were not significant (P>0.05). However, the values between the metal adsorption were significant (P<0.05) indicating dependence of % removal and adsorption on concentration of metal ion.

The effect of concentration on uptake and adsorption showed high correlation between Pb and Cr (r = 0.8994), Ni (r = 0.8838), Cu (r = 0.8788); Cr and Ni (r = 0.9961), Cu (r = =0.9695); Ni and Cu (r = 0.9547). However high but negative correlations were observed between concentrations and Pb (r = -0.9552).

The application of the biomass to waste water showed high percentage removal of the metal ions viz 73% Pb, 78% Cr, 87% Ni and 82% Cu. The percentage removal indicates that Pb which was highest in concentration was the least removed and Ni which was least in concentration was the highest removed (Ayawei et al., 2005). This implies that metals with low concentration are more highly removed.

Effect of contact time

Effects of contact time (Figures 3 and 6) showed that the amount of metal ions removed by the biomass at different contact times did not show regular pattern. The amount of sorption observed increased to a certain point (Soudani et al., 2011) and decreased later with increasing contact time. This observation indicates that

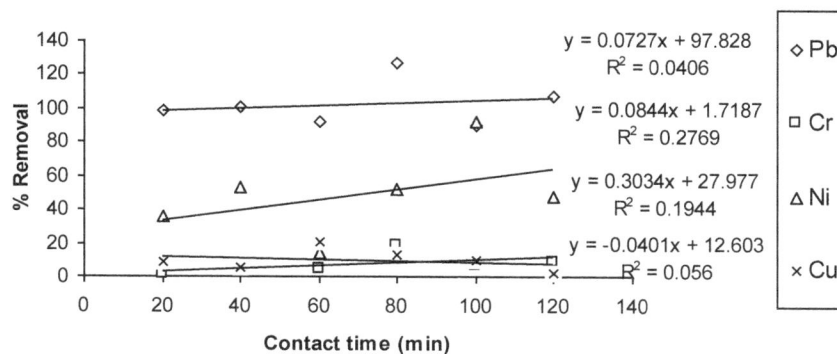

Figure 6. Variation in % removal of metal ion with biomass contact.

the removal of the metal ion by biomass does not depend on the contact time. This observation agrees with the report of Ayawei et al. (2005). The results also indicate that the metal ion removal by the biomass is relatively rapid for all metals investigated. The optimum removal of each metal occurred between 60-100 min. The non uniform pattern of sorption observed could be attributed to non uniform surface area or binding of the biomass.

The regression of metal uptake with contact time was fitted into linear regression equation:

$$y = mx + c$$

The slopes of the regression equations on variation of metal uptake with contact time of the biomass were positive and less than unity for all metals except Cu which is negative. This indicate low rate of variation of % removal as contact time increases except for Cu whose effect is different but negligible. The squared correlation coefficients, R^2 shown in variation of metal uptake with biomass contact time were not significant (P>0.05) (Figure 6). This implies that sorption is not feasible and the metal uptake does not depend on the contact time rather, on ionic radii (sizes) and concentration of the metal ion in solution.

High positive correlation exists between contact time and Ni (r = 0.6289), Cr (r = 0.5260); Pb and Cr (r = 0.7274). The trend of the mean amount of metal removed from the solution by the biomass followed the order, Pb > Ni > Cu > Cr implying that Pb was highly adsorbed and Cr was least adsorbed. This could be attributed to differences in ionic radii (sizes) and binding of biomass. This observation agrees with the order reported by Chiban et al. (2011a, b) and Soudani et al. (2011) in industrial wastewater samples.

Analysis of variance (ANOVA) showed no significant difference (P>0.05) between biomass capacity and the metal ions. However, the differences between percentage removal, amount adsorbed and uptake were significant (P<0.05). This is in contrast with the outcome of the effect

of concentration indicating that contact time does not have major influence on % removal of metal ions.

Effect of pH

The in effect of pH on the sorption of metal ion by palm fruit fiber biomass showed that the amount of sorption of metal ions initially decreased except for Cu but later increased with increasing pH. The highest sorption of all the metal ions except Ni was observed at pH 10-12. The highest value of 69.71% observed at pH 2 for Ni could be attributed to binding and irregular measurement of pH. However, the results showed that the amount of sorption fairly depend on the pH of the metal ion solution. This observation implies that high pH favours high sorption of metal ion. This observation agrees with the report of Hashem (2007) who found that in the adsorption of lead ions in aqueous solution by okra waste the pH increased with sorption capacity and lead removal was pH dependent.

High positive correlation was observed between pH and Cu (r = 0.7972) and between Ni and Cr (r = 0.7857). ANOVA showed significant difference (P<0.05) between the values of metal ions adsorbed indicating non dependence of adsorption on pH. The trend of the mean amount of metal removed from the solution by the biomass followed the order, Pb > Cu > Ni > Cr.

The graph (Figure 7) of variation of metal uptake with pH indicates positive slopes and the regression equations were positive for all metals except Ni whose % removal was high at low pH. Furthermore the slopes of Pb and Ni were less than unity indicating low rate of variation of % removal at higher pH (high % removal at low pH). The slopes of Cr and Cu were greater than unity indicating that the % removal increased with increased pH (higher variation of pH than % removal of metal ions).The squared correlation coefficient, R^2 for Cu (0.6356) was significant while the R^2 for other metals were low and not significant. This implies that the sorption of these metals do not solely depend on pH.

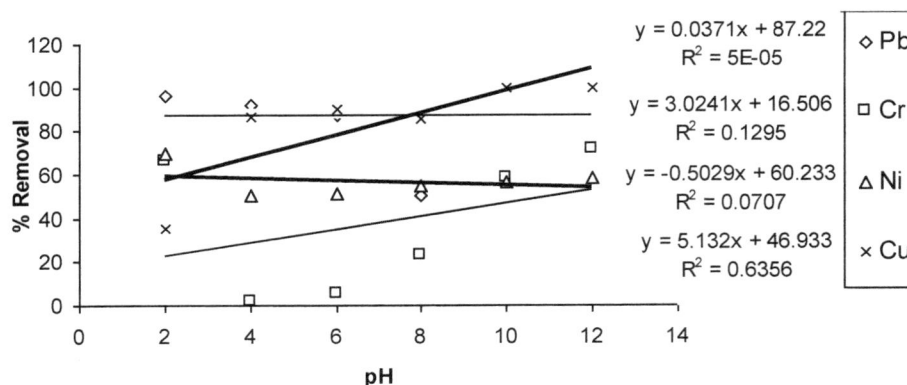

Figure 7. Variation in % removal of metal ion with pH of metal solutions.

It is observed that the % removal of the metal ions followed same order of Pb > Cu > Ni > Cr at concentrations of 0.2 and 1 M contact time of 60 min and pH 4. Also the same order of Pb > Ni > Cr > Cu at concentration of 0.4 M, contact times of 40, 80 and 120 min and pH 2.

Conclusion

The findings of this study have shown that in testing the sorption capacity of palm oil fruit fiber, some metals had 100% removal by this biomass. Therefore it can be concluded that this biomass is environmentally friendly, has great adsorption potential and can be used in the removal or recovery of heavy metal ions from aqueous solutions at low concentrations (treatment of wastewater). Metals with low concentration were more highly removed (high % removal) than metals of high concentration. The sorption capacity of this biomass decreased with increase in concentration. Although the sorption of metal ion did not show any particular trend with contact time and pH, higher sorption occurred at higher contact time and pH. Chemical modification of the adsorbent increased the percentage removal of the metals. The maximum sorption capacities depend on the type, nature and size of the metal ion. The use of palm fruit fiber for the removal of heavy metals has the advantages of availability, efficient and easy disposal. Based on the observed order of frequency of % removal of the metal ions a range of concentrations, 0.2 – 0.4 M, contact times, 60 – 80 min and pH 2 – 4 is recommended for use in the removal of heavy metals from solutions and polluted water.

REFERENCES

Ayawei N, Horsfall M, Spiff A (2005). *Rhizophora Mangle* waste as adsorbent for metal ions removal from aqueous solution. Eur. J. Sci. Res., 9(1): 6-21.

Babel S, Kurniawan TA (2003). Low-cost adsorbents for heavy metals uptake from contaminated water: A review. J. Hazard. Mater., 97: 219-243.

Benhima H, Chiban M, Sinan F, Seta P, Persin M (2008). Removal of lead and cadmium ions from aqueous sodium by adsorption onto micro-particles of dry plants. Colloids Surf.B: Biointerfaces, 61: 10-16.

Blanco AB, Sunz B, Liama MJ, Serra JL (1999). Biosorption of heavy metals to immobilized biomass. J. Biotechnol., 69, 227-240.

Chiban M, Soudani A, Sinan F, Persin M (2011a). Wastewater treatment by batch adsorption method onto micro-particles of dried *Withania frutescens* plant as a new adsorbent. J. Environ. Manage., In press, doi:10.1016/j.jenvman.,06.044.

Chiban M, Soudani A, Sinan F, Persin M (2011b). Characterization and application of dried plants to remove heavy metals, nitrate and phosphate ions from industrial wastewaters in a batch system. CLEAN - Soil Air Water, 39(4): 376-283.

Chiban M, Soudani A, Sinan F, Persin M (2011c). Single, binary and multi-component adsorption of some anions and heavy metals on environmentally friendly Carpobrotus edulis plant. Colloids Surf. B: Biointerf., 82: 267-276.

Davies OA, Allision ME, Uyi HS (2006). Bioaccumulation of heavy metals in water, sediment and Periwinkle from Elechi Creek. Niger Delta. Afr. J. Biotechnol., 5(10): 968-933.

Deans JR, Dixon BG (1992). Uptake of lead and copper (ii) ions by novel biopolymers. Water Res., 26(4): 469-472.

Fatoki OS, Lujizan O, Ogunfowokan AO (2002). Trace metal pollution in Umata River. Water Res., 28(2): 183-189.

Hashem MA (2007). Adsorption of lead ions from aqueous solution by Okra Waste. Int. J. Phys. Sci., 2(7): 178-184.

Igwe JC, Abia AA (2003). Maize cob and husk as adsorbents for the removal of heavy metals from waste water. Phys. Sci., 2: 83-92.

Juang RS, Shiau RC (2000). Metal removal from aqueous solution using chitosan enhanced membrane filtration. J. Membr. Filtration, 165(2): 159-167.

Laws EA (1993). Aquatic Pollution. 2nd edition, London: John Willey and Sons Publishers, pp. 351-415.

Mohan D, Pittman Jr CU (2007). Arsenic removal from water/wastewater using adsorbents - A critical review. J. HAZARD. Mater., 142: 1-53.

Okoye PAC, Eboatu LO, Okerulu RE, Onochie CC (2001). The use of organic wastes in decontamination of industrial effluents. Afr. J. Sci., 2: 200–228.

Osinowo FAO, Olayinka K (1998). Treatment of waste effluents on the electroplating plant using different diatomaceous materials. 21st International Annual Conference of chemical society of Nigeria held at Ibadan 23rd – 27th September 1998.

Raji C, Anirudhan TS (1997). Chromium (VI) adsorption by saw dust kinetics and equilibrium. Indian J. Chem. Technol., 4(5), 227 – 236.

Rau UN, Amit P (2002). Health hazards and Heavy metals. Int. Soc. Environ. Bot., 8(1): 1-5.

Saifuddin MN, Kumaran P (2005). Removal of heavy metal from industrial waste water using Chitosan coated oil palm shell charcoal. Environ. Biol., 8(1): 1-13.

Soudani A, Chiban M, Zerbet M, Sinan F (2011a). Use of of Mediterranean plant as potential adsorbent for municipal and industrial wastewater treatment. J. Environ. Chem. Ecotoxicol., 3(8): 199-205.

Soudani A, Chiban M, Zerbet M, Sinan F (2011b). Bensergao wastewater treatment by adsorption on dried *Launea arborescens* plant as an environmentally friendly material. Phys. Chem. News, 58. 17-19.

23

Chelate-assisted phytoextraction of metals from chromated copper arsenate (CCA) contaminated soil

Gift O. Tsetimi and Felix E. Okieimen*

Geo Environmental and Climate Change Adaptation Research Centre, University of Benin, Benin City, Edo State, Nigeria.

This study examined the effect of citric acid and ethylenediaminetetraacetic acid (EDTA) application on As, Cr and Cu phytoextraction of maize (*Zea mays* L.) plant in chromated copper arsenate (CCA) contaminated soil. The soluble and available metal pools in the contaminated soil were determined to be 6.20 ± 0.12 mg/kg and 9.80 ± 0.39 mg/kg As; 34.70 ± 1.10 mg/kg and 69.80 ± 1.44 mg/kg Cr and 18.40 ± 0.70 mg/kg and 46.30 ± 1.16 mg/kg Cu respectively, while the pseudo-total metal contents were 31.70 ± 0.29 mg/kg As, 241.40 ± 1.28 mg/kg Cr and 152.90 ± 1.82 mg/kg Cu. Maize seedlings grown on contaminated soil samples were treated with 100 ml of citric acid and EDTA solutions of various concentrations (0, 2, 4, 6 8 and 10 mM), 15 days after germination. The plants were harvested five days after amendment application and the levels of As, Cr and Cu in the roots and shoots (mg/kg dw) were determined by AAS. Post-harvest mobilization and redistribution of As, Cr and Cu in the soil samples were examined using the BCR sequential extraction method. The total levels of metal uptake (root + shoot) from the unamended soil sample were 6.8 mg/kg As, 5.7 mg/kg Cr and 38.8 mg/kg Cu representing 35, 9 and 37%, respectively, of the potentially available metals. It was found that citric acid and EDTA application markedly enhanced As, Cr and Cu extraction by maize plant, with uptake varying in the order Cu > Cr > As.

Key words: Phytoextraction, citric acid, ethylenediaminetetraacetic acid (EDTA), heavy metals, chromated copper arsenate, maize plant.

INTRODUCTION

Phytoremediation, the use of plants for the containment and/or absorption of xenobiotics from soil and water offers an economic, ecofriendly and non-invasive alternative remediation technology for heavy metal(s) contaminated soils. The success of phytoremediation technology whereby metals are effectively removed from soil is dependent on adequate plant yield and on efficient transfer of metals from plant roots to shoots (Evangelou et al., 2007). Some plants such as *Thalpsi, Urtica, Chenopodium, Polygonum, Sachalasse, Alyssum* etc, which are known metal hyperaccumulators have shown the ability to extract, accumulate and tolerate high levels of heavy metals (McGrath and Zhao, 2003). However, most of these plants are innately slow-growing and have small biomass and these tend to limit their application in remediation of heavy metals contaminated soils (Mulligan

et al., 2001). More recent research efforts in phyto-extraction have focused on fast growing crop species such as maize, tobacco, etc, which though are not metal hyperaccumulating, have high biomass yields (Robinson et al., 2000; Luo et al., 2005; Komarek et al., 2004; Meers et al., 2005; Tandy et al., 2006). For example, Komarek et al. (2004) compared phytoextraction efficiency of maize (*Zea mays*) to hybrid poplar (*Polulus nigra* × *Populus maximoviczika*) after application of EDTA to Pb contaminated soils and reported that phytoextraction efficiency was pH dependent: maize exhibited better results than poplar in more acidic (pH ≈4) soils, while poplar proved more efficient at near-neutral pH (about 6).

However, in many soils, only a fraction of heavy metals is readily available for plant uptake, thus limiting the level of metal uptake and the practical field application of phytoextraction. To increase metal availability and extend practical field application of phytoextraction in the remediation of soils contaminated with heavy metals, the use of complexing agents such as amino polycarboxylic

---
*Corresponding author. E-mail: fexokieimen@yahoo.com.

acids for example ethylenediaminetetraacetic acid (EDTA), chelating organic acids (for example citric acid) have been used to desorb metals from soil matrix into soil solution to facilitate uptake by plants (Wu et al., 1999; Blaylock and Huang, 2000; Jiang et al., 2003).

The objectives of this study were to examine the effects of citric acid and EDTA application on the phytoextraction of As, Cr and Cu by maize (*Zea mays* L.) plant in CCA contaminated soil and on the mobilization and redistribution of the metals in post-harvest maize soil fractions.

MATERIALS AND METHODS

Soil samples were collected from ten locations within the premises of an active wood treatment factory located in the south-west peri-urban area in Benin City, Nigeria, bounded by latitude 6°06' and 6°30 N and longitude 5°30' and 5°45' E in the geomorphic unit referred to as the Benin low-lands. The samples were pooled, sieved through a 2 mm screen and air-dried.

The physico-chemical properties and As, Cr and Cu levels in the soil sample are given in Table 1 (Uwumarongie and Okieimen, 2010). Analar grade chemicals viz: acetic acid, ammonium acetate, citric acid, hydroxylamine hydrochloride and EDTA were obtained from BDH Ltd. and used without further purification.

Determination of water soluble and bioavailable metal fractions in the contaminated soil

The water soluble fractions of As, Cr and Cu in the soil samples were determined by agitating 5 g-portion of the soil in 25 mL of distilled water for 6 h followed by centrifugation at 1500 rpm. The amount of the metals in the supernatant determined by AAS (Buck Scientific VGF model 210A) is reported as the water soluble fractions. The bioavailable fractions of As, Cr and Cu in the soil was determined by extraction of an aliquot of the soil sample with 0.01 M $CaCl_2$ solution following the method described by Oliver et al. (1999).

Pot experiments

Air-dried soil samples (1 kg) were placed in plastic pots and maintained at 60% field water capacity by adding deionised water. Four grains of viable maize were sown in each pot. Fifteen days after germination, sub-samples of the pots were treated with 100 mL of 0, 20, 40, 80 and 100 mM solutions of citric acid and EDTA. Amendments application was performed by applying the solution to the top of the pots. Post-germination treatment as opposed to pre-sow treatment was adopted to preclude phytotoxic growth depressions (Meers et al., 2004).

The maize plants were harvested five days after application of the amendments by cutting the shoots 0.5 cm above the surface of the soil, and the roots were carefully removed. The roots were steeped in 0.01 M $CaCl_2$ for 30 min to remove any exogenous metals and were thereafter washed free of salt solution. The roots and shoots were washed and rinsed thoroughly with deionised water and were thereafter dried at 70°C until constant weight. The dried plant materials were ground using agate mill.

Subsamples of the ground shoots (200 mg) and roots (100 mg) were digested in a mixture of concentrated HNO_3 and $HClO_4$ (4:1 by volume) and the As, Cr and Cu in the digestate solutions were determined by AAS. Reagent blank and analytical duplicates were used to ensure accuracy and precision of analyses. The data

reported in this paper are the mean values of triplicate determinations.

Heavy metal fractionation in the contaminated soil

The heavy metal fractionation in the contaminated soil sample was determined by the BCR (European Communities Bureau of Reference) sequential procedure (Golia et al., 2007; Tokalioglu et al., 2006). The protocol operationally defined the metal distribution into the following pools: B_1, extractable; B_2, reducible; B_3, organic bound and; R, residual fractions. The results given in Table 2 indicate that As is fairly evenly distributed among the operationally defined pools, while relatively larger proportions of Cr and Cu are associated with the intransigent soil phases

RESULTS AND DISCUSSION

Physico-chemical properties of contaminated soil

Textural analysis showed a preponderance of sand fraction (73.10%) followed by clay (24.80%) and silt (2.10%) thus classifying the soil as sandy loam. Although sandy soils are known to have poor retention capacity of water and metals, the relatively large proportion of clay (24.80%) in the contaminated soil sample suggests that the soil will drain poorly with implications for potential deleterious impact of retained pollutants on environmental receptors.

The acidic pH 5.92 of the soil is generally within the range for soil in the region. Soil pH plays a major function in the sorption of heavy metals as it controls the solubility and hydrolysis of metal hydroxides, carbonates and phosphates. It also influences ion-pair formation, solubility of organic matter, as well as surface charge of Fe, Mn and Al-oxides, organic matter and clay edges (Tokalioglu et al., 2006). The soil had moderate organic matter content (2.15%) and relatively high cation exchange capacity (CEC) (47.84 meq/100 g). CEC measures the ability of soils to allow for easy exchange of cations between its surface and solutions. The relatively high level of clay and CEC indicate low permeability and leachability of metals in the soil.

Contamination status of the soil

The pseudototal levels of As, Cr and Cu given in Table 1 were used to estimate their intervention levels in the contaminated soil via the Department of Petroleum Resources (DPR, 2002) method which considers the organic matter and clay contents of the soil. Intervention levels of metal contaminants in soil give indications of quality for which the functionality of the soil for human, animal and plant lives is considered to be impaired. The intervention levels of As, Cr and Cu are given in Table 3. The results show that the pseudo-total levels of As, Cr and Cu are markedly higher than the intervention values

Table 1. Physico-chemical properties of CCA contaminated soil.

Physico-chemical property	Value
pH	5.92 ± 0.10
Clay (%)	24. 80 ± 0.00
Silt (%)	2.10 ± 0.00
Sand (%)	73.10 ± 0.00
Nitrogen (%)	0.34 ± 0.08
Carbon (%)	1.22 ± 0.30
Organic matter (%)	2.15 ± 0. 40
Phosphorus (mg/kg)	44.74 ± 3.73
Calcium (meq/100 g)	5.68 ± 0.40
Magnesium (meq/100 g)	1.96 ± 0.30
Sodium (meq/100 g)	0.19 ± 0.10
Potassium (meq/100 g)	0.57 ± 0. 10
CEC (meq/100 g)	48.74 ± 0.10
As (mg/kg)	31.70 ± 2.90
Cr (mg/kg)	241.40 ± 12.80
Cu (mg/kg)	152.90 ± 18.20

Table 2. BCR sequential fractionation of As, Cr and Cu in CCA contaminated soil.

Step	Extractant	As (%)	Cr %	Cu %
B_1 - Extractable	40 mL of 0.1 M CH_3COOH, 16 h at room temperature	23.13	18.98	20.36
B_2 - Reducible	40 mL of 0.5 M $NH_2OH.HCl$ (pH 2) 16 h at room temperature	24.15	21.84	23.35
B_3 - Organic-bound	10 ml of 8.8 M H_2O_2, 1 h at room temperature; then 1 h at 85°C; cool, add 50 mL of 1 M CH_3COONH_4 (pH 2) 16 h at room temperature	25.17	25.44	25.78
R - Residual	Aqua regia digestion (21 mL conc. HCl + 7 mL conc. HNO_3) 16 h at 180°C	27.55	33.74	30.51

Table 3. Contamination status of soil sample from active wood treatment site.

Metal	Intervention value (mg.kg^{-1})	Contamination factor (M_{contam}/M_{ref})	Contamination/Pollution index (M_{Target})
As	8.81 ± 1.26	90.57[a]; 21.13[b]	1.09
Cr	90.52 ± 9.28	689.71[a]; 2.41[b]	2.41
Cu	76.51 ± 6.08	16.09[a]; 3.06[b]	4.25
C_D = degree of contamination = Σc_f		796.39[a]; 26.60[b]	

$C_f = M_{contam}/M_{ref}$; a is with reference to the control soil sample (0.35 mg.kg^{-1} As, 0.35 mg.kg^{-1} Cr and 9.50 mg.kg^{-1} Cu) and b is with reference to uncontaminated soil (1.50 mg.kg^{-1} As, 100.0 mg.kg^{-1} Cr and 50.0 mg.kg^{-1} Cu (Sparks, 2000).

and these values correspond to moderate (with respect to As) and high (with respect to Cr and Cu) levels of contamination of the soil (Uwumarongie and Okieimen, 2010).

The contamination status of the soil measured in terms of contamination factor, C_f, a ratio of the metal concentration in the contaminated soil to that in uncontaminated soil, and degree of contamination, C_D, is given in Table 3. The values of these contamination indices classify the soil as moderate – to – highly contaminated (Hakanson, 1980). On the basis of the C/P index and the

total levels of As, Cr and Cu in the contaminated soil sample, the soil in the wood preservation site may be classified as slightly contaminated with respect to As, moderately polluted with respect to Cr and severely polluted with respected to Cu.

Contaminant solubility in soil solution and mobility are increasingly being used as key indicators of potential risk to environmental receptors. Table 4 gives the water-soluble, bioavailable and mobile pools of As, Cr and Cu in the contaminated soil. The mobile pool of the metals is

Table 4. Soluble, available and mobile pools of As, Cr and Cu in CCA contaminated soil.

Metal	Soluble fraction (mg.kg^{-1})	Available fraction (mg.kg^{-1})	Mobile fraction (mg.kg^{-1})
As	6.20 ± 0.12	9.80 ± 0.39	6.80 ± 0.27
Cr	34.70 ± 1.10	69.80 ± 1.44	43.20 ± 1.63
Cu	18.40 ± 0.70	46.30 ± 1.15	29.30 ± 0.80

given as the B_1 fraction from the BCR sequential extraction scheme.

These results when compared with the data in Table 3 suggest that with the exception of Cr for which the available fraction exceeds the regulated target value, the contaminated soil in the wood preservation site may not present imminent risk to environmental receptors. However, the relatively low levels of soluble and available forms of the metals may be bioaccumulated and become associated with long-term deleterious effects on human and environmental receptors.

Metal accumulation by maize plant

Figure 1 shows the amount of As, Cr and Cu in the roots of maize plant. It can be seen that: (i) citric acid and EDTA application markedly enhanced the concentrations of the metals in the roots; (ii) that the enhancements in metal uptake in EDTA amended soils were more marked than for citric acid amended soils and increased with increased level of amendment application and; (iii) that the amounts of the metals taken up by the plant varied in the order Cu > Cr > As. The results indicate that at relatively high levels of amendment application, more than the bioavailable pool of As and Cu were taken up in the roots of maize with the apparent order of the levels of metal uptake being Cu > Cr > As and did not correlate with the bioavailable (0.01 M CaCl$_2$ extractable) pool of the metals in the contaminated soil i.e. Cr > Cu > As. Copper being an essential element to plants may have been taken up by the plant actively, while As and Cr which would exhibit phytotoxic effects at relatively lower concentrations than Cu, may have being taken up by the plant *via* a passive mechanism.

The processes involved in chelate-assisted phytoextraction include: (a) desorption/dissolution of metal from the soil matrix; (b) transport to the roots by diffusion and mass flux; (c) adsorption and uptake by roots and; (d) transfer to xylem and translocation to shoots. The predominant theory for metal-chelate uptake is the split-uptake (free metal ion) mechanism, by which only free metals are absorbed by plant roots (Marschner et al., 1986; Samet et al., 2001).

According to Schowanek et al. (1997), complexing agents may be divided into three categories depending on their metal complex formation constant: weak (for example, zeolites, polycarboxylate and citrate); moderate

and; high (for example EDTA). It would therefore be expected that metal complexes formed with EDTA will less readily yield the free metal ion (the form in which it was thought that plant uptake of metals mainly occurred) than metal-citrate complexes; and should result to lower enhancement in metal uptake. The results of more recent studies suggest that metal-chelate uptake occurs simultaneously with free metal ion by plant roots in chelate-assisted phytoextraction (Wenger et al., 2008).

Figure 2 shows the amounts of As, Cr and Cu in the shoot of maize plant. The results show that as with the roots, the amount of As, Cr and Cu accumulated in the maize shoots were similarly enhanced by citric acid and EDTA application. In comparison with metal accumulation in maize shoots in the unamended soil sample, the increase in the amounts of the metals accumulated in maize shoots in the amended soils were generally about the same order of magnitude (about 5-fold for As, 11-fold and 13-fold for Cr and 7-fold and 8-fold for Cu in citric acid and EDTA amended soils, respectively). The relatively small increase in metal accumulation in the plant shoots may suggest a low phytotoxic threshold of the plant for containment and that severally cropping cycles may be required to effectively phytoremediate the contaminated soil.

Metal transfer coefficient

Soil-to-plant transfer ratio is an important component of phytoextraction. For an amendment to be considered effective in enhancing phytoextraction of metals, it must not only enhance mobilization of metals from soil matrix into soil solution, it should in addition facilitate metal uptake by plant roots and their translocation into plant shoots. Metal transfer coefficients, T_c, given as the ratio of the metal concentration in plant shoots to the pseudototal concentration in soil obtained for As, Cr and Cu are shown in Figures 3 and 4 given in for citric acid and EDTA amended soils, respectively. It can be seen from the results that: (a) amendment application markedly enhanced the values of T_c and; (b) the values of T_c are generally about the same order of magnitude in citric acid and EDTA amended soils. It has been suggested that values of T_c based on the potentially available fraction of metals rather than on the pseudototal amounts, may provide a more reliable assessment of the effectiveness of amendment application in phytoextraction (Okieimen et

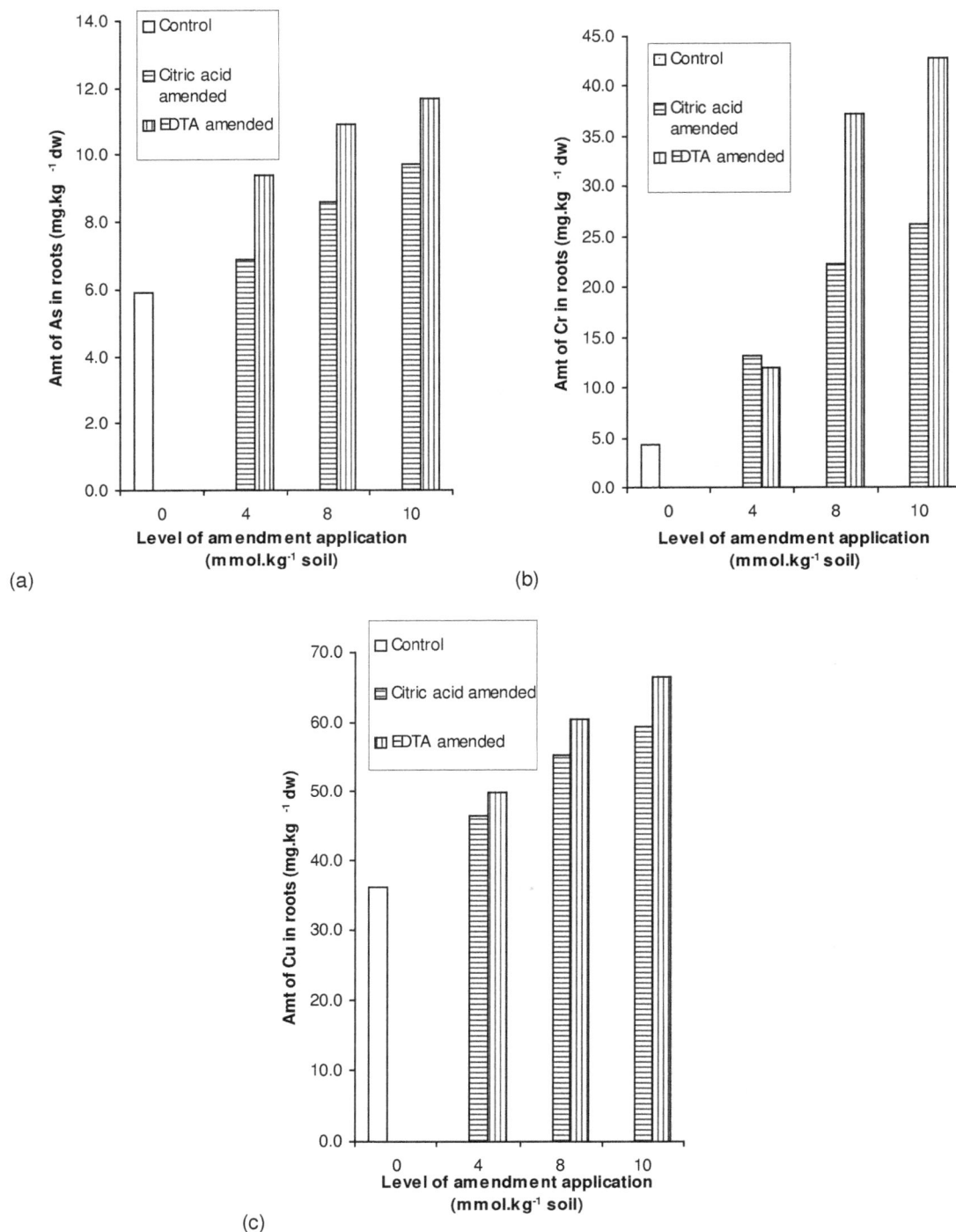

(a)

(b)

(c)

Figure 1. Amount of metals: (a) As, (b) Cr and, (c) Cu in the roots of maize (*Zea mays* L.) grown on CCA contaminated soil amended with citric acid and EDTA.

al., 2010).

Metal translocation factor

Translocation factor, (TF) defined as the ratio of a metal

(loid) concentration in plant shoots to that in the roots, may be used to evaluate the effectiveness of a chelating agent in enhancing the capacity of plant to transfer metals from roots to shoots. The effect of citric acid and EDTA application on the values of TF of As, Cr and Cu in maize plant is shown in Figures 5 and 6. The results

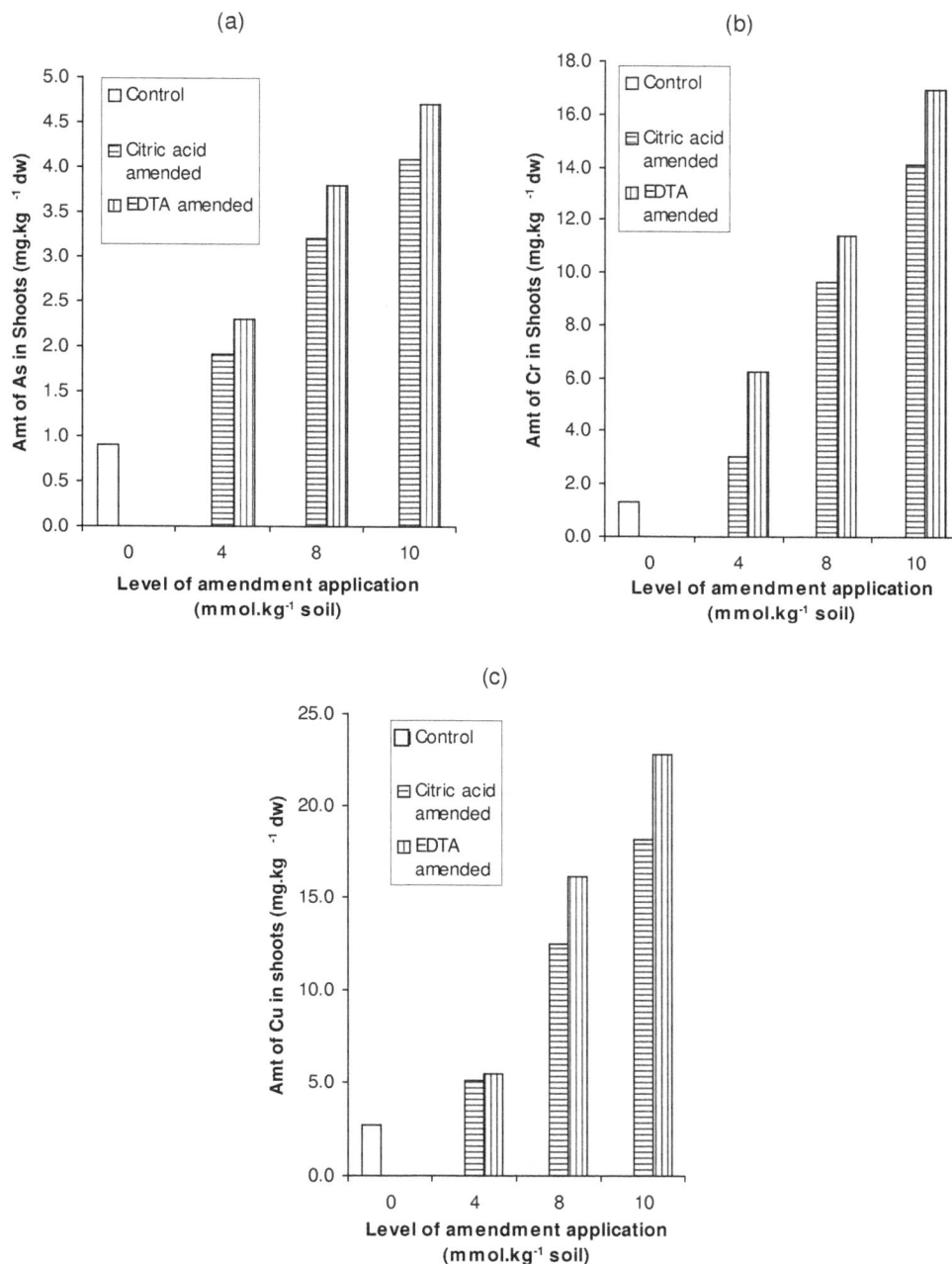

Figure 2. Amount of metals: (a) As, (b) Cr and, (c) Cu in the shoots of maize (*Zea mays* L) grown on CCA contaminated soil amended with citric acid and EDTA.

show marked improvements in the translocation of As, Cr and Cu from the roots of maize plants in the CCA contaminated soil by the application of citric acid and EDTA, with increases in the values of TF generally 100% higher than the corresponding values for contaminated soil. The substantial increase in the values of TF is consistent with the report of Luo et al. (2005) for EDTA and EDSS enhanced metal phytoextraction of metals from contaminated soil. As with transfer coefficient, the levels of improvement in metal translocation in citric acid

amended soils are about the same order of magnitude but are somewhat lower than in EDTA amended soil.

The results in Figures 5 and 6 indicate that less than 40% of the metals absorbed in the roots of the maize plant in the unamended soil was translocated to the shoot. Adsorption by carboxylic groups and other cationic exchanging moieties within the apoplasmic cell wall reportedly contributed to net accumulation in plant roots (Huang and van Steveninck, 1989) and may explain the disproportionate amounts of the metals in the shoots

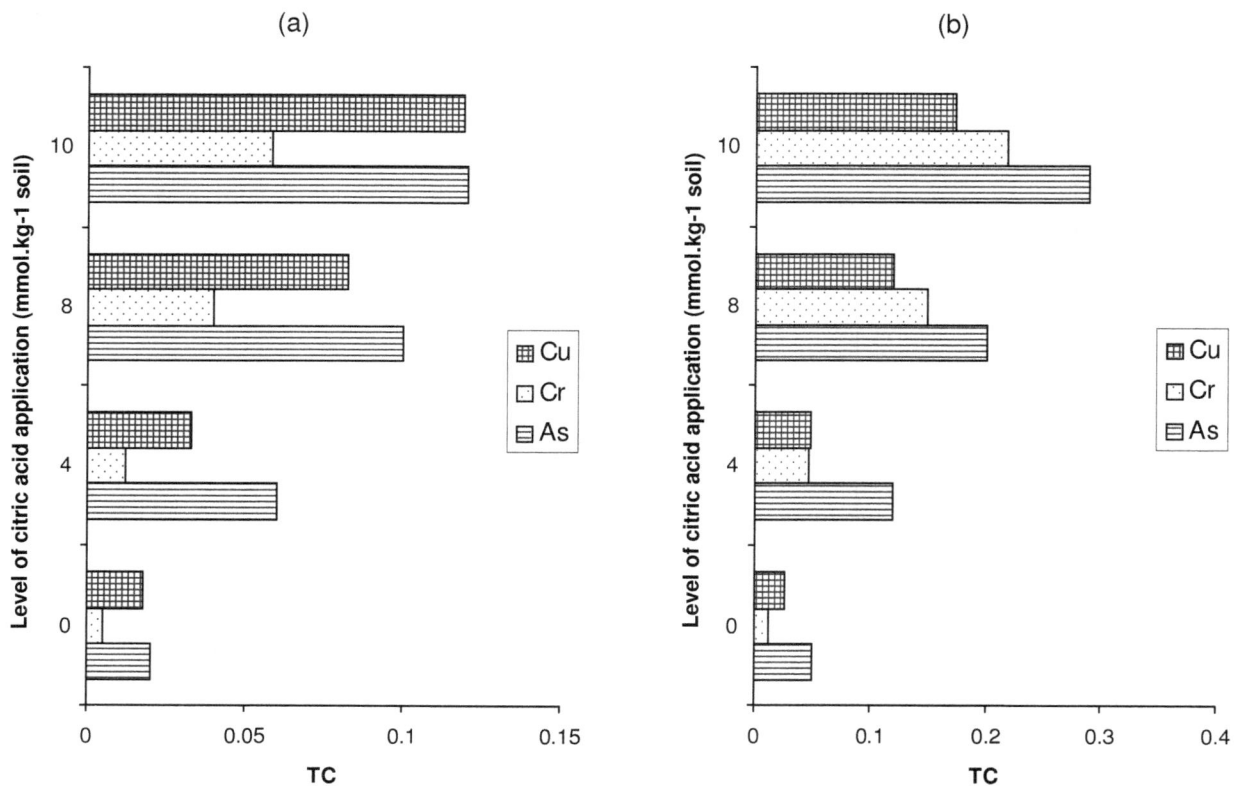

Figure 3. Effect of citric acid application on transfer coefficients of As, Cr and Cu with reference to: (a) pseudototal metal levels and; (b) plant available metal pools to maize in CCA contaminated soil.

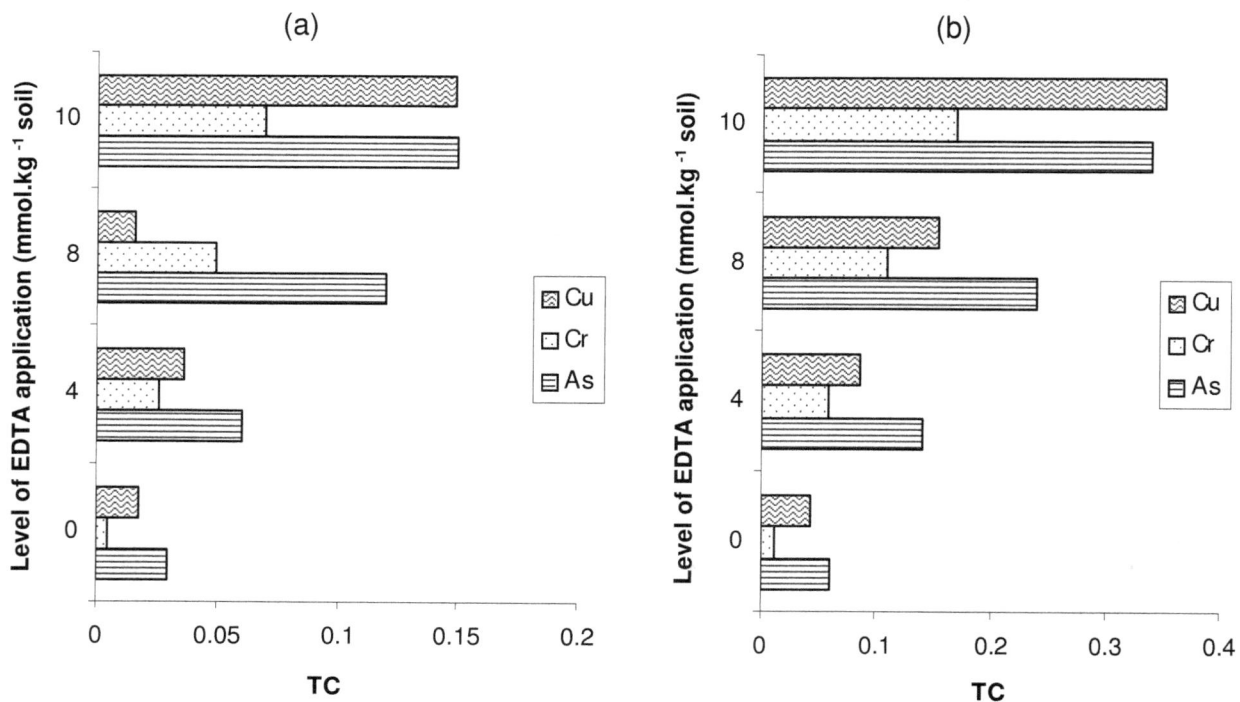

Figure 4. Effect of EDTA application on transfer coefficients of As, Cr and Cu with reference to: (a) pseudototal metal levels and; (b) plant available metal pools to maize in CCA contaminated soil.

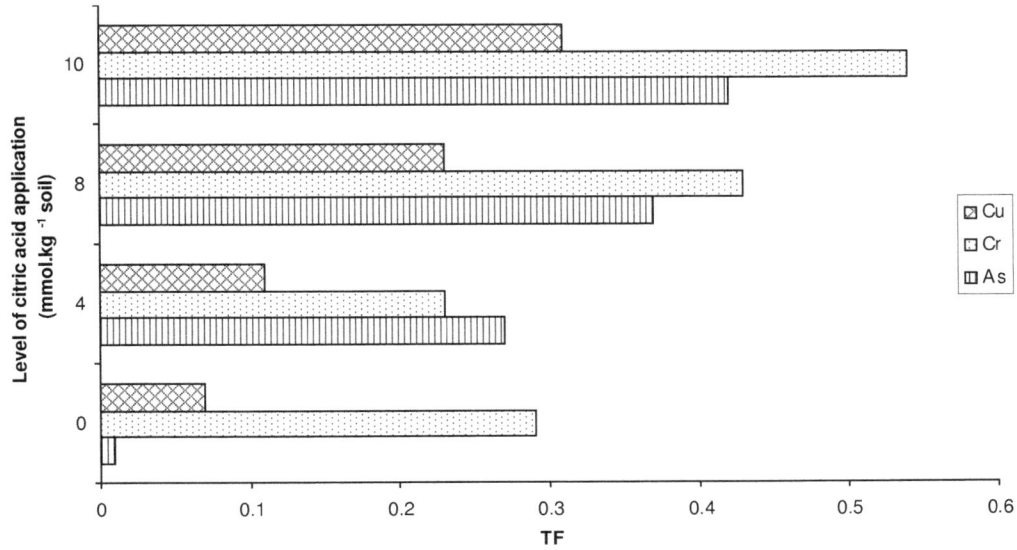

Figure 5. Effect of citric acid application on translocation factor of As, Cr and Cu from root to shoot of maize in CCA contaminated soil.

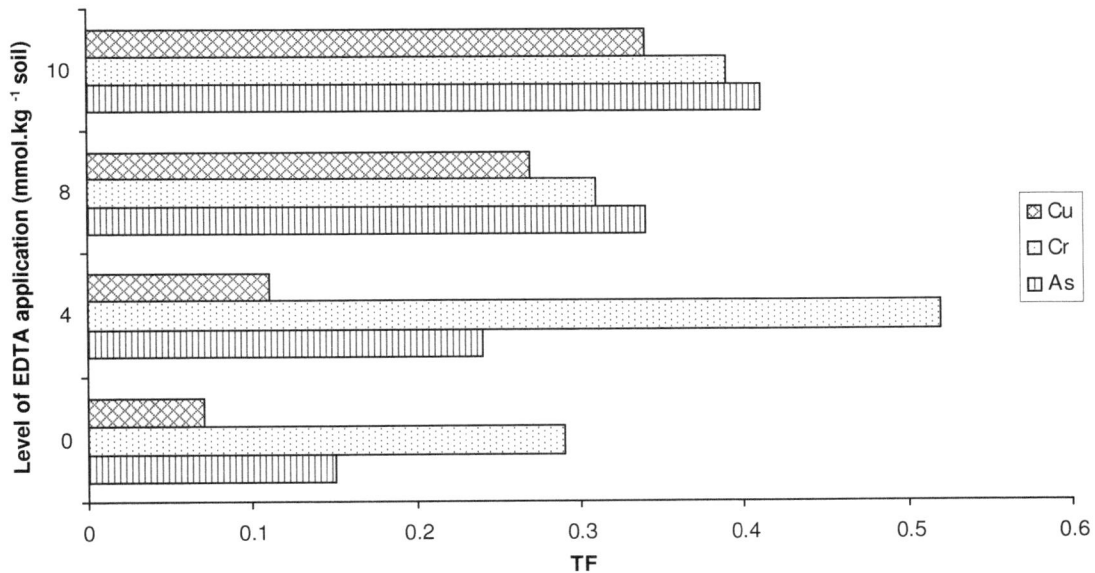

Figure 6. Effect of EDTA application on translocation factor of As, Cr and Cu from roots to shoots of maize in CCA contaminated soil.

relative to corresponding amounts in the roots of the maize plant grown on the CCA contaminated soil (Figures 5 and 6).

Post-harvest distribution of As, Cr and Cu in the contaminated soil

A major environmental concern in chelate-assisted phytoextraction of metals from contaminated soil is the residual pools of mobilized metals which may not be assimilated by the plant. Prolonged mobilization of heavy metals after harvest is undesirable as the absence of an actively transpiring plant may result in percolation of mobilized metals with implications for ground water impactation. To examine these effects, BCR sequential extraction procedures were applied to the soil samples 3 days after harvesting the maize plants. The distribution patterns of As, Cr and Cu in the post harvest soil samples are shown in Figures 7 and 8 for citric acid and EDTA

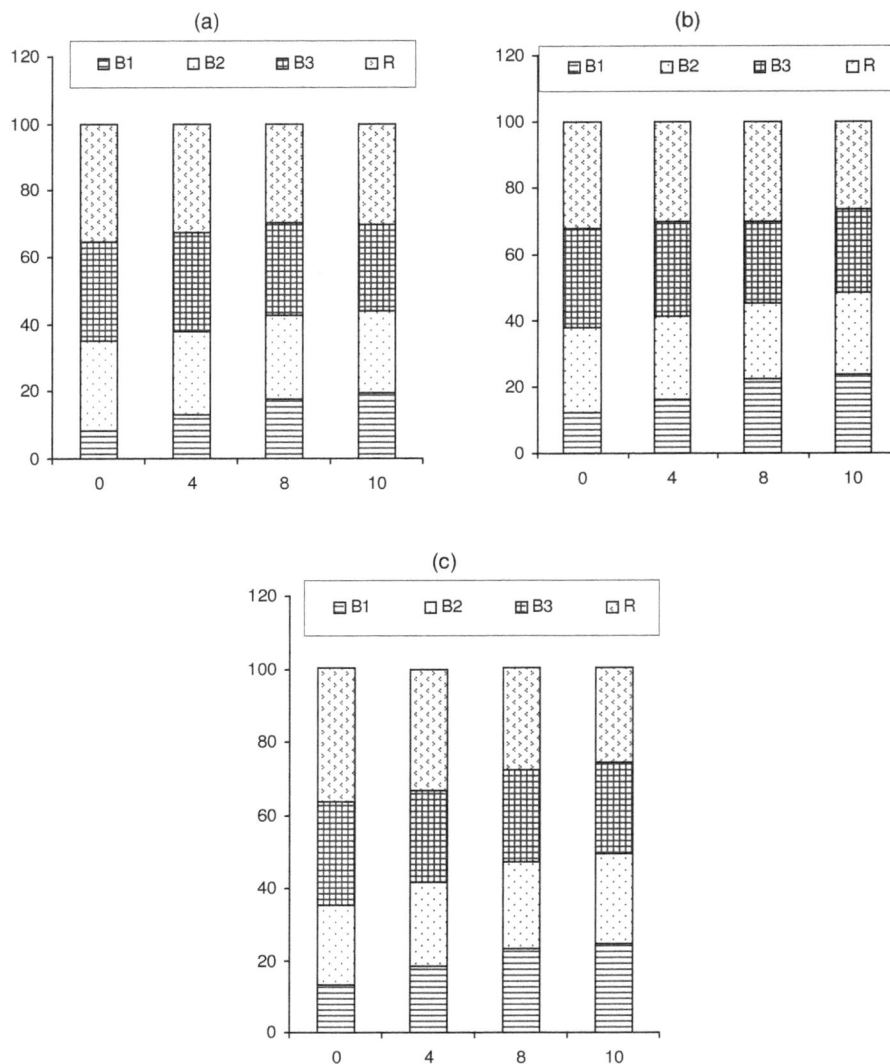

Figure 7. Metal mobilization and redistribution in post-harvest soil samples amended with citric acid (a) As, (b) Cr and (c) Cu.

amended soils, respectively. The result show that amendment application increased post harvest mobile pools of the metals in the soil; As from 1.90 ± 0.88 mg.kg^{-1} in the unamended soil to 3.10 ± 0.11 mg.kg^{-1} and 3.70 ± 0.17 mg.kg^{-1} in EDTA – and citric acid amended soils, respectively; Cr from 26.30 ± 1.11 mg.kg^{-1} in the unamended soil to 44.60 ± 2.10 mg.kg^{-1} and 45.20 ± 3.80 mg.kg^{-1} in EDTA- and citric acid amended soils, respectively, and Cu from 16.90 ± 1.00 mg.kg^{-1} in the unamended soil to 17.60 ± 3.50 mg.kg^{-1} and 21.40 ± 4.6 mg.kg^{-1} in EDTA- and citric acid amended soils, respectively. Residual leachable pools of metal in assisted post-harvest, chelate-assisted phytoextraction soils are not uncommon (Schmidt, 2003; Lai and Chen, 2005) but the relatively high levels of leachable pools of As, Cr and Cu obtained in this study may be connected with the rather short duration of the study.

Conclusion

This study demonstrated that citric acid could be regarded as a good candidate chelate for environmentally safe phytoextraction of heavy metals in contaminated soils. Its enhancement of As, Cr and Cu uptake by maize and of the translocation of the metals from the roots to the shoots of the plant is comparable with that of EDTA. In the short duration of this study, increases in transfer coefficient and translocation factor following citric acid applications provide an impetus for further detailed studies on the capacity of maize to accumulate heavy metals in the presence of citric acid. Being biodegradable, increase in leachable metal pools in post-harvest, citric acid amended soil would present less environmental concern than the post-harvest leachable metal pools in EDTA-amended soil.

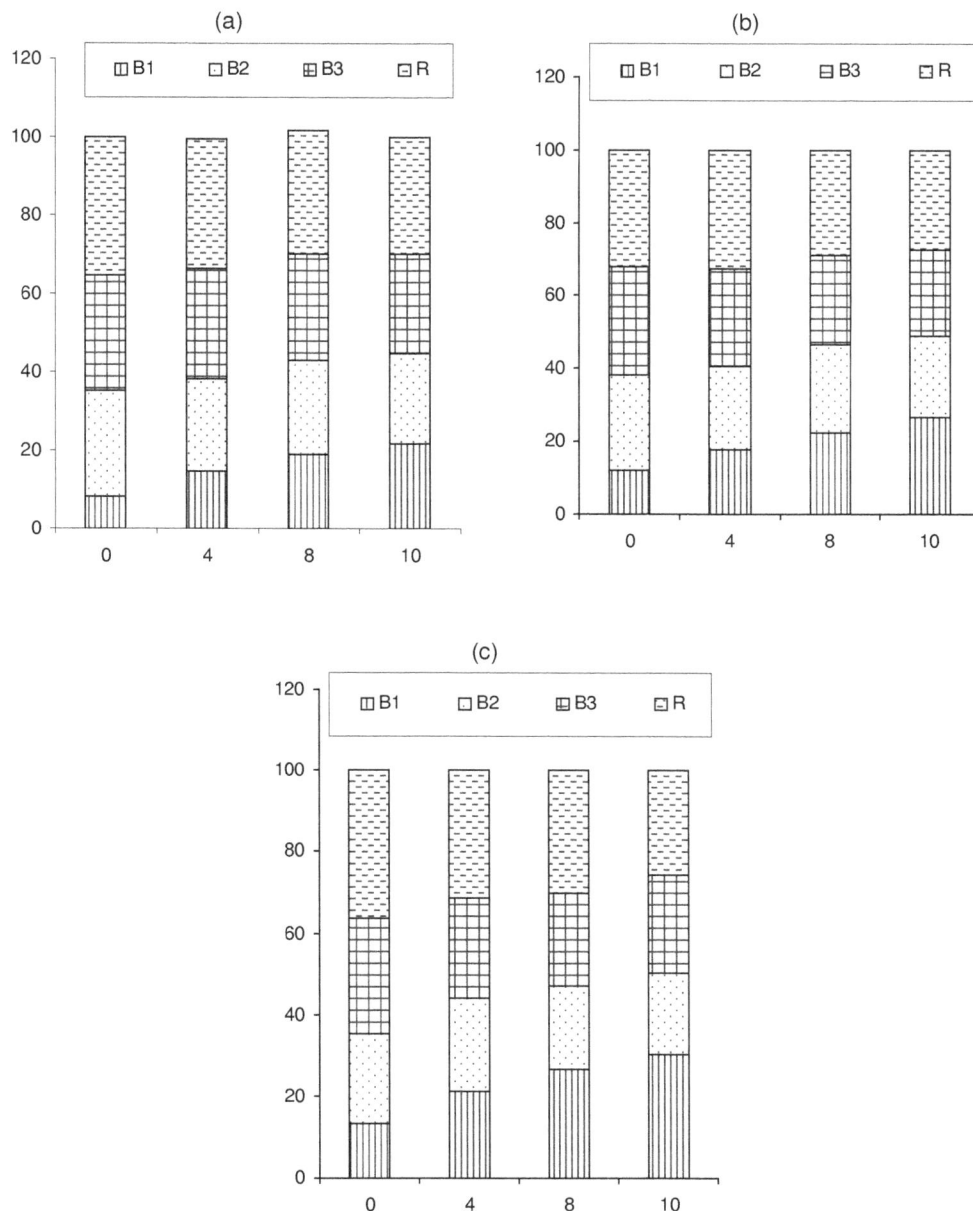

Figure 8. Metal mobilization and redistribution in post-harvest soil samples amended with EDTA (a) As, (b) Cr and (c) Cu.

REFERENCES

Blaylock MJ, Huang JW (2000). Phytoremediation of heavy metals. In: Using Plants to Clean Up the Environment. (Raskin I and Ensley, B.D. Eds). John Wiley & Sons, New York, pp. 53-69.

DPR (2002). Environmental Guidelines and Standards for the Petroleum Industry, (EGASPIN), Department of Petroleum Resources, Lagos, pp. 278-281.

Evangelou MHW, Nauer U, Ebel M, Schaeffer A (2007). The influence of EDDS and EDTA on the uptake of heavy metals of Cd and Cu from soil with tobacco Nicotianna tabacum. Chemosphere, 68: 345-353.

Golia EE, Tsiropolous NG, Dimikrou A, Mitsios A (2007). Distribution of heavy metals in agricultural soils in Greece using the modified BCR sequential extraction method. Int. J. Environ. Anal. Chem.,

87(13/14): 1055-1063.

Hakanson I (1980). Ecological risk index for aquatic pollution control. A sedimentologial approach. Water Res., 14(5): 975-1001.

Huang CX, van Steveninck RFM (1989). The role particular pericyle cells in the apoplastic transport in root meristems of barley. J. Plant Physiol., 133: 554-558.

Jiang XJ, Luo YM, Zhao QG, Baker AJM, Christie P, Wong MH (2003). Soil Cd availability to Indian mustard and environment risk following EDTA addition to CD contaminated soil. Chemosphere, 50: 813-818

Komarek M, Klustos P, Szakova J, Chrasiny V, Ettler V (2004). The use of maize and poplar in chelant-enhanced phytoextraction of lead from contaminated agricultural soils. Chemosphere, 57: 640-651.

Lai HY, Chen ZS (2005). The EDTA effect on phytoextraction of single and combined metals contaminated soil using rainbow pink (Dianthus chinensis). Chemosphere, 60: 1062-1071.

Luo C, Shen Z, Li X (2005). Enhanced phytoextraction of Cu, Pb, Zn and Cu with EDTA and EDSS. Chemosphere, 59: 1-11

Marschner H, Romheld V, Kiessel M (1986). Different strategies in higher plants in mobilization and uptake of iron. J. Plant Nutr., 6: 429-438.

McGrath SP, Zhao FJ (2003). Phytoextraction of metals and metalloids from contaminated soil. Curr. Opin. Biotech., 14: 277-282.

Meers E, Hopgood M, Lesage E, Vervaeke P, Tack FGM, Verloo M (2004). Enhanced phytoremediation: in search of EDTA alternatives. Int. J. Phytorem., 6: 95-100.

Meers E, Ruttens A, Hopgood M, Lesage E, Vervaeke P, Tack FMG (2005). Comparison of EDTA and EDSS as potential soil amendments for enhanced phytoextraction of heavy metals. Chemosphere, 58: 1011-1023.

Mulligan CN, Young RN, Gibbs BF (2001). Remediation technologies for metal-contaminated soil and groundwater. An evaluation. Eng. Geol., 60: 193-207

Oliver DP, Tiller KG, Alston AM, Naidu R, Cozens GD (1999). A comparison of three soil tests for assessing Cd accumulation in wheat grain. Aust. Soil Res., 37: 1123-1138.

Okieimen FE, Tsetimi GO, Egbuchunam TO (2010). The effect of citric application on phytoextraction of As, Cr and Cu by maize from CCA contaminated soil. Int. J. Appl. Environ. Sci., 5(6): 831-844.

Robinson BH, Millis TM, Petit D, Fung LE, Green SR, Clothier BE (2000). Natural and induced cadmium accumulation in poplar and willow: implication for phytoremediation. Plant Soil, 227: 301-306.

Samet G, Vangronsveld J, Manceau A, Musso A, Ditean J, Menthonnex JJ, Hazemann JI (2001). Accumulation forms of Zn and Pb in *Phaseolus vulgaris* in the presence and absence of EDTA. Environ. Sci. Technol., 35: 2854-2859.

Schmidt U (2003). Enhancing phytoextraction: the effect of chemical soil manipulation on mobility, plant accumulation and leaching of heavy metals. J. Environ. Qual., 23: 1939-1954

Schowanek D, Feijtel TCJ, Perkins CM, Hartman FA, Federle TW, Larson RJ (1997). Biodegradation of [S, S], [R, R] and mixed stereoisomers of ethylene diaminedisussinic acid (EDSS), a transition metal chelator. Chemosphere, 34: 2375-2391.

Tokalioglu S, Kavtal S, Gultekin A (2006). Investigation of heavy metal uptake by vegetables growing in contaminated soil using the modified BCR sequential extraction method. Int. J. Environ. Anal. Chem., 88(6): 417-430.

Tandy S, Schulin R, Nowack B (2006). Uptake of metals during chelant-assisted phytoextraction with EDDS related to solubilised metal concentration. Environ. Sci. Technol., 40: 2753-2758.

Uwumarongie EG, Okieimen FE (2010). Extractive decontamination of CCA contaminated soil with organic acids. Afr. J. Environ. Sci. Technol., 4(9): 567-576.

Wenger SK, Gupta SK, Schulin R (2008). The value of nitrilotriacetate in chelate-assisted phytoremediation. Dev. Soil Sci., 32: 679-695

Wu J, Hsu FC, Cunninghan SD (1999). Chelate-assisted Pb phytoextraction: Pb availability, uptake and translocation constraints. Environ. Sci. Technol., 33: 1898-1904.

24

Chemical fractionation of heavy metals in soils around the vicinity of automobile mechanic workshops in Kaduna Metropolis, Nigeria

M. M. Achi[1]*, A. Uzairu[1], C. E. Gimba[1] and O. J. Okunola[2]

[1]Department of Chemistry, Ahmadu Bello University, Zaria, Kaduna State, Nigeria.
[2]National Research Institute for Chemical Technology, P. M. B. 1052, Basawa, Zaria, Kaduna State, Nigeria.

The study of heavy metals in environmental niches is essential, especially with their potential toxicity to human life. Metal toxicity depends on chemical associations in soils. For this reason, determining the chemical form of a metal in soils is important to evaluate its mobility and bioavailability. In this manner sequential extraction was used to fractionate seven heavy metals (Cd, Cu, Ni, Zn, Cr, Fe and Pb) from ten contaminated soils into six operationally defined groups: water soluble, exchangeable, carbonate, Fe -Mn oxide, organic, and residual. The residual fraction was the most abundant pool for all the metals examined. A significant amount (2.00 to 73.47, 1.77 to 17.78, 9.76 to 58.54 and 1.71 to 54.11% respectively) of Cu, Cd, Cr and Pb was present in the potentially available fraction: non residual fraction. Contamination of Zn, Ni and Fe in these soils was not as severe as Cu, Cd, Cr and Pb. Overall, the order of contamination was Cu > Cr > Pb > Cd > Ni > Zn > Fe. The study indicated that the possible pollution of soils of this environment was as a result of activities carried out within these areas.

Key words: Fractionation, heavy metals, mechanic, workshops, Kaduna.

INTRODUCTION

The problem of soil and vegetation pollution due to toxic metals in spent oil is beginning to cause concern now in most major metropolitan cities (Vwioko et al., 2006). These toxic heavy metals entering the ecosystem may lead to geoaccumulaton, bioaccumulation and biomagnification (Wong et al., 2002; Lokeshwari and Chandrappa, 2006). Spent oil soil pollution leads to the build-up of essential and non-essential elements in the soil and eventual translocation in plant tissues (Vwioko et al., 2006). Soil pollution by spent lubricating oil has been reported to cause growth retardation/reduction in plants and this has been attributed to the presence of heavy metals at toxic concentrations in the soil (Anoliefo and Vwioko, 1995).

Due to the ever-increasing population and Industrialization, most environments are to some extent unnaturally polluted. Nriagu (1978) reported that we may be experiencing a silent epidemic of environmental metal poisoning from the increasing amounts of metals released into the biosphere. Waste engine oil pollution is responsible for several environmental problems, including disruption of plant water relations, direct toxicity and indirect effects on plant metabolism (Racine, 1994). Edebiri and Nwanokwale (1981) found that metals present in spent lubricating oil are not necessarily the same as those present in unused lubricant. Heavy metals such as vanadium, lead, aluminum, nickel and iron which are low in unused engine oil gave higher concentrations in spent engine oil. Hall (2002) observed that heavy metals such as copper and zinc are essential for normal plant growth, elevated concentrations of both essential and non essential metals can result in growth inhibition and toxicity symptoms.

There are researches in the Southern part of Nigeria on oil pollution of soil and vegetation, but data is scarce on oil pollution in the Northern region of the country.

*Corresponding author. E-mail: mmachi42005@yahoo.com

Consequently, this work was initiated to study and evaluate soil and vegetation pollution due to heavy metals in spent lubricating oil in the vicinity of mechanic workshops in Kaduna metropolis of Northern Nigeria.

Kaduna was selected as the regional headquarters of Northern Nigeria in 1912 by the British colonial officials led by Lord Frederick Lugard. "Kaduna" is a Hausa word that means "crocodiles". This name was given to the town because of the presence of crocodiles in the river (River Kaduna) that now divides the town into two. The growth of Kaduna into an urban area was stimulated by its choice as an administrative centre and a garrison town by the colonialist because of its central location in the North and its now the capital city of Kaduna State. The development of industrial outfits and the growth of commercial activities led to an increased influx of people searching for jobs. By 1963, the population of Kaduna was less than 250,000 but the 1991 Census results put the population at 1,307,311. The population in 2001 was estimated to be 2,466, 760 (Africa Atlases, 2002) while in 2008, it was estimated to be over 4,000,000.

MATERIALS AND METHODS

Kaduna lies at latitude 10°28N and at longitude 7°25E. It is located in the central area of what used to be called the Northern Region of Nigeria. The mean annual rainfall in the area ranges from 924.3 to 1,543.6 mm. Annual temperature varies between 29 to 38.6°C. Presently, Kaduna is the capital of Kaduna State which was created in 1987. The creation of states did not affect the growth of Kaduna as an administrative, industrial, a veritable commercial center and a functional urban area. Kaduna today ranks second only to Kano in Northern Nigeria in terms of population (about 4,000,000 residents), industrial and commercial activities.

Ten mechanic workshops: Old Artillery Barracks (OA), Inland Container (IC), Sabon Tasha (ST), Ungwan Boro (UB), Television (TV), Kurmi Mashi (KM), Badiko (BD), Ungwan Muazu (UM), Kakuri (KK) and Gonin Gora (GG): were selected based on availability of mechanic workshops and geographical location for soil and plant samplings from the study area (Kaduna) excluding control (Figure 1). Surface soil samples were collected from the immediate vicinity of the plant samples with the aid of a stainless spoon, washed with soap and rinsed with distilled water after each sampling (Awofolu, 2005). Twelve soil samples from each mechanic workshop (sampling location), were collected and mixed properly to give a composite sample mixture. Three control samples (A, B, and C) were collected to validate the heavy metal concentration in soil and plant in areas where no human activities such as those of auto mechanics have taken or are taking place.

Soil and plant (*Sida acuta* Brum. f commonly known as brown weed) samples were collected in April 2008 according to Chimuka et al. (2005) and Awofolu (2005) respectively. The plant and soil samples collected were pretreated and digested according to Awofolu (2005).

The pH and electrical conductivity of soil samples were measured with a soil: water ratio of 1:2 using pH and electrical conductivity meters (HANNA Model No 111991000)) as described by Hendershot et al. (1993); organic carbon was determined by the method of Walkley and Black (1934); Cation Exchange Capacity (CEC) was determined by Herdershot et al. (1993); while particle size distribution was determined by the hydrometer method as described by Bougoucos (1951). Extractable sulphate (SO_4^{2-}) and phosphate (PO_4^{3-}) were determined by methods described by Allen et al. (1974).

Sequential extraction was carried out on the principle of selective extraction, proposed by Tessier et al. (1979), with the following modifications: Water soluble was first of all extracted (Ma and Rao, 1997) and then, $Mg(NO_3)_2$ was used instead of $MgCl_2$ to extract exchangeable fraction because the chloride ion can complex metals (Shrivastava and Benerjee, 2004) and increase the solubility of several heavy metals within the soil. Also, the extraction of the oxidisable phase was undertaken after the extraction of the exchangeable phase. This method allows the destruction of organic matter, which entraps the mineral materials and then provides a better extraction of the following phases (Shrivastava and Banerjee, 2004). For the residual fraction, a combination of aqua regia/hydrofluoric acid ($HCl/HNO_3/HF$) was used (Shrivastava and Banerjee, 2004).

Quality control test was performed on soil and plant samples in order to validate the experimental procedures. This was done by spiking the pre-digested soil and plant samples with Multielement metal Standard Solution, MESS (0.5 mgl[-1] of Cd and Cr and 5 mgl[-1] for Cu, Fe, Ni, Pb and Zn) according to Awofolu (2005).

The validity of the extraction procedure and the precision and accuracy of the Atomic Absorption Spectrophotometer were tested by spiking experiment. The results of the 7 heavy metals present in the soil and plant samples as shown in Table 1 varied between 89.67 to 108.67 and 89.00 to 103.00 for the soil and plant respectively. The highest percentage recovery of 108.67 and 103.00 were recorded for Zn in soil and plant respectively while the lowest percentage recovery of 89.67 and 89.00 were recorded for Ni in soil and plant respectively. The pattern of recovery efficiency for both soil and plant were found to follow the decreasing orders; Zn > Cu > Fe > Cr = Pb > Cd > Ni and Zn > Cr = Pb > Fe > Cu > Cd > Ni.

RESULTS AND DISCUSSION

Tables 2 and 3 show the results of the physico-chemical parameters of the surface soil of Kaduna in the study and control areas. The percentage of particle size distribution in Kaduna soil was in the range 22.0 to 34, 12.0 to 30.0 and 42.0 to 64% for clay, silt and sand respectively. Generally, the textural class of the soils was mainly loamy sand with a few (that is soil samples from Sabon Tasha, Badiko and Gonin Gora mechanic workshops) being loamy clay.

As shown in Table 2, soil pH varied from 7.00 to 8.00 and 7.30 to 7.60 for the study and control sites, respectively. This is consistent with the results of Maiz et al. (2000) and Osakwe and Otuya (2008). The soil pH serves as a useful index of availability of nutrients, the potency of toxic substances present in the soil and the physical properties of the soil. Several studies have shown that availability of heavy metals is pH dependent (Iwegbue et al., 2006; Gonzalez-Fernandez et al., 2008). Most of the pH values of the study areas did not indicate the tendency for availability of these metals and this could indicate low metal uptake by plants. Electrical Conductivity (EC) values ranged from 0.30 to 9.50 μscm[-1] and 0.30 to 0.35 μscm[-1] for study and control sites, respectively. Also, variation observed in electrical conductivity between sites could be due to the content of

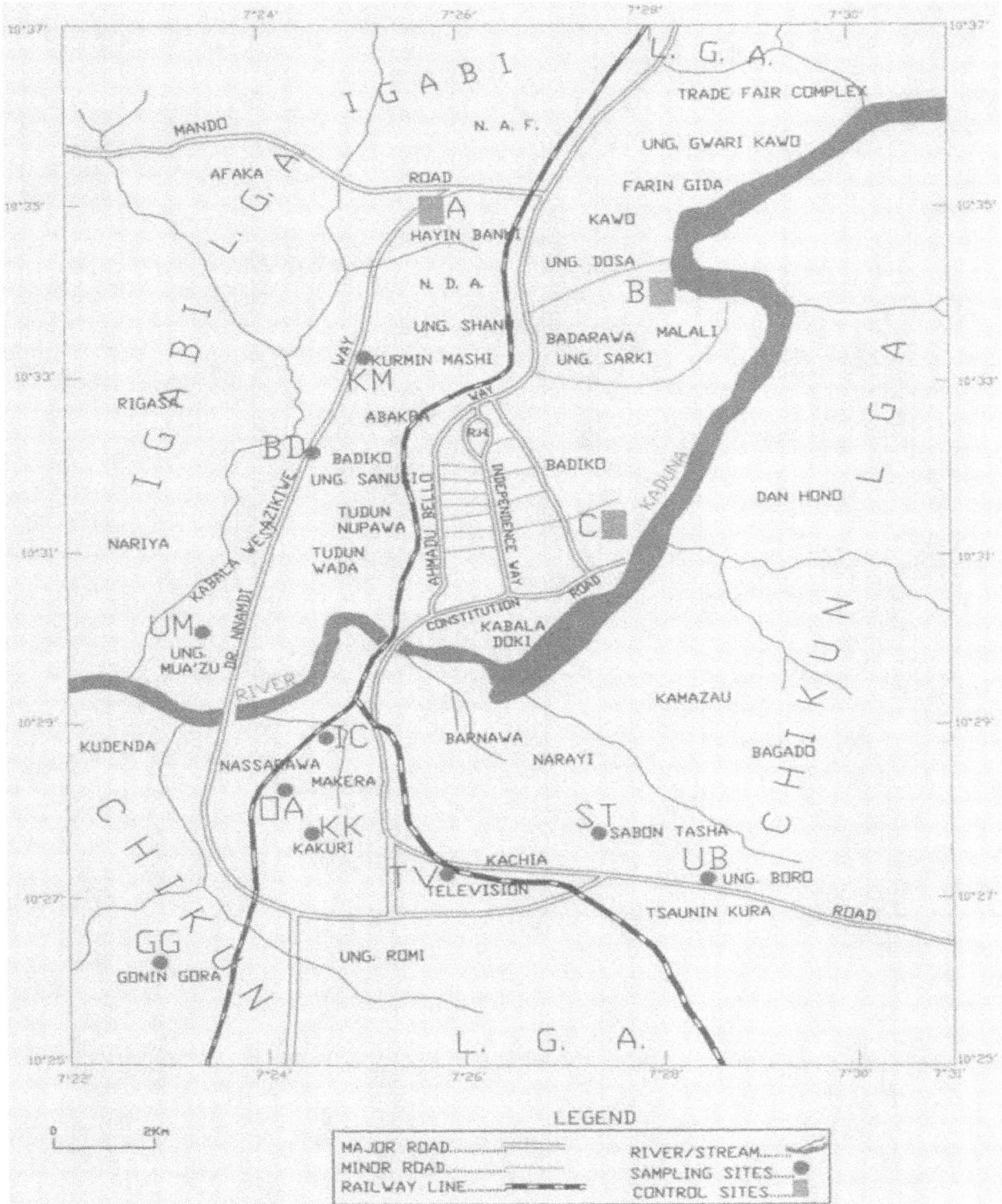

Figure 1. Kaduna Metropolis showing sampling sites. Source: Kaduna State Environmental Protection Authority

available soluble salts. The electrical conductivity values were higher than those obtained by Iwegbue et al. (2006) but much lower than Osakwe and Otuya (2008). The percentage organic carbon content in the soil samples ranged between 25.90 (Inland Container) - 43.10% (Sabon Tasha) for the study soils and 13.20 to 16.40 for

Table 1. Percentage recovery of heavy metals in soil and plant samples.

Metals	Soil sample	Plant sample
Cd	90.67 ± 1.00	90.67 ± 1.20
Cr	93.00 ± 1.20	97.00 ± 2.00
Zn	108.67 ± 1.00	103.00 ± 2.10
Cu	98.00 ± 2.00	93.33 ± 3.00
Fe	93.33 ± 3.00	95.67 ± 4.00
Ni	89.67 ± 1.00	89.00 ± 1.00
Pb	93.00 ± 1.30	97.00 ± 2.50

Table 2. Physicochemical parameters of the soil in the study areas.

Site	Clay (%)	Silt (%)	Sand (%)	pH	EC (μScm^{-1})	OC (%)	CEC (Meq/100g of soil)	*Ext. PO_4^{3-} ($mgkg^{-1}$)	*Ext. SO_4^{2-} ($mgkg^{-1}$)
Old Artillery Barracks	24.00±0.00	12.00±0.00	64.00±0.00	7.00±0.21	0.30±0.01	27.50±0.02	19.20±0.40	55.00±1.20	12.00±0.30
Inland Container	22.00±0.00	14.00±0.00	64.00±0.10	8.00±0.34	0.75±0.01	25.90±0.10	18.70±0.40	45.00±1.00	16.00±0.10
Sabon Tasha	30.00±0.00	26.00±0.00	44.00±0.21	7.40±0.11	1.10±0.00	43.10±0.45	21.10±0.98	25.00±1.40	14.00±0.30
Ungwan Boro	26.00±0.00	20.00±0.00	54.00±0.00	7.50±0.01	0.40±0.01	30.30±0.30	15.20±0.65	22.00±1.00	11.00±0.33
Television	24.00±0.00	22.00±0.00	54.00±0.00	7.30±0.10	1.00±0.02	41.50±0.10	16.40±0.38	38.00±2.00	18.00±0.45
Kurmi Mashi	34.00±0.00	30.00±0.00	46.00±0.00	7.40±1.00	0.80±0.01	28.70±0.30	20.00±0.67	22.00±1.00	20.00±0.90
Badiko	32.00±0.00	28.00±0.00	40.00±0.00	7.40±0.00	0.85±0.00	40.70±0.11	23.10±0.10	30.00±2.00	16.00±0.14
Ungwan Muazu	26.00±1.00	22.00±0.00	52.00±0.00	7.00±0.13	9.50±0.00	38.30±0.10	20.20±0.39	30.00±1.00	18.00±0.20
Kakuri	24.00±0.00	20.00±0.00	56.00±0.00	7.50±0.00	0.75±0.23	31.90±0.01	17.30±0.37	28.00±1.00	22.00±0.56
Gonin Gora	30.00±0.00	28.00±0.00	42.00±0.00	7.40±0.00	0.50±0.10	35.90±0.50	15.20±0.20	25.00±1.00	22.00±0.50
Mean	27.20	22.20	51.60	7.39	1.60	34.38	18.64	32.00	16.90

*Ext. = Extractable.

Table 3. Physicochemical parameters of the soil in the control area.

Site	Clay (%)	Silt (%)	Sand (%)	pH	EC (μScm^{-1})	OC (%)	CEC (Meq/100 g of soil)	*Ext. PO_4^{3-} ($mgkg^{-1}$)	*Ext. SO_4^{2-} ($mgkg^{-1}$)
A	18.00±0.00	28.00±0.10	54.00±0.00	7.30±0.50	0.30±0.01	13.20±1.20	12.80±0.20	12.00±1.00	4.00±0.50
B	20.00±0.00	26.00±0.00	54.00±1.00	7.60±0.30	0.30±0.20	16.40±0.57	9.20±0.40	14.00±0.50	6.00±0.00
C	24.00±0.00	14.00±0.00	62.00±0.00	7.50±0.56	0.35±0.10	14.40±0.60	17.20±0.80	13.50±0.90	4.00±0.00
Mean	20.67	22.67	56.67	7.47	0.32	14.67	13.07	13.17	4.67

*Ext. = Extractable.

Table 4. Chemical fractionation of Cd (mgkg⁻¹) in surface soils of some selected mechanic workshops in the study areas.

Fractions	Sites									
	Old Artillery Barracks	Inland Container	Sabon Tasha	Ungwan Boro	Television	Kurmi Mashi	Badiko	Ungwan Muazu	Kakuri	Gonin Gora
Water soluble	0.00±0.00	1.85±0.10	1.85±0.20	1.85±0.01	3.70±0.02	0.00±0.00	1.85±0.00	3.70±0.01	1.85±0.00	1.85±0.16
% Water soluble	0.00	1.79	1.78	1.77	3.57	0.00	1.78	3.58	1.77	1.77
Exchangeable	7.41±0.01	7.41±0.34	7.41±0.15	7.41±0.24	9.26±0.44	7.41±0.21	7.41±0.11	9.26±0.02	7.41±0.10	7.41±0.13
% Exch.	7.17	7.16	7.11	7.08	8.92	7.21	7.11	8.96	7.08	7.08
CB	3.70±0.10	5.56±0.12	11.11±0.24	5.56±0.50	7.41±0.15	9.26±0.15	3.70±0.23	7.41±0.34	3.70±0.11	9.26±0.15
%CB	3.59	5.37	10.67	5.31	7.13	9.01	3.56	7.16	3.54	8.85
Fe-MnO	11.11±0.21	12.96±0.22	0.00±0.00	12.96±0.22	11.11±0.26	9.26±0.15	9.26±0.12	11.11±0.23	9.26±0.01	9.26±0.29
% Fe-MnO	10.76	12.54	0.00	12.39	10.70	9.01	8.89	10.75	8.85	8.85
Org. B	11.11±0.20	11.11±0.14	18.52±0.36	12.96±0.21	11.11±0.26	16.67±0.11	9.26±0.13	16.67±0.56	9.26±0.10	12.96±0.33
% Org. B	10.76	10.75	17.78	12.39	10.70	16.22	8.89	16.12	8.85	12.39
Residual	69.91±0.16	64.51±0.21	65.28±0.82	63.89±0.50	61.27±0.21	60.19±0.21	72.69±0.45	55.25±0.07	73.15±0.11	63.89±0.67
% Residual	67.71	62.39	62.67	61.06	58.99	58.56	69.78	53.43	69.91	61.06
∑$_T$	103.24	103.40	104.17	104.63	103.86	102.78	104.17	103.40	104.63	104.63

the control soils. The high percentage organic carbon content in the study sites as compared to the control sites could be due to automechanic activities carried out within these sites, which release hydrocarbon waste. Hence, the relatively high total organic carbon content in samples could be due to the high organic content of used lubricants that were discharged at the sites (Akoto et al., 2008). The CEC of the sites varied between 15.20 (Ungwan Boro and Gonin Gora) to 23.10 mmolkg⁻¹ (Badiko) in study sites and 9.20 to 17.20 mmolkg⁻¹ in control sites. The CEC of soils is related to the nature and quality of clay and organic carbon contents. Hence, high percentage organic carbon from these sites could be responsible for the CEC. According to Awode et al. (2008), CEC of soil is more greatly influenced by organic matter than by the concentration of clays, hence CEC tends to be higher in the study sites than in the control sites. All study sites had

comparatively higher mean cation exchange capacity than the mean CEC of 8.14 mmolkg⁻¹ reported by Isirimah (1987).

In the study area, extractable sulphate content ranged from 11.00 to 22.00 mgkg⁻¹ while the sulphate in samples from the control area ranged between 4.00 to 6.00 mgkg⁻¹. Significant variation was observed between the sulphate level in the study and control areas. This could be due to the activities of battery chargers within the study area, an indication of soil contamination within this area. The level of extractable PO_4^{3-} in soil samples ranged from 22.00 – 55.00 mgkg⁻¹ in the study area while in the control area, the PO_4^{3-} ranged from 12.00 – 14.00 mgkg⁻¹. Similar to sulphate, phosphate level in the study area is very much greater than the level in the control area. For similar reasons, sulphate could be responsible for high phosphate within these sites. Extractable phosphates in the study area had a mean of 32.00

mg/kg which is similar to reports by Iwegbue et al. (2006).

The mobility and availability of heavy metals in soils depend on how the metals are associated with the components of the soil, and the measure of the mobility and availability of metals serves to predict the behavior of heavy metals in the soil. The metals are bound to the solid component of the soil through phases such as water soluble, exchangeable, carbonate, oxides and hydroxides, and organics. The selective sequential extraction of heavy metals provides detailed insight on how the metals are associated with the components of the soil. Tables 4 to 10 present total and sequentially fractionated metal (Cd, Cr, Zn, Cu, Fe, Ni and Pb) concentrations in soil, and the heavy metal percentage for each fraction calculated for Kaduna soil.

Analysis of variance between heavy metals across sites revealed significant differences ($p <$

Table 5. Chemical fractionation of Cr (mgkg⁻¹) in surface soils of some selected mechanic workshops in the study areas.

Fractions	Old Artillery Barracks	Inland Container	Sabon Tasha	Ungwan Boro	Television	Kurmi Mashi	Badiko	Ungwan Muazu	Kakuri	Gonin Gora
Water soluble	0.00±0.00	0.00±0.00	0.00±0.00	0.00±0.00	0.00±0.00	0.00±0.00	0.00±0.00	0.00±0.00	0.00±0.00	0.00±0.00
% Water soluble	0.00	0.00	0.00	0.00	0.00	0.00	0.00	0.00	0.00	0.00
Exchangeable	0.00±0.00	0.00±0.00	0.00±0.00	0.00±0.00	0.00±0.00	0.00±0.00	0.00±0.00	0.00±0.00	0.00±0.00	0.00±0.00
% Exch.	0.00	0.00	0.00	0.00	0.00	0.00	0.00	0.00	0.00	0.00
CB	0.00±0.00	0.00±0.00	0.00±0.00	0.00±0.00	0.00±0.00	0.00±0.00	0.00±0.00	0.00±0.00	0.00±0.00	0.00±0.12
%CB	0.00	0.00	0.00	0.00	0.00	0.00	0.00	0.00	0.00	0.00
Fe-MnO	9.76±0.23	9.76±0.11	0.00±0.00	0.00±0.00	19.51±0.34	0.00±0.00	9.76±0.34	9.76±0.00	9.76±0.23	9.76±0.34
% Fe-MnO	15.69	12.31	0.00	0.00	26.23	0.00	13.56	14.37	14.04	11.94
Org. B	0.00±0.00	29.27±0.34	29.27±1.20	58.54±0.23	39.02±0.22	29.27±0.67	29.27±0.56	39.02±0.45	29.27±0.45	29.27±0.98
% Org. B	0.00	36.92	33.80	78.69	52.46	39.34	40.68	57.49	42.11	35.82
Residual	52.44±0.34	40.24±0.19	57.32±0.45	15.85±0.11	15.85±0.12	45.12±0.45	32.93±0.67	19.11±0.78	30.49±0.22	42.68±0.34
% Residual	84.31	50.77	66.20	21.31	21.31	60.66	45.76	28.14	43.86	52.24
$\sum_T$	62.20	79.27	86.59	74.39	74.39	74.39	71.95	67.89	69.51	81.71

Table 6. Chemical fractionation of Zn (mgkg⁻¹) in surface soils of some selected mechanic workshops in the study areas.

Fractions	Old Artillery Barracks	Inland Container	Sabon Tasha	Ungwan Boro	Television	Kurmi Mashi	Badiko	Ungwan Muazu	Kakuri	Gonin Gora
Water soluble	7.50±0.023	12.50±0.34	0.00±0.00	2.50±0.01	7.50±0.23	0.00±0.00	2.50±0.00	12.50±0.01	7.50±0.01	7.50±0.10
% Water soluble	1.39	2.07	0.00	0.43	1.129	0.00	0.41	1.343	1.23	1.09
Exchangeable	25.00±0.45	50.00±0.01	22.50±0.34	10.00±0.19	20.00±0.56	32.50±0.67	20.00±0.120	27.50±0.20	22.50±0.10	25.00±1.20
% Exch.	4.62	8.28	1.919	1.70	3.011	4.88	3.31	2.954	3.70	3.65
CB	142.50±0.56	172.50±1.56	250.00±2.10	115.00±2.00	225.00±0.33	142.50±1.00	150.00±2.10	250.00±2.00	137.50±2.00	225.00±1.33
%CB	26.33	28.57	21.322	19.59	33.877	21.39	24.84	26.858	22.62	32.85
Fe-MnO	42.50±0.78	35.00±0.38	47.50±0.67	42.50±0.30	40.00±0.20	40.00±1.11	35.00±0.20	50.00±0.00	40.00±1.00	40.00±0.10
% Fe-MnO	7.85	5.80	4.051	7.24	6.023	6.00	5.80	5.372	6.58	5.84
Org. B	87.50±0.37	140.00±5.10	145.00±1.00	100.00±1.00	180.00±2.10	117.50±1.20	125.00±3.00	145.00±0.10	115.00±1.00	150.00±1.00
% Org. B	16.17	23.19	12.367	17.03	27.102	17.64	20.70	15.577	18.92	21.90
Residual	236.25±0.23	193.75±0.90	707.50±1.00	317.08±1.00	191.67±1.00	333.75±0.99	271.25±1.90	445.83±1.20	285.42±1.00	237.50±1.00
% Residual	43.65	32.09	60.341	54.01	28.858	50.09	44.93	46.95	46.95	34.67
$\sum_T$	541.25	603.75	1172.500	587.08	664.167	666.25	603.75	930.833	607.92	685.00

Table 7. Chemical fractionation of Cu (mgkg⁻¹) in surface soils of some selected mechanic workshops in the study areas.

Fractions	Sites									
	Old Artillery Barracks	Inland Container	Sabon Tasha	Ungwan Boro	Television	Kurmi Mashi	Badiko	Ungwan Muazu	Kakuri	Gonin Gora
Water soluble	2.00±0.00	0.00±0.00	2.00±0.10	0.00±0.00	2.00±0.00	2.00±0.19	0.00±0.00	2.00±0.45	0.00±0.00	2.00±0.56
% Water soluble	12.25	0.00	8.70	0.00	2.00	3.46	0.00	2.45	0.00	2.00
Exchangeable	0.00±0.00	0.00±0.00	0.00±0.00	4.00±0.16	2.00±0.00	2.00±0.34	0.00±0.00	2.00±0.00	0.00±0.00	2.00±0.78
% Exch.	0.00	0.00	0.00	13.71	2.00	3.46	0.00	2.45	0.00	2.00
CB	6.00±0.10	14.00±0.45	10.00±0.20	12.00±0.56	72.00±1.00	10.00±0.67	14.00±0.23	60.00±0.45	14.00±1.00	14.00±0.23
%CB	36.74	38.89	43.48	41.14	72.00	17.29	42.42	73.47	36.84	13.98
Fe-MnO	0.00±0.00	0.00±0.00	0.00±0.00	2.00±0.77	2.00±0.00	2.00±0.77	0.00±0.00	2.00±0.10	0.00±0.00	2.00±0.00
% Fe-MnO	0.00	0.00	0.00	6.86	2.00	3.46	0.00	2.45	0.00	2.00
Org. B	2.00±0.34	2.00±0.20	2.00±0.40	2.00±0.11	2.00±0.00	2.00±0.26	2.00±0.67	2.00±0.10	2.00±0.11	2.00±0.01
% Org. B	12.25	5.56	8.70	6.86	2.00	3.46	6.06	2.45	5.26	2.00
Residual	6.33±0.50	20.00±0.30	9.00±0.27	9.17±0.00	20.00±1.00	39.83±0.87	17.00±0.56	13.67±0.30	22.00±0.45	78.17±0.90
% Residual	38.77	55.56	39.13	31.43	20.00	68.88	51.52	16.74	57.89	78.04
∑T	16.33	36.00	23.00	29.17	100.00	57.83	33.00	81.67	38.00	100.17

Table 8. Chemical fractionation of Fe (mgkg⁻¹) in surface soils of some selected mechanic workshops in the study areas.

Fractions	Sites									
	Old Artillery Barracks	Inland Container	Sabon Tasha	Ungwan Boro	Television	Kurmi Mashi	Badiko	Ungwan Muazu	Kakuri	Gonin Gora
Water soluble	14.71±0.56	17.65±0.10	14.71±0.10	14.71±0.20	14.71±0.13	11.76±0.87	14.71±0.10	17.65±0.00	14.71±0.00	17.65±0.11
% water soluble	3.28	3.64	3.02	3.08	3.06	2.46	3.08	3.64	3.13	3.64
Exchangeable	29.41±0.01	11.76±0.34	17.65±0.10	11.76±0.10	20.59±0.56	14.71±0.33	17.65±0.30	29.41±0.10	14.71±0.00	14.71±0.21
% Exch.	6.56	2.42	3.62	2.46	4.29	3.08	3.69	6.06	3.13	3.03
CB	23.53±0.00	20.59±0.45	26.47±0.34	35.29±0.22	29.41±0.56	20.59±0.29	23.53±0.10	29.41±0.10	32.35±0.10	29.41±0.10
%CB	5.25	4.24	5.43	7.38	6.12	4.31	4.92	6.06	6.88	6.06
Fe-MnO	29.41±0.34	14.71±0.10	14.71±0.44	17.65±0.11	14.71±0.89	17.65±0.33	17.65±0.34	20.59±0.30	14.71±0.10	20.59±0.11
% Fe-MnO	6.56	3.03	3.02	3.69	3.06	3.69	3.69	4.24	3.13	4.24
Org. B	14.71±0.20	26.47±0.20	11.76±0.56	17.65±0.34	17.65±0.20	14.71±0.67	17.65±0.20	20.59±0.10	17.65±0.30	17.65±0.20
% Org. B	3.28	5.45	2.41	3.69	3.67	3.08	3.69	4.24	3.75	3.64
Residual	336.76±0.11	394.12±1.90	402.45±1.00	380.88±1.11	383.33±1.67	398.53±5.10	386.76±1.30	367.65±1.89	376.47±2.10	385.29±1.90
% Residual	75.08	81.21	82.51	79.69	79.80	83.38	80.92	75.76	80.00	79.39
∑T	448.53	485.29	487.75	477.94	480.39	477.94	477.94	485.29	470.59	485.29

Table 9. Chemical fractionation of Ni (mgkg^{-1}) in surface soils of some selected mechanic workshops in the study areas.

Fractions	Sites									
	Old Artillery Barracks	Inland Container	Sabon Tasha	Ungwan Boro	Television	Kurmi Mashi	Badiko	Ungwan Muazu	Kakuri	Gonin Gora
Water soluble	26.67±0.10	26.67±0.76	26.67±0.11	20.00±0.20	20.00±0.30	20.00±0.90	26.67±0.10	26.67±0.12	20.00±0.00	26.67±0.00
% Water soluble	13.01	12.28	12.60	9.84	9.68	9.52	12.70	12.70	9.52	12.73
Exchangeable	20.00±0.30	13.33±0.30	20.00±0.00	6.67±0.34	13.33±0.20	13.33±0.78	13.33±0.50	20.00±0.34	13.33±0.43	13.33±0.12
% Exch.	9.76	6.14	9.45	3.28	6.45	6.35	6.35	9.52	6.35	6.37
CB	20.00±030	13.33±0.11	20.00±0.10	13.33±0.67	6.67±0.56	6.67±0.28	20.00±0.10	20.00±0.56	13.33±0.10	13.33±0.12
%CB	9.76	6.14	9.45	6.56	3.23	3.17	9.52	9.52	6.35	6.37
Fe-MnO	33.33±0.10	26.67±0.21	26.67±0.23	33.33±0.78	33.33±0.78	26.67±0.89	26.67±0.00	33.33±0.89	33.33±0.00	33.33±0.00
% Fe-MnO	16.26	12.28	12.60	16.39	16.13	12.70	12.70	15.87	15.87	15.92
Org. B	40.00±0.11	26.67±0.20	26.67±0.20	26.67±0.98	40.00±0.20	33.33±0.87	26.67±0.00	33.33±0.44	33.33±0.20	33.33±0.00
% Org. B	19.51	12.28	12.60	13.11	19.35	15.87	12.70	15.87	15.87	15.92
Residual	65.00±0.45	110.56±1.00	91.67±0.00	103.33±1.90	93.33±0.57	110.00±1.00	96.67±1.90	76.67±0.10	96.67±0.56	89.44±0.87
% Residual	31.71	50.90	43.31	50.82	45.16	52.38	46.03	36.51	46.03	42.71
$\sum_T$	205.00	217.22	211.67	203.33	206.67	210.00	210.00	210.00	210.00	209.44

Table 10. Chemical fractionation of Pb (mgkg^{-1}) in surface soils of some selected mechanic workshops in the study areas.

Fractions	Sites									
	Old Artillery Barracks	Inland Container	Sabon Tasha	Ungwan Boro	Television	Kurmi Mashi	Badiko	Ungwan Muazu	Kakuri	Gonin Gora
Water soluble	18.80±0.20	19.64±0.45	20.00±0.20	22.00±0.12	16.80±0.56	17.60±0.00	16.00±0.18	21.20±0.10	20.40±0.10	18.00±0.50
% Water soluble	2.63	2.85	3.05	3.80	2.76	2.59	2.40	3.40	2.96	2.56
Exchangeable	12.40±0.10	10.44±0.34	12.40±0.30	12.00±0.56	10.40±0.20	12.00±0.10	10.00±0.65	10.00±0.00	10.40±0.19	13.20±0.10
% Exch.	1.74	1.52	1.89	2.07	1.71	1.77	1.50	1.60	1.51	1.87
CB	172.00±1.20	173.20±2.10	194.40±1.20	213.20±1.29	184.80±2.11	208.00±0.80	178.00±1.50	163.60±8.10	173.60±0.70	200.00±2.00
%CB	24.10	25.17	29.63	36.80	30.38	30.62	26.67	26.24	25.18	28.40
Fe-MnO	15.20±0.10	15.60±0.19	17.20±0.20	15.60±0.45	16.40±0.50	16.80±0.22	15.20±0.20	17.20±0.30	15.60±1.00	17.60±0.20
% Fe-MnO	2.13	2.27	2.62	2.69	2.70	2.47	2.28	2.76	2.26	2.50
Org. B	324.40±1.20	302.00±2.10	304.40±3.00	302.40±2.00	329.20±1.00	303.60±1.99	304.00±1.00	308.80±1.10	326.00±2.13	296.00±1.20
% Org. B	45.46	43.89	46.40	52.20	54.11	44.69	45.55	49.52	47.28	42.04
Residual	170.83±1.00	167.19±1.80	107.63±1.12	14.10±0.30	50.77±0.40	121.30±0.90	144.23±0.50	102.73±1.00	143.53±1.50	159.37±1.11
% Residual	23.94	24.30	16.41	2.43	8.34	17.86	21.61	16.48	20.82	22.63
$\sum_T$	713.63	688.07	656.03	579.30	608.37	679.30	667.43	623.53	689.53	704.17

Table 11. Correlation matrix among physicochemical parameters and heavy metals in surface soils of mechanic workshops in Kaduna.

Parameter	Clay	Silt	Sand	pH	EC	OC	CEC	Ext. P	Ext. SO$_4^{2-}$	Cd	Cr	Zn	Cu	Fe	Ni	Pb
Cd	-0.054	0.162	-0.307	0.186	-0.266	0.352	-0.436	-0.426	0.075	1						
Cr	0.314	0.450	-0.461	0.557	-0.271	0.328	-0.098	-0.461	0.113	0.315	1					
Zn	0.655*	0.759*	-0.663*	0.332	-0.214	0.096	-0.138	-0.655*	0.653*	0.157	0.654*	1				
Cu	0.035	0.286	-0.324	0.129	-0.095	0.225	-0.564	-0.199	0.561	0.515	0.378	0.554	1			
Fe	0.267	0.540	-0.503	0.422	0.284	0.462	-0.005	-0.601	0.302	0.211	0.764*	0.559	0.367	1		
Ni	0.002	0.050	-0.013	0.615	0.105	-0.045	0.380	0.048	0.348	-0.225	0.460	0.344	0.060	0.514	1	
Pb	0.109	-0.089	0.066	0.091	-0.314	-0.357	0.219	0.364	0.349	-0.208	-0.059	0.266	0.156	-0.373	0.410	1

* . Correlation is significant at the 0.05 level (2-tailed). Ext.= Extractable.

0.05). A correlation matrix for heavy metals in the soil of the different sites was calculated to see if some metals were interrelated with each other and the results are presented in Table 11. Negative correlation was especially found between the metals in the study areas, Cd: Ni and Pb; Pb: Cr; and Pb: Cd, though not significant (p > 0.05).

However, others showed significant (p < 0.05) positive correlation between Cr: Zn and Fe. Also, Table 11 shows the correlation between heavy metals in soils and the soil physico-chemical parameters revealed significant (p < 0.05) positive correlation between Zn: clay silt and extractable SO$_4^{2-}$ and Cu: extractable SO$_4^{2-}$. Hence, positive correlation between metal concentration and physico-chemical parameters could imply a significant effect on the amount of heavy metals in the soil, since the mobility and bioavailability of metals present in soils depend on physico-chemical properties of both the metal and the soil (McEldowney et al., 1993).

The distribution pattern of metals in the six geo-chemical phases of the soil showed differences in the study sites. As shown in the results, the highest percentage of metals was found in the residual fractions while the water soluble fraction had the least percentage. However, the chemical partitioning trends were found to be different for each metal. Majority of the metals were found to be concentrated in the residual fraction with the exception of Pb. The highest percentage fractions obtained by the sequential extraction procedure shows the following metal distribution pattern:

Cd: Residual > Organic > Fe-MnO > Exchangeable > Carbonate > Water Soluble
Cr: Residual > Organic > Fe-MnO > Exchangeable = Carbonate = Water Soluble
Zn: Residual > Carbonate > Organic > Fe-MnO > Exchangeable > Water Soluble
Cu: Residual > Carbonate > Organic > Water Soluble > Exchangeable > Fe-MnO
Fe: Residual > Carbonate > Exchangeable > Fe-MnO > Organic > Water Soluble
Ni: Residual > Fe-MnO > Organic > Water Soluble > Carbonate = Exchangeable
Pb: Organic > Carbonate > Residual > Water Soluble > Fe-MnO > Exchangeable.

As observed from the results, the highest percentage fraction of 69.91 and 44.50 were associated with the residual fraction for Cd in Kakuri. This situation is however different from Tessier et al. (1979), where insignificant amounts

of Cd are associated with Fe-MnO fraction, whereas in the other fractions, Cd concentrations were below the limit of detection. Association of Cd to the residual fraction does not generally constitute an environmental risk. This is due to the stable nature of the compound and the fact that the metals are bonded firmly within a mineral lattice that restricts the bioavailability of this metal (Coetzee, 1993; Abu-Kukati, 2001).

The distribution of Cr, Zn, Cu, Fe, Ni and Pb were similarly dominated by residual and organic fractions. The variations across these fractions between the study areas could be due to different anthropogenic sources of these metals. For Cr, the highest value of 83.31% was associated with residual and organic fractions in Old Artillery. There is no water soluble, exchangeable and carbonate fractions of Cr noticeable in all the sites; this was also similar to Venkateswaran et al. (2007) where no Cr was detected in the first three fractions.

According to Venkateswaran et al. (2007), the leaching of Cr to the environment from these samples may not occur readily Cr(VI) is a highly toxic metal that has been linked to cancer in humans following prolonged inhalation, and is toxic to plants at relatively low concentrations

(USEPA, 1998).

The predominant form of Zn (60.34% in Sabon Tasha) was in the residual and carbonate fractions, followed by organic and Fe-MnO fractions. There was no significant variation in absolute mobile fractions (water soluble exchangeable and carbonate bound fraction). Higher concentrations of Zn in Fe-MnO fraction can be attributed to diffusion mechanism (Backes et al., 1995). This metal can be released into the environment under extremely reducible conditions (Venkateswaran et al., 2007).

Also, the predominant form of Cu available in the entire fractions was residual fractions. The mobile fraction of Cu contributed over 40%; the high percentage of Cu in the organic fraction was due to higher stability constant of Cu complexes with organic matter (Venkateswaran et al., 2007). Infact, the highest value of 87.69 of Cu was found in the residual fraction, indicating that it is mainly of lithogenic origin (Xuelu et al., 2008). This coincided with the researches carried out on soils in China (Liu et al., 2003; Yuan et al., 2004).

Though Fe is not a toxic heavy metal because it serves as a micronutrient, it was also analysed in the present study and the results discussed since it was the predominant metal available in all sites. The highest value of 76.47% was available in the residual fractions in the studied soil. Since the Fe concentration was very high in the residual fraction, it could be converted to reducible fraction by plant roots (Otte et al., 1995).

Fe-MnO, residual and organic bound fractions were the predominant forms of Ni in all the samples. About 15 - 33.33% was in the exchangeable and carbonate fractions which can cause environmental toxicity during mobility (Karbassi and Shankar, 2005). Similar to this study, Behnar and Catherine (2006) showed that Ni in soil was concentrated in residual fractions. Pb in the soil was found mainly in the organic and residual fractions and then in carbonate and exchangeable forms in all the sites with few exceptions. In a smaller amount, it was bound to water soluble and Fe-MnO fractions. A similar distribution of Pb forms among fractions was reported for the fluvial deposits by Sobczynski and Siepak (2001).

Conclusion

The distribution pattern of metals in the six geochemical phases of the soil showed differences in metal content across the sites. The highest percentages of metals were found in the residual fractions while the water soluble fraction had the least percentages.

However, in most soils, a significant percentage of the total metals were associated with the no residual fractions. The metal fraction associated with the residual fraction can only be mobilized as a result of weathering and as such, this is only associated with long-term effects. Generally, the study indicated possible soil pollution in this environment by activities carried out within these areas.

REFERENCES

Abu-Kukati Y (2001). Heavy metal distribution and speciation in sediments from Ziqlab Dam – Jordan. Geological Eng., 25(1): 33 – 40.

Africa Atlases Nigeria (2002). Les Editions, J.A. pp. 142 – 144.

Akoto O, Ephraim JH, Darko G (2008).Heavy metals pollution in surface soils in the vicinity of abundant railway servicing workshop in Kumasi, Ghana. Int. J. Environ. Res., 2(4):359-364.

Allen SE (1974). Chemical analysis of ecological materials, Backwell scientific Publications, Oxford. pp. 13 – 106.

Anoliefo GO, Vwioko DE (1995). Effects of spent lubrication oil on the growth of Capsicum anum L. and Hycopersicon esculentum Miller. Environ. Pollut., 88(3): 361-364.

Awode UA, Uzairu A, Balarabe ML, Harrison GFS, Okunola OJ (2008). Assessment of Peppers and Soils for some Heavy metals from Irrigated Farmlands on the banks of River Challawa, Nigeria. Pakistan J. Nutr., 7(2): 244 – 248 2008.

Awofolu OR (2005). A survey of trace metals in vegetation, soil and lower Animals along some selected major roads in metropolitan city of Lagos. Environ. Monit. Assess., 105: 413-447.

Backes CA, McLaren RG, Rate AW, Swift RS (1995). Kinetics of Cd and Co desorption from iron and Manganese Oxides. Soil Sci. Soc. Am. J., 59: 778 – 785.

Bougoucos GH (1951). Determination of particle sizes. Soil Agron. J., 43: 434 – 438.

Chimuka L, Odyo O, Bapela J, Maboladisoro H, Mugwedi R (2005). Metals in environmental media: A study of trace and platinum group metals in Thohoyandou, South Africa. Water S.A., 31: 4

Coetzee PP (1993). Determination of speciation of heavy metals in sediments of the Hortbees port: dam by sequential chemical extraction. Water SA, 9(4): 291 – 300.

Edebiri RAO, Nwanokwale E (1981). Control of pollution from internal combustion engine used lubricant. Proc. Int. Seminar Petroleum Industry and the Nigeria Environment, pp. 9-12.

Gonzalez – Fernandez O, Hidalgo M, Margui E, Carvalho ML, Queralt I (2008). Heavy metals' content of automotive shredder residues (ASR): Evaluation of environmental risk. Environ. Pollut., 153: 476 – 482.

Hall JL (2002). Cellular mechanisms for heavy metal detoxification and tolerance. J. Exp. Bot., 53: 1 -11.

Hendershot WH, Lalande H, Duguette M (1993). Soil reaction and exchangeable acidity. In: Carter, M.R(Ed), Soil sampling and methods of Analysis for Canadian Society of Soil Science, Lewis, Boca Raton, FL, pp. 141 – 145.

Isirimah NO (1987). An inventory of some chemical prosperities of selected surface soils of Rivers state of Nigeria. In proceeding of 15th Annual Conference of Soil Science Association of Nigeria, Kaduna, 217 – 233.

Iwegbue CMA, Isirimah NO, Igwe C, Williams ES (2006). Characteristic levels of heavy metal in soil profiles of automobile mechanic waste dumps in Nigeria. Environmentalist, 26: 123 – 128.

Karbassi AR, Shankar R (2005). Geochemistry of two sediment cores from the west coast of India. Int. J. Environ. Sci. Tech., 1: 307 -316.

Liu WX, Lia XD, Shen ZG, Wang DC, Wai OWH, Li YS (2003). Multivariate statistical study of heavy metal enrichment in sediments of the Pearl River Estuary. Environ. Pollut., 121: 377 – 388.

Lokeshwari H, Chandrappa GT (2006). Impact of heavy metal contamination of Bellandur Lake on soil and cultivated vegetation. Curr. Sci., 91(5): 622 – 627.

Ma LQ, Rao GN (1997). Chemical fractional of cadmium, copper nickel and zinc in contaminated soils. J. Environ. Qual., 26: 259-264.

Maiz I, Arambarri I, Garcia R, Millan E (2000). Evaluation of heavy metal availability in polluted soils by two sequential extraction procedures using factor analysis. Environ. Pollut., 110: 3-9.

McEldowney S, Hardman DJ, Waite S (1993). Pollution, Ecology, and Biotreatment. Longman Scientific and Technical, pp. 6 – 8, 251 – 258.

Nriagu JO (1978). Lead in the atmosphere, in the Biogeo-chemistry of lead in the environment, Part A, Elsevier, North Holland, Amsterdam, pp. 138 – 164.

Osakwe SA, Otuya OB (2008). Elemental composition of soils in some mechanic dumpsites in Agbor, Delta state, Nigeria. Chemical Society of Nigeria. 31st Annual International Conference and Exhibition. Deltachem 2008 Conference Proceedings, pp. 557 – 559.

Otte ML, Kearns CC, Doyle MO (1995). Accumulation of arsenic and zinc in the rhizosphere of wetland plants. Bull. Environ. Contam. Toxicol., 55: 154 – 161.

Racine CH (1994). Long-term recovery of vegetation on two experimental crude oil spills in interior Alaska Black Spruce Taiga. Can. J. Bot., 72: 1711 – 1177

Shrivastava SK, Banerjee DK (2004). Speciation of metals in sewage sludge amended soils. Water Air Soil Pollut., 152: 219 – 232.

Sobczynski T, Siepak J (2001). Speciation of heavy metals in bottom sediments of lakes in the area of Wielkopolski National Park. Polish J. Environ. Stud., 10(6): 463 – 474.

Tessier A, Campell PGC, Bisson M (1979). Sequential extraction procedure for the speciation of particulate trace metals. Anal. Chem., 51: 844-851.

USEPA (1998). Toxicological review of hexavalent chromium. Environmental Protection Agency.

Venkateswaran P, Vellaichamy S, Palanivelu K (2007). Speciation of heavy metals in electroplating industry sludge and wastewater residue using inductive coupled plasma. Int. J. Environ. Sci. Tech., 4(4): 497 – 504.

Vwioko DE, Anoliefo GO, Fashemi SD (2006). Metal contamination in plant tissues of *Ricinus communis* L. (Castor Oil) grown in Soil contaminated with spent lubricating oil. J. Appl. Sci. Environ. Manage., 10(3): 127 – 134.

Walkley A, Black IA (1934). An examination of the Detjare method for Determining soil organic matter and a proposed modification of the chromic Acid titration. Soil Sci., 37: 29-36.

Wong SC, Li XD, Zhang G, Qi SH, Min YS (2002). Heavy metnals in agriocultural soils of the pearl River Delta, South China. Enviorn. Pollut., 119: 33-44.

Xuelu G, Shaoyong C, Aimin L (2008). Chemical speciation of 12 metals in surface sediments from the Northern South China Sea under natural grain size. Baseline /Marine Pollut. Bull., 56: 770 – 797.

Yuan C, Shi J, He B, Liu J, Liang L, Jiang G (2004). Speciation of heavy metals in marine sediments from the East China Sea by ICP-MS with sequential extraction. Environ. Int., 30: 769 – 783.

Permissions

All chapters in this book were first published in JECE, by Academic Journals; hereby published with permission under the Creative Commons Attribution License or equivalent. Every chapter published in this book has been scrutinized by our experts. Their significance has been extensively debated. The topics covered herein carry significant findings which will fuel the growth of the discipline. They may even be implemented as practical applications or may be referred to as a beginning point for another development.

The contributors of this book come from diverse backgrounds, making this book a truly international effort. This book will bring forth new frontiers with its revolutionizing research information and detailed analysis of the nascent developments around the world.

We would like to thank all the contributing authors for lending their expertise to make the book truly unique. They have played a crucial role in the development of this book. Without their invaluable contributions this book wouldn't have been possible. They have made vital efforts to compile up to date information on the varied aspects of this subject to make this book a valuable addition to the collection of many professionals and students.

This book was conceptualized with the vision of imparting up-to-date information and advanced data in this field. To ensure the same, a matchless editorial board was set up. Every individual on the board went through rigorous rounds of assessment to prove their worth. After which they invested a large part of their time researching and compiling the most relevant data for our readers.

The editorial board has been involved in producing this book since its inception. They have spent rigorous hours researching and exploring the diverse topics which have resulted in the successful publishing of this book. They have passed on their knowledge of decades through this book. To expedite this challenging task, the publisher supported the team at every step. A small team of assistant editors was also appointed to further simplify the editing procedure and attain best results for the readers.

Apart from the editorial board, the designing team has also invested a significant amount of their time in understanding the subject and creating the most relevant covers. They scrutinized every image to scout for the most suitable representation of the subject and create an appropriate cover for the book.

The publishing team has been an ardent support to the editorial, designing and production team. Their endless efforts to recruit the best for this project, has resulted in the accomplishment of this book. They are a veteran in the field of academics and their pool of knowledge is as vast as their experience in printing. Their expertise and guidance has proved useful at every step. Their uncompromising quality standards have made this book an exceptional effort. Their encouragement from time to time has been an inspiration for everyone.

The publisher and the editorial board hope that this book will prove to be a valuable piece of knowledge for researchers, students, practitioners and scholars across the globe.

List of Contributors

Ayşe Bahar Yılmaz
Water Pollution Laboratory, Faculty of Fisheries, Mustafa Kemal University, 31200, skenderun-Hatay, Turkey

Cemal Turan
Fisheries Genetics Laboratory, Faculty of Fisheries, Mustafa Kemal University, 31200, skenderun-Hatay, Turkey

Tahsin Toker
Water Pollution Laboratory, Faculty of Fisheries, Mustafa Kemal University, 31200, skenderun-Hatay, Turkey

E. S. Abechi
Department of Chemistry, Ahmadu Bello University, Zaria, Kaduna State, Nigeria

O. J. Okunola
National Research Institute for Chemical Technology, Basawa, Zaria, Kaduna State, Nigeria

S. M. J. Zubairu
Department of Chemistry, Ahmadu Bello University, Zaria, Kaduna State, Nigeria

A. A. Usman
Salem University, Lokoja, Kogi State, Nigeria

E. Apene
Federal College of Forestry Mechanization P. M. B. 2273, Afaka, Kaduna, Kaduna State, Nigeria

C. Turra
Centro de Energia Nuclear na Agricultura, Universidade de São Paulo, Piracicaba, SP, Brazil

E. A. N. Fernandes
Centro de Energia Nuclear na Agricultura, Universidade de São Paulo, Piracicaba, SP, Brazil

M. A. Bacchi
Centro de Energia Nuclear na Agricultura, Universidade de São Paulo, Piracicaba, SP, Brazil

L. O. Odokuma
University of Port-Harcourt, Port-Harcourt, Rivers State, Nigeria

E. Akponah
Delta State University, Abraka, Delta State, Nigeria

R. K. Meindinyo
Department of Physics, Bayelsa State College of Education, Okpoama, Brass, Nigeria

E. O. Agbalagba
Department of Physics, University of Port Harcourt, Rivers state, Nigeria

M. A. Hamed
National Institute of Oceanography and Fisheries, Red Sea branch, Egypt

Kh. S. Abou El-Sherbini
Department of Inorganic Chemistry, National Research Center, Dokki, Giza, Egypt

Y. A. Soliman
National Institute of Oceanography and Fisheries, Red Sea branch, Egypt

M. S. El-Deek
National Institute of Oceanography and Fisheries, Red Sea branch, Egypt

M. M. Emara
Department of Chemistry, Faculty of Science, Al-Azhar University Egypt

M. A. El-Sawy
National Institute of Oceanography and Fisheries, Red Sea branch, Egypt

Abdulmojeed O. Lawal
Nuclear Technology Centre, Nigeria Atomic Energy Commission, Sheda, P. M. B. 007, Gwagwalada, Abuja, FCT, Nigeria

Abdulrahman A. Audu
Department of Chemistry, Bayero University, Kano. P. M. B. 3011, Kano, Nigeria

Mark Fungayi Zaranyika
Chemistry Department, University of Zimbabwe, P. O. Box MP 167, Mount Pleasant, Harare, Zimbabwe

Tsitsi Chirinda
Chemistry Department, University of Zimbabwe, P. O. Box MP 167, Mount Pleasant, Harare, Zimbabwe

S. Singh
Division of Environmental Sciences, Insian Agricultural Research Institute, New Delhi 110012, India

M. Zacharias
Division of Environmental Sciences, Insian Agricultural Research Institute, New Delhi 110012, India

S. Kalpana
Division of Environmental Sciences, Insian Agricultural Research Institute, New Delhi 110012, India

S. Mishra
Division of Environmental Sciences, Insian Agricultural Research Institute, New Delhi 110012, India

Swati Pattnaik
Department of Ecology and Environmental Sciences, Pondicherry University, Puducherry – 605014, India

M. Vikram Reddy
Department of Ecology and Environmental Sciences, Pondicherry University, Puducherry – 605014, India

O. J. Okunola
National Research Institute for Chemical Technology, Zaria, Kaduna State, Nigeria

Y. Alhassan
National Research Institute for Chemical Technology, Zaria, Kaduna State, Nigeria

G. G. Yebpella
National Research Institute for Chemical Technology, Zaria, Kaduna State, Nigeria

A. Uzairu
Department of Chemistry, Ahmadu Bello University, Zaria, Kaduna State, Nigeria

A. I. Tsafe
Usmanu Danfodiyo University, Sokoto, Sokoto State, Nigeria

E. S. Abechi
Department of Chemistry, Ahmadu Bello University, Zaria, Kaduna State, Nigeria

E. Apene
Federal College of Forestry Mechanization, P. M. B. 2273, Afaka, Kaduna, Kaduna State, Nigeria

T. Jimoh
Department of Chemistry, Federal University of Technology, Minna, Nigeria

J. N. Egila
Department of Chemistry, University of Jos, Plateau State, Nigeria

B. E. N. Dauda
Department of Chemistry, Federal University of Technology, Minna, Nigeria

Y. A. Iyaka
Department of Chemistry, Federal University of Technology, Minna, Nigeria

M. K. Ladipo
Yaba College of Technology, Lagos, Nigeria

V. O. Ajibola
Department of Chemistry, Ahmadu Bello University, Zaria, Nigeria

S. J. Oniye
Department of Biological Sciences, Ahmadu Bello University, Zaria, Nigeria

S. A. Abdallah
Hussaini Adamu Federal Polytechnic, Kazaure, Jigawa State, Nigeria

A Uzairu
Department of Chemistry, Ahmadu Bello University, Zaria, Nigeria

J. A Kagbu
Department of Chemistry, Ahmadu Bello University, Zaria, Nigeria

O. J Okunola
National Research Institute for Chemical Technology, Basawa, Zaria, Nigeria

O. Ukabiala Chinwe
Department of Chemistry, School of Natural and Applied Sciences, College of Science and Technology, Covenant University, Km 10 Idiroko Road, P. M. B. 1023, Ota, Ogun State, Nigeria

C. Nwinyi Obinna
Department of Biological sciences, School of Natural and Applied sciences, College of Science and Technology, Covenant University, Km 10 Idiroko Road, P. M. B. 1023, Ota, Ogun State, Nigeria

Abayomi Akeem
Department of Chemistry, University of Lagos, Akoka, Yaba, Lagos State, Nigeria

B. I. Alo
Department of Chemistry, University of Lagos, Akoka, Yaba, Lagos State, Nigeria

J. C. Akan
Department of Chemistry, University of Maiduguri, P. M. B. 1069, Maiduguri, Borno State, Nigeria. Maiduguri, Borno State, Nigeria

F. I. Abdulrahman
Department of Chemistry, University of Maiduguri, P. M. B. 1069, Maiduguri, Borno State, Nigeria. Maiduguri, Borno State, Nigeria

O. A. Sodipo
Department of Clinical Pharmacology and Therapeutics, College of Medical Sciences, University of Maiduguri, Maiduguri, Borno State, Nigeria

A. E. Ochanya
Department of Chemistry, University of Maiduguri, P. M. B. 1069, Maiduguri, Borno State, Nigeria. Maiduguri, Borno State, Nigeria

Y. K. Askira
Department of Chemistry, University of Maiduguri, P. M. B. 1069, Maiduguri, Borno State, Nigeria. Maiduguri, Borno State, Nigeria

Chimezie Anyakora
Department of Pharmaceutical Chemistry, University of Lagos, Nigeria

Kenneth Nwaeze
Department of Pharmaceutical Chemistry, University of Lagos, Nigeria

Olufunsho Awodele
Department of Pharmacology, University of Lagos, Nigeria

Chinwe Nwadike
Department of Pharmaceutical Chemistry, University of Lagos, Nigeria

Mohsen Arbabi
Department of Environmental Health Engineering, School of Health, Shahrekord University of Medical Sciences, Iran

Herbert Coker
Department of Pharmaceutical Chemistry, University of Lagos, Nigeria

Rohani Ambo-Rappe
Department of Marine Sciences, Faculty of Marine Sciences and Fisheries, Hasanuddin University, Tamalanrea Km. 10 Makassar, 90245, Indonesia

Dmitry L. Lajus
Department of Ichthyology and Hydrobiology, Faculty of Biology and Soil Sciences, St. Petersburg State University 16 line V.O. 29, 199178, St. Petersburg, Russia

Maria J. Schreider
School of Environmental and Life Sciences, Ourimbah Campus, Newcastle University, Brush Road, Ourimbah, NSW, 2258, Australia

Elnor C. Roa
Department of Chemistry, College of Science and Mathematics, Mindanao State University-Iligan Institute of Technology (MSU-IIT), 9200 Iligan City, Philippines Institute of Fisheries Research and Development, Mindanao State University-Naawan, 9023 Naawan, Misamis Oriental, Philippines

Mario B. Capangpangan
Department of Chemistry, College of Science and Mathematics, Mindanao State University-Iligan Institute of Technology (MSU-IIT), 9200 Iligan City, Philippines

Madeleine Schultz
School of Physical and Chemical Sciences, Queensland University of Technology, Brisbane, Queensland 4001, Australia

Shuaibu Tela Garba
Department of Chemistry, P. M. B. 1069. University of Maiduguri, Borno State.Nigeria

Akuewanbhor Sunday Osemeahon
Department of Chemistry, P. M. B. 2076. Federal University of Technology Yola (FUTY), Adamawa State, Nigeria

Humphrey Manji Maina
Department of Chemistry, P. M. B. 2076. Federal University of Technology Yola (FUTY), Adamawa State, Nigeria

Jeffry Tsaware Barminas
Department of Chemistry, P. M. B. 2076. Federal University of Technology Yola (FUTY), Adamawa State, Nigeria

T. J. K. IDERIAH
Institute of Pollution Studies, Rivers State University of Science and Technology, Rivers State, Nigeria

O. D. DAVID
Department of Environmental and Applied Biology, Rivers State University of Science and Technology, Port Harcourt Rivers State, Nigeria

D. N. OGBONNA
Institute of Geosciences and Space Technology, Rivers State University of Science and Technology, Rivers State, Nigeria

Gift O. Tsetimi
Geo Environmental and Climate Change Adaptation Research Centre, University of Benin, Benin City, Edo State, Nigeria

Felix E. Okieimen
Geo Environmental and Climate Change Adaptation
Research Centre, University of Benin, Benin City, Edo
State, Nigeria

M. M. Achi
Department of Chemistry, Ahmadu Bello University,
Zaria, Kaduna State, Nigeria

A. Uzairu
Department of Chemistry, Ahmadu Bello University,
Zaria, Kaduna State, Nigeria

C. E. Gimba
Department of Chemistry, Ahmadu Bello University,
Zaria, Kaduna State, Nigeria

O. J. Okunola
National Research Institute for Chemical Technology, P.
M. B. 1052, Basawa, Zaria, Kaduna State, Nigeria